Preface

The authors of this book have recognized the difficulties that African University students encounter whenever they try to read and understand Probability Theory and Mathematical Statistics textbooks. Generally, Mathematics is perceived to be a very difficult subject, and therefore we submit Probability Theory and Mathematical Statistics as difficult disciplines, which students still grapple with. The facilities that render these subjects easy for teaching (i.e. simple textbooks, well-equipped statistics libraries, and bookshops, over and above specialized teaching staff) are usually absent. Compounding the problems further, most textbooks in Probability Theory and Statistics require advanced knowledge of Mathematics, particularly in the areas of Measure and Integration theory. For these reasons, the authors of this book have attempted to present Probability Theory by moving away from the Measure and Integration theory approach, which would otherwise limit the reading of the textbook only to students of Mathematics. The principal subjects of this textbook are Probability Theory and Stochastic Processes.

The importance of Probability Theory and Stochastic Processes to the Social, Engineering and Biological sciences has long been recognized. Statistical methods are now widely used in Physics, Biology, Medicine, Economics, Industry, Agriculture, Fisheries, and Meteorology and in Communications. Areas of application of Probability Theory and Mathematical Statistics are increasing more and more. It is therefore necessary to develop the student's ability to relate Mathematics to modern-day problems in Biology, Social Sciences, Law, Engineering and Medicine. These are the aims of this book. We have attempted to write a flexible book that will serve the extremely varied needs and backgrounds of the students for whom it is intended. To this end, prerequisites are kept to a minimum.

Students with some background in Calculus, Algebra and Differential equations can read the entire text of the book with ease. No advanced knowledge in these fields or knowledge of Measure and Integration Theory is required. This book is also suitable for non-mathematicians as far as concepts, theorems, and methods of application are concerned. The main purpose of this book is to give a systematic introduction to Probability theory and Stochastic Processes, present an outline of possible applications of these theories accompanied by descriptive concrete examples; and provide extensive (not exhaustive) references of other books in these areas, thereby giving the reader opportunity to complete his knowledge of the subjects considered.

This content has been designed so that it can be used as a textbook for an undergraduate Mathematics and Statistics course. For this reason the proofs and solutions to the problems have been carried out by the most elementary methods possible. The exposition is indeed very pedagogic. The textbook has been written in such a way that each chapter is independent of the others although there are many cross-references, which unify the subjects at hand.

The material in the textbook proceeds from simple set theory to simple random experiments, from concepts of probability and probability distributions to functions of random variables, and generating functions to simple stochastic processes, in what we regard as the logical approach to probability Theory and Stochastic Processes. While definitions, theorems and proofs are the essence of this subject, we have tried to avoid the definition — theorem — proof format insofar as this seemed to be feasible.

The definitions and theorems are numbered sequentially in each chapter, the designation denoting the section and number of each definition or theorem. For reference purposes, the section is indicated only if the definition or theorem occurs in a different chapter. For example, in Chapter 6, any reference to Theorem 6.1.8 of chapter 6 would be simply to Theorem 6.1.8 but, if the same theorem is referred to in some other chapter, the reference would be to Theorem 6.1.8 of section 6.1. Same thing for the definitions. All worked examples and diagrams are numbered, in sequence for each chapter, but with no reference to the section in which one is located. For example, the fourth worked example in chapter 2 is designated as Example 2.4, but this particular worked example happens to be in section 2.2. Similarly, the second diagram in chapter 2 is designated as Figure 2.2 but this particular diagram happens to be in section 2.1. It would have been easy to use a more elaborate and definitive system of numeration for definitions, theorems, worked examples and diagrams, but we felt that the disadvantages would outweigh the advantages.

The Notes, a common feature of the book, are an important part in their aspects. Firstly, they illuminate points made in the text. Secondly, they present material that does not quite fit anywhere else and thirdly, they discuss details whose exposition would disturb the flow of argument.

In this book more than 140 worked examples have been provided and it is essential that the student go through all of them. We have also provided a set of questions at the end of each chapter. All in all, more than 150 problems have been provided, and it is essential that the student go through most of them. At the rear of the book, answers have been provided to over 90 percent of the questions. We have not included answers for some of the questions, since we regard these problems as best suited for class discussions. Statistical tables are also not included since most of them are available on the Web The questions at the end of each chapter range from trivial numerical exercises and elementary problems, intended to familiarize the students with the concepts, to material more difficult than that worked out in the book. They are included both as a check on the student's mastery of the material and as pointers to the wealth of ideas and results that for obvious reasons of space could not be put into the body of the text.

Various notational conventions and abbreviations are used in this textbook. A list indicating where they are used is given at the end of the text.

J.A.L. Kamara.

Anani Lotsi

F.O. Mettle

A brief tour may now be taken through this book

Chapter 1 is introductory and is needed for the material that follows. It includes Set Theory, functions and Relations.

Chapters 2 to 4, on discrete probability in one and two dimensions, include all probability theory that is standard in courses in finite Mathematics. These courses include simple random experiments, concepts and definition of probability, conditional probability, Bayes Theorem, Expected value and Variance, Conditional expectation and variance, permutations and combinations, one and two dimensional probability distributions, marginal probabilities, functions of random variables, and some important discrete random variables such as: Binomial, Hypergeometric, Geometric, Poisson distributions, which are so useful in Biological applications.

Chapters 5 and 6, on continuous random variables, capture most of the topics of chapters 2 to 4 except that they are now treated as continuous processes (the integration sign replacing the summation sign). Additional topics have also been included viz:- Normal (or Gaussian) distribution, Bivariate and Multivariate transformations (Jacobian matrix), Uniformly distributed random variables, Central limit theorem, Chebyshev's inequality, Law of large numbers, Exponential distribution, the Gamma and Beta distributions, the Chi-square, F and T distributions.

Chapter 7 deals with Generating Functions: Moment, Factorial, and Probability Generating functions, Characteristic functions, Multivariate Generating functions and Negative Binomial distributions, and Multinomial distributions.

Finally, **Chapter 8** deals with Introductory Stochastic processes:- Discrete parameter cases: Simple random walks and Gambler's ruin, Stochastic Matrices, transition probabilities, recurrence and transience, classification and decomposition of states of a Markov chain, Periodicity, and Stationary processes. The chapter also deals with Continuous parameter cases: the simple Poisson process, multivariate and compound Poisson processes, Waiting time between successive events, Pure and Simple Birth processes, Pure Death process, Immigration-Birth process, and the Birth-Death process.

We believe that chapter 8 of this book could be used as a revision for students starting their Master's degree programmme.

Acknowledgements

"God created the universe and gave knowledge to the creatures, The rest is the work of man. To God be the glory and praise."

First and foremost, we express our sincere thanks to the Almighty God for giving us the knowledge and the stamina to complete this book.

There are many people to whom we owe a debt of gratitude for their assistance in bringing this book to publication. Most of all, we are indebted to many students who have studied Probability Theory in our classes over the years and who have contributed unwittingly to the evolution of the manuscript material. Several anonymous reviewers provided us with very encouraging comments on the manuscript, but we would like to identify Professor Nicholas. N.N. Nsowah-Nuamah of the Kumasi Polytechnic, Ghana, as one whose comments and suggestions were particularly appreciated and very useful to us. He made careful reading of the entire manuscript, and pointed out many ways in which it could be improved.

We also wish to thank all persons who have worked so expeditiously on the various stages of the publication process. Last and most important we would like to thank Mrs Emelia Lotsi (wife of Dr Anani Lotsi), Mrs. Deborah Naa Dzama Odotei-Mettle (wife of Dr F.O. Mettle), Mrs Antonia Hawa Kamara (wife of Professor J.A. Lawrence Kamara) and our families for their support, encouragement, and active participation in an enterprise that at times seemed endless. They provided us with the necessary inspiration, and we are grateful to these gurus.

If there are only few technical errors left in this book, then it is because many professional scholars assisted us in reading the preliminary manuscript. We are indeed indebted to such scholars.

Any remaining errors in the book are of course our responsibility, including any shortcomings in the text. We appreciate comments on any flaws or inadequacies that might be detected, and we suggest that these be sent to us at the following email addresses:

(i) Lawrence_kamara@yahoo.co.uk,
(ii) alotsi@ug.edu.gh.

J.A.L. Kamara
Anani Lotsi
F.O. Mettle

Contents

CHAPTER 1
SET THEORY

In order to study elementary Probability Theory it is best to first discuss some ideas and concepts of Mathematical Set Theory. The subject of Set Theory is a very extensive one, and much has been written about it. However, we will present a few basic notions.

1.1.   Definition:
A set is a well-defined collection of objects.

1.2.   Description of a Set
(a) Sets are usually represented by capital letters, say, S, S_1, S_2, P, H, etc. The individual objects belonging to a set S are called members (or elements) of S. When an object denoted by say "r", is a member of a set S we write r $\in$ S. The symbol "$\in$" means "is a member of". Similarly, when the object "2" is a member of the set P we write 2$\in$P. When "r" is not a member of S we write r $\notin$ S and when 2 is not a member of P we write 2 $\notin$ P. The symbol "$\notin$" means "is not a member of".

The set S, for example, can be represented as S = { •, •, •, •, •}, where the dots represent the members of S; each member being separated from the other by a comma; and all the members are enclosed within brackets. For instance, the set
S = {1, 2, 3, 4, r, t} comprises six members 1, 2, 3, 4, r and t. Suppose another set P has members 6, 7 and g. We may write P = {6, 7, g}. The following statements are now true about sets S and P mentioned above.

(i)   6 $\notin$ S, i.e. element 6 does not belong to S

(ii)   g $\notin$ S, i.e. element g does not belong to S

(iii)   2 $\notin$ P, i.e. element 2 does not belong to P.

(iv)   r $\notin$ P, i.e. element r does not belong to P.

(b) Sets may also be described in words. For example, the set S_1 may consist of all real numbers between 1 and 2 inclusive. To describe the above set we can simply write
S_1 = {x : 1 $\leq$ x $\leq$ 2}; that is, S is the set of all members x, where x is a real number between 1 and 2 inclusive.

1.3.   Types of Sets

1.1.1.   Finite sets

In most of the situations we are dealing with, we will be concerned with the study of finite sets, which have a definite number of objects. Examples of such sets are:- the set of the real numbers, the set of pupils in a school, the set of cars produced by a factory each month, etc. We usually describe each of these situations in relation to a much bigger set called the Universal set. The Universal set can be defined as the set of all objects under consideration. This set is usually denoted by U.

1.1.2.   Null or empty Set

A null or empty set contains no members at all. It is usually represented by the Greek Symbol ϕ (Phi). Thus, we can write $\phi = \{ \ \}$. Note that there is no dot within the brackets. It is important to note that, the empty set $\phi = \{ \ \}$ which contains no element, is not the same as the set $P = \{0\}$ which has 0 (zero) as its only member.

1.1.3.   Subsets and Equal Sets

It may happen that when two sets A and B are considered, being a member of A implies being a member of B or vice versa. In that case we say that A is a subset of B or B is a subset of A and we write $A \subset B$ or $B \subset A$.

Two sets are said to be equal if they contain the same number of identical elements, in that case we write A=B.

EXAMPLE 1.1

If we let the set A represent all students of Fourah Bay College, and let the set B represent only those students of Fourah Bay College, who are registered in the Faculty of Pure and Applied Sciences, then we can say that B = {students of Fourah Bay College studying Pure and Applied Sciences} is a subset of A = {all students of Fourah Bay College}, i.e. B $\subset$ A, which can also be written as

$$A \supset B.$$

EXAMPLE 1.2

Let A = {1, 2, 3} and B = {1, 2, 3, 5, 6}, then A is a subset of B, i.e. A ⊂ B.

EXAMPLE 1.3

Let P represent the pupils of Christ the King College (C.K.C), and let R represent pupils of C.K.C who are in the science stream. Then it is true that,
R = {pupils of C.K.C. who are in the science stream} is a subset of P = {all the pupils of C.K.C.}.
Two sets A and B are said to be equal (denoted A = B), if and only if A is a subset of B and at the same time B is a subset of A; A ⊂ B and B ⊂ A. That is, the two sets are equal if and only if they contain the same members.

EXAMPLE 1.4

If A = {1, 2, 3, 4, 5} and B = {1, 3, 5, 4, 2}, we say that the set A is the same as the set B, since they both contain the same members 1, 2, 3, 4, 5. The order in which the members appear within the brackets is immaterial.
Two sets A and B are unequal if they have different elements (i.e. they do not have exactly the same members). For example, if
A = {Δ, r, 1, 2, 3} and B = {1, 2, 3, 4, 5} we can write A ≠ B (A is not equal to B), since all the elements of set A are not the same as those of set B.
We can now make the following important statements about sets:-
(i)   The empty set φ, is a subset of itself. In fact, every set is a subset
         of itself.

(ii)   Once the Universal set, U, has been agreed upon, then for every other set, A, considered in the context of U, we have A ⊂ U. As an illustration, suppose we agree on students of Prince of Wales School as our universal set. Then the various groups within the school, for example, the football team, the cricket team, the science club, the quiz team etc. are subsets of the set of students of Prince of Wales School.

1.4.   Algebra of sets

We consider the idea of combining given sets in order to form a new set. Two basic operations are considered, the operation of addition and multiplication of numbers. Let A and B be any two given sets.

1.1.4.   Definition (Union of Sets)

We define the set C as the union of A and B (sometimes called the sum of A and B) as follows:-
$C = \cup B = \{: x \in A \text{ or } x \in B \text{ (or } x \in \text{ both A and B)}\}$, where $A \cup B$ means the union of set A and set B. Thus, C consists of all members x which are in A, or in B or in both (A and B). The Venn diagram of Figure 1.1 below shows the universal set $\mho$, and the shaded area $A \cup B$.

Figure 1.1

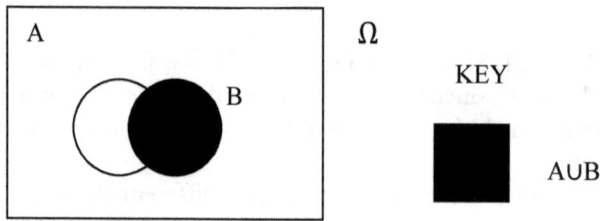

EXAMPLE 1.5
Suppose $\mho = \{1, 2, 3, 4, 5, 6, 7, 8, 9, 10\}$,
$A = \{1. 2, 3, 4\}$, $B = \{3, 4, 5, 6\}$.
We find that
$C = A \cup B = \{1, 2, 3, 4\} \cup \{3, 4, 5, 6\}$
i.e. $C = A \cup B = \{1, 2, 3, 4, 5, 6\}$. Note that in the set $C = A \cup B$, a particular member is listed exactly once.

1.1.5.   Definition: (Intersection of Sets)

We define the set D as the intersection of A and B (sometimes called the product of A and B) as follows:
$D = A \cap B = \{x: x \in A \text{ and } x \in B\}$, where $A \cap B$ means the intersection of set A and set B. Thus D consists of all members, which are in A and in B (members common to both A and B). The Venn

diagram of Figure 1.2 below shows the Universal set $\mho$, and the shaded area

$A \cap B$.

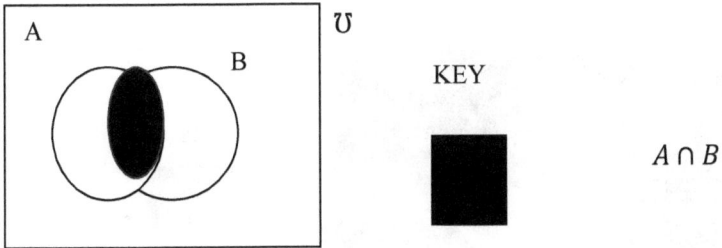

Figure 1.2

EXAMPLE 1.6
Suppose $\mho = \{1, 2, 3, 4, 5, 6, 7, 8, 9, 10\}$,
$A = \{1, 2, 3, 4\}$, $B = \{3, 4, 5, 6\}$

We find that $D = A \cap B = \{1,2,3,4\} \cap \{3,4,5,6\}$

i.e. $D = A \cap B = \{3,4\}$, again a member of D is listed exactly once.

1.1.6.   Definition (Complement of a Set)
The set denoted by A^c, consisting of all members not in A (but one in the Universal set $\mho$) is called the complement of A.
Thus $A^c = \{x: x \notin A\} = \{$the set of all members x such that x is not a member of A$\}$.
The Venn diagram of Figure 1.3 below shows the Universal set, $\mho$, and the shaded area A^c.

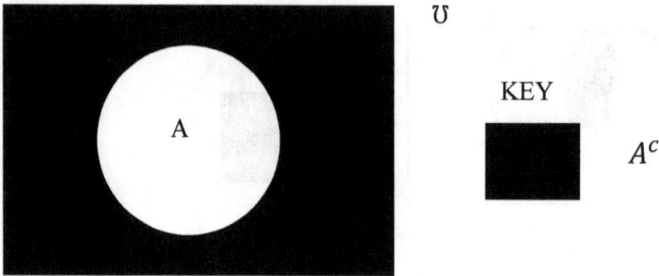

U

KEY

A

A^c

EXAMPLE 1. Figure 1.3

Suppose that $U = \{1, 2, 3, 4, 5, 6, 7, 8, 9, 10\}$, and
$A = \{1, 2, 3, 4\}$, we find that the complement of set A is given as

$$A^c = \{5, 6, 7, 8, 9, 10\}$$

The above operations of union and intersection defined for just two sets may be extended in an obvious way to any finite number of sets. Thus, for any three sets A, B, C we define A ∪ B ∪ C as A ∪(B ∪C) or (A ∪ B) ∪ C which are the same, by associativity as can easily be checked. Similarly, we define $A \cap B \cap C$ as $A \cap (B \cap C)$ or $(A \cap B) \cap C$ (A∩B) ∩C, which can be proven to be the same.

Also

$$\bigcup_{i=1}^{n} A_i = A_1 \cup A_2 \cup ... \cup A_n = \{x : x \in A_1 \text{ or } x \in A_2 \text{ or } ... \ x \in A_n\}$$

The above represents the events that at least one of the A_i's occurred. Similarly

$$\bigcap_{i=1}^{n} A_i = A_1 \cap A_2 \cap ... \cap A_n$$

$$= \{x : x \in A_1 \text{ and } x \in A_2 \text{ and } ... \ x \in A_n\}$$

This represents the event that all of the A_i's occur.

EXAMPLE 1.8

$\mho$ = {All doctors in a particular hospital}
A = {All female doctors in the hospital}
We find that A^c = {*All male doctors in the hospital*}

EXAMPLE 1.9

Given that S = { 5,6,7,8,9,10,11,12,13,14,15} and sets A, B, and C are subsets of $\mho$ given by
$$A = \{6,8,10,12\}, B = \{5,7,9,11\}, C = 13,15,6,11\}$$

Find (a) A^c (b) B^c (c) C^c

We find that:

(a) A^c= {5, 7, 9, 11, 13, 14, 15} i.e. all elements that are in the Sample space $\mho$ but are not in A.
(b) B^c= {6, 8, 10, 12, 13, 14, 15} i.e. all elements that are in the Sample space $\mho$ but are not in B.
(c) C^c ={5, 7, 8, 9, 10, 12, 14}   i.e. all elements that are in the Sample space $\mho$ but are not in C.

EXAMPLE 1.10
Let
A = {5, 10, -3, 6, 8, 9}
C = {4, 1, 2}
B = {4, 3, 6, 8}
Find (a) $A \cup B$ (b) $B \cup C$ (c) $A \cap \{B \cup C\}$

We find that:

(a) $A \cup B$ = {5, 10, -3, 6, 8, 9, 4, 3}
(b) $B \cup C$ = {4, 3, 6, 8, 1, 2}
(c) $A \cap \{B \cup C\}$ = {6, 8}

1.1.7.   Definition (Cartesian Product of Set)

The Cartesian product of two sets A and B, written A x B, is the set of pairs of members (a, b) such that 'a' is a member of set A and 'b' is a member of B. (i.e. the set {(a, b): a $\in$ A and b $\in$ B}. The pair (a, b) is obtained in the order a $\in$ A followed by b $\in$ B.

The above notion can be extended as follows:

If $A_1; A_2; A_3, \ldots A_n$ are sets, then
$A_1 \times A_2 \times A_3 \times \ldots \times A_n = \{(a_1, a_2, \ldots a_n), a_i \in A_i, i=1, 2, \ldots, n \}$,
that is the set of ordered n – tuples.

An important special case arises when we take the Cartesian product of a set with itself, that is A x A or A x A x A.

EXAMPLE 1.11

Let A = {1, 2},
A x A = {1, 2} x (1, 2}
A x A is obtained by pairing (or combining) each member in the first A in turn, with every member of the second A. Thus, A x A = {(1,1), (1, 2), (2, 1), (2, 2)}, where
(1, 1) means: '1' from the first A is combined with '1' from the second A,
(1, 2) means: '1' from the first A is combined with '2' from the second A,
(2, 1) means:' 2' from the first A is combined with '1' from the second A, and
(2, 2) means: '2 'from the first A is combined with' 2' from the second A.

EXAMPLE 1.12

If A = {1, 2},
A x A x A = {1, 2} x {1, 2} x {1, 2}

A x A x A is obtained by combining each member of the first A in turn, with each member of the other two A's. Thus,

A x A x A is given by:

First		Second		Third	(A x A x A)
A	X	A	X	A	
1		1		1	(1, 1, 1)
1		1		2	(1, 1, 2)
1		2		1	(1, 2, 1)
1		2		2	(1, 2, 2)
2		1		1	(2, 1, 1)
2		1		2	(2, 1, 2)
2		2		1	(2, 2, 1)
2		2		2	(2, 2, 2)

EXAMPLE 1.13

Let A = {1, 2}, B = {r, s, t, u, v, w}, A× B is given by:

A	B	(A x B)
1	r	(1, r)
1	s	(1, s)
1	t	(1, t)
1	u	(1, u)
1	v	(1, v)
1	w	(1, w)
2	r	(2, r)
2	s	(2, s)
2	t	(2, t)
2	u	(2, u)
2	v	(2, v)
2	w	(2, w)

The idea of Cartesian product is applied whenever the same experiment is repeated, or two separate experiments are performed. We wish to pair the results of the first experiment with that of the second. Examples of such experiments are: the tossing of a coin two or more times, the throwing of a die two or more times, tossing a coin and throwing a die simultaneously etc.

1.5.   Relations

A set, S, consisting of five science students at Fourah Bay College: Khadija, Michael, Sei, Khalil and Dumbuya, were asked to say which subjects they liked best out of the following: Mathematics, Statistics, Physics, Geology and Chemistry. Their replies were recorded as follows:

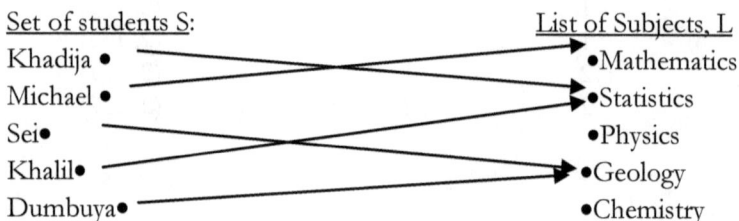

Set of students S: List of Subjects, L
Khadija •   •Mathematics
Michael •   •Statistics
Sei•   •Physics
Khalil•   •Geology
Dumbuya•   •Chemistry

The arrows indicate, for instance, that Khadija and Khalil like Statistics. The facts could have been listed thus: {(Khadija, Statistics), (Michael, Mathematics), (Sei, Geology), (Khalil Statistics), (Dumbuya, Geology)}. We say that there is a relation between the set of students S, and the list of subjects, L. Here the relation is "like best". (Khadija, Statistics) is called an ordered pair. The order is important for if we wrote (Statistics, Khadija) with the same relation it would mean that Statistics likes Khadija best. The first member of the ordered pair usually called the first co-ordinate, must be a member of the set from which the arrows emanate and the second member of the pair, called the second co-ordinate, must be a member of the set to which the arrows are directed. The members of the second set are referred to as "images". The set of students, S, is called the domain of the relation. The subset of the list of subjects, which consists only of those members, that are images, is called the range of the relation. Thus, in the example above, the domain is {Khadija, Michael, Sei, Khalil, Dumbuya} and the range is {Mathematics, Statistics, Geology}.

If every student of the first set liked one and only one subject (i.e. if every member of the domain had one and only one image), this kind of relation is called a one-to-one mapping. The example we illustrated previously (i.e. the example of students and subjects) is not a one-to-one mapping. It is called a many-to-one mapping. This is because both Khadija and Khalil like Statistics. In other words, Statistics is the image of more than one member. Similarly, Geology is the image of more than one member. There are some elements of the second set which are images of more than one element of the first set – hence the word "many", but the elements of the first set have only one image – hence the word "one".

1.1.8. Definition: (Relation)

Any set of ordered pairs is a relation.

The set of all first components of the ordered pairs in a relation is the domain of the relation, and the set of all second components of the ordered pairs is the range of the relation.
A special kind of relation with which we shall be concerned is called a function.

1.1.9. Definition: (Function)

A function is a relation in which no two ordered pairs have the same first components. Thus the relation $\{(1, 3), (1, 4), (2, 3), (2, 5), (3, 6)\}$ shown in
Figure 1.4 below is not a function.

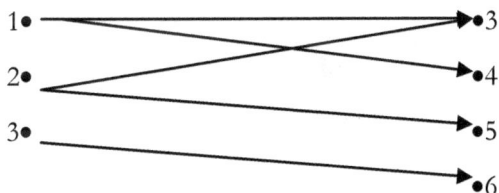

Figure 1.4

Several ordered pairs have the same first components, that is, one or more elements in the domain are mapped onto more than one element in the range.

On the other hand, the relation $\{(1, 3), (2, 4), (3, 5)\}$ and $\{(1, 3), (2, 3), (3, 4)\}$ shown in figures 1.5 and 1.6 respectively are functions.

Figure 1.5

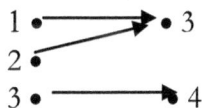

Figure 1.6

The concepts introduced above, although representing only a snapshot treatment of the theory of sets, are fairly adequate for the purposes of this book and can describe with sufficient rigour and precision, the basic ideas of Probability Theory.

EXERCISE 1

1.1 A and B are given sets. Express in words the following combinations of the sets, and illustrate them by use of Venn diagrams.

(a) $A \cap B$ (b) $(A \cap B)^c$ (c) $A \cap B^c$ (d) $A^c \cap B^c$

(e) $A \cup B$ (f) $(A \cup B)^c$ (g) $A \cup B^c$ (h) $A^c \cup B^c$

1.2   Simplify the following combinations of sets:-

(a) $A \cap A^c$ (b) $A \cap A$ (c) $A \cup A^c$ (d) $A \cup A$

(e) $A \cap (A \cap B)$ (f) $A \cap (A \cup B)$ (g) $A \cup (A \cap B)$ (h) $A \cup (A \cup B)$

(i) $(A \cap B) \cup (A \cap B)^c$ (j) $(A^c \cap B) \cup (A^c \cap B^c)$

1.3   Show by use of Venn diagrams or otherwise, the indicated properties hold:

(a) $A \cup B = B \cup A$ –Commutative Law

(b) $A \cap B = B \cap A$ – Commutative Law

(c) $A \cup (B \cup C) = (A \cup B) \cup C$ Associative Law

(d) $A \cap (B \cap C) = (A \cap B) \cap C$ - Associative Law

(e) $A \cup (B \cap C) = (A \cup B) \cap (A \cup C)$—Distributive Law

(f) $A \cap (B \cup C) = (A \cap B) \cup (A \cap C)$--- Distributive Law

(g) $A \cap \phi = \phi$

(h) $A \cup \phi = A$

(i) $(A \cap B)^c = A^c \cup B^c$ De Morgan's Law

(j) $(A \cup B)^c = A^c \cap B^c$ De Morgan's Law

(k) $(A^c)^c = A$

1.4   Suppose that the Universal set consists of the positive integers from 1 through 10. Let A = {2, 3, 4}, B = {3, 4, 5}, C = {5, 6, 7}. List the members of the following sets

(a) $A^c \cap B$ (b) $A^c \cup B$ (c) $(A^c \cap B)^c$
(d) $(A \cap (B \cap C)^c)^c$ (e) $(A \cap (B \cup C))^c$

1.5   Let $A = \{H, T\}$, $B = \{1, 2, 3, 4, 5, 6\}$. List the members of the following sets:
(i) $A \times A$ (ii) $A \times A \times A$ (iii) $B \times B$ (iv) $A \times B$

1.6. Consider the number of customers in a queue at a particular bank waiting to perform a transaction. We can model the number of customers in the queue with the sample space N^+ the set of positive intergers (including 0). Let
$$A_1 = \{n: 2 \le n \le 15\}$$

$$\text{and } A_2 = \{n: 5 \le n \le 155\}$$

Find the followings:
(i) $A_1 \cup A_2$ (ii) $A_1 \cap A_2$ (iii) A_1^c (iv) A_1^c (v) A_2^c
(vi) $A_2^c \cap A_1$

1.7.   The following diagram shows a relation. State the domain and the range.

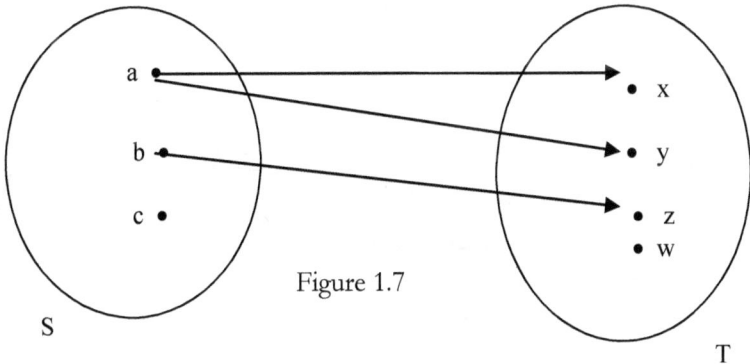

Figure 1.7

1.8.   Given that $S = \{1, 2, 3\}$ and $T = \{1, 2, 3, 4, 5, 6\}$ list the members of the relation R, where $R = \{(x, y)\}$: $x \in S$, $y \in T$ and $y = 2$ $x\}$. Show this relation on a diagram with arrows going from the members of S to their appropriate images in T.

1.9. Which of the following diagrams represent mapping?

(i)

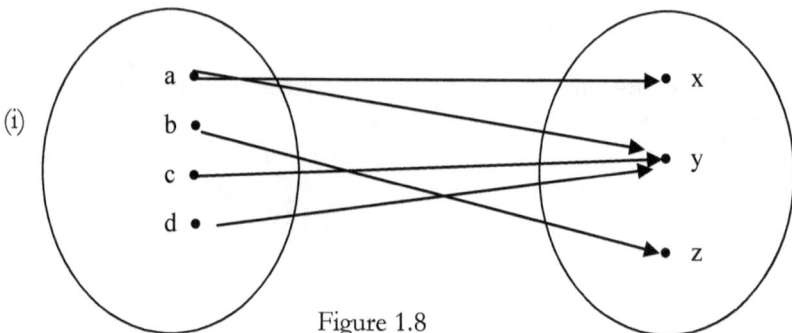

Figure 1.8

(ii)

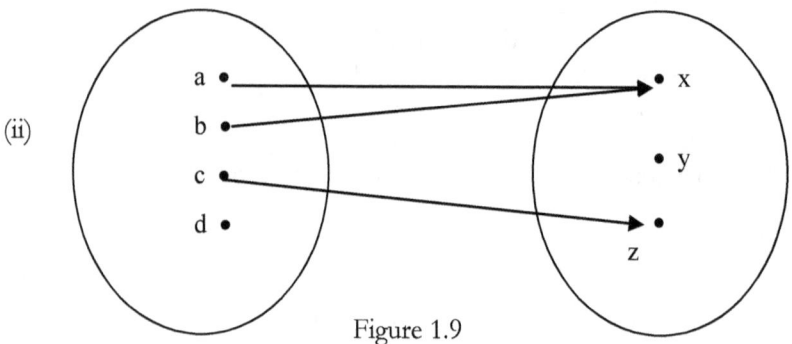

Figure 1.9

(iii)

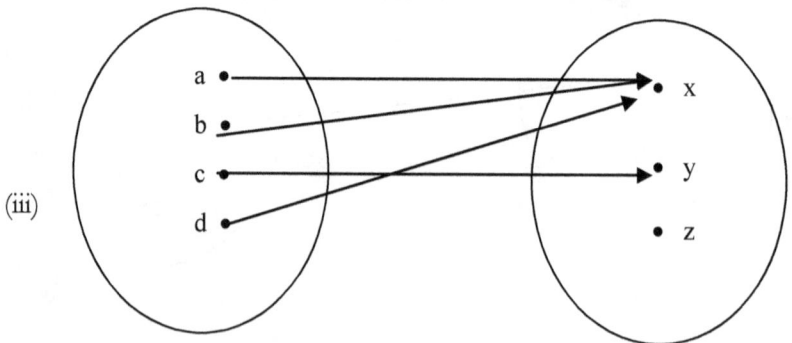

Figure 1.10

CHAPTER 2
SIMPLE RANDOM EXPERIMENTS

An experiment to begin with, is a process of observation and a random experiment is any process of observation or activity in which the results cannot be predicted with certainty. Stated differently, an experiment or random experiment is any specified set of actions the results of which cannot be predicted with certainty. Each repetition of an experiment is called a trial.

In the study of Probability Theory we usually talk about random experiments. Below are a few examples of such experiments.

(a)   Throw a die and count the number of points that show up.
(b)   Toss a fair coin three times and observe the total number of tails obtained.
(c)   A box of 10 marbles contains 4 red marbles and 6 white marbles. One marble is chosen after another (without replacing the chosen marble) until the last red marble is obtained. The total number of marbles removed is counted.
(d)   Items are manufactured until 12 non-defective items are produced. The total number of manufactured items is counted.

The above experiments have certain features in common: namely:

(i)   Each experiment is capable of being repeated indefinitely under essentially unchanged conditions.
(ii)   Prior to performing the experiment, we are not able to state what a particular result (or outcome) will be; we are, however, able to list the set of all possible outcomes of the experiment.
(iii)   As the experiment is performed a few times, the individual outcomes seem to occur in a haphazard manner. However, as the process is repeated a large number of times, the outcomes appear to have a definite or regular pattern. For example, in the repeated tossing of a fair coin, although heads and tails will appear, successively, in an almost arbitrary fashion, it is a well-known empirical fact that after a large number of tosses, the proportion of heads and tails will be approximately equal. All the experiments described in 2(a) to (d) above satisfy these general characteristics (i) to (iii), (of course, the last mentioned characteristic, i.e. (iii), can only be verified by experimentation; we all leave it to the reader's intuition to believe that if the experiment were repeated a large number of times, the regularity referred to would be evident).

In describing the various experiments, we have specified not only the procedure, which is being performed but, also, what we are interested in observing. This is a very important point to which we shall refer again and again when discussing random experiments.

2.1. The Sample Space

In the previous section, we mentioned that although we are, in general, not able to state, a priori, what a particular outcome of a random experiment will be: we are, however, able to list the set of all the possible outcomes of the experiment.

2.1.1 Definition (Sample Space)
An outcome is a result of an experiment. With each random experiment E we are considering, we define the sample space as the set of all possible outcomes of E. We usually designate this set by S. (In our present context, S represents the Universal set described previously).

We will now consider two random experiments and describe a sample space for each. The sample spaces S_1 and S_2 will relate to the experiments E_1 and E_2 respectively.

2.1.2 Experiment (Tossing a Fair Coin Once)

The experiment E_1 represents the tossing of a fair coin once. As we all know, a fair coin has a head on one side and a tail (T) on the other. The sample space S_1 associated with the experiment E_1 is the list of all possible outcomes. In this case, there are only two possible outcomes; a Head (H) or a Tail (T), as shown in Figure 2.1 below:

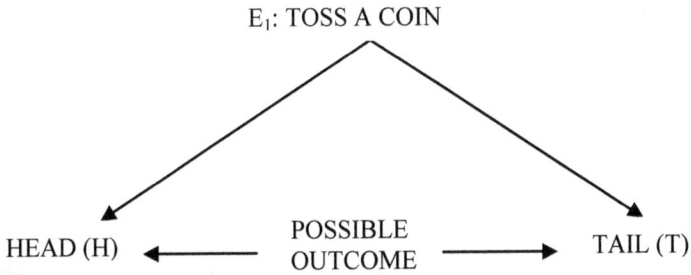

E_1: TOSS A COIN

HEAD (H) ← POSSIBLE OUTCOME → TAIL (T)

Figure 2.1

So the sample space $S_1 = \{H, T\}$

2.1.3 Experiment (Throwing a Ludo Die Once)

The experiment E_2 represents the throwing of a fair die once. As we know, a fair Ludo die has six faces numbered 1, to 6 respectively. The sample space S_2 associated with the experiment E_2 is the list of all possible outcomes 1, 2, 3, 4, 5, 6 as shown in Figure 2.2 below:

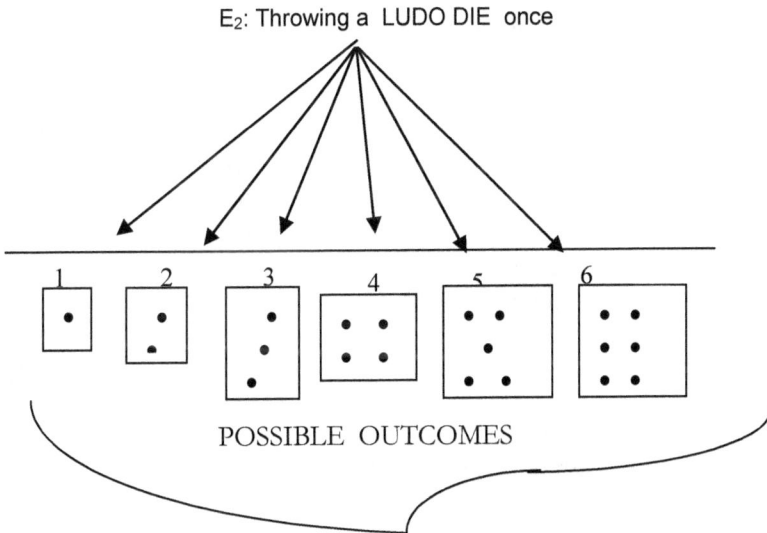

E_2: Throwing a LUDO DIE once

POSSIBLE OUTCOMES

Figure 2.2

So the sample space $S_2 = \{1, 2, 3, 4, 5, 6\}$.

For the moment, we will restrict ourselves to the two examples given above. In this book we will not concern ourselves with very complex sample spaces. However, such sample spaces do arise, but require more advanced Mathematics for their study than we are presupposing.

2.2. Events

Whenever we perform an experiment E, and a sample space S associated with the experiment has been obtained, we may further be interested in observing certain points or outcomes within the sample space. Such points or outcomes are referred to as events (of interest) and are subsets of the sample space S. An event associated with a sample space S associated with an experiment E, is usually designated with the letter A. For example, in the experiment of tossing a fair coin once, the event A_1 of observing a head only, is given as $A_1 = \{H\}$ and is of course a subset of the sample space $S = \{H, T\}$. Similarly, the event of observing a tail only, is given as $A_2 = \{T\}$, and is of course a subset of the sample space $S = \{H, T\}$. In view of this discussion, the sample space itself is an event and so is the empty set ϕ (phi). An individual outcome may also be viewed as an event.

If a_1 is an element of S (or belong to S), then we write $a_1 \in S$. If a_1 is not an element of S (or does not belong to S), then we write $a_1 \notin S$. A set A is called a subset of B, denoted $A \subset B$; if every element of A is also an element of B. Any subset of the sample space S is called an event. Note that a sample point of S is often referred as an elementary event. The sample space S is the subset of itself, that is $S \subset S$. Since S is the set of all possible outcomes of the experiment, it is often referred to as certain event.

The following are some examples of events associated with the sample space S, associated with the experiment E of throwing a fair dice once.

EXAMPLE 2.1

Experiment E_1: Throw a fair die once.
Sample space S: S = {1, 2, 3, 4, 5, 6}
Event of interest: An even number occurs

That is, A_1 = {2, 4, 6}
Note that A_1 = {2, 4, 6} is a subset of S = {1, 2, 3, 4, 5, 6}. We write
$A_1 \subset S$.

EXAMPLE 2.2:

Experiment E_2: Throw a fair die once.
Sample space S: S = {1, 2, 3, 4, 5, 6}
Event of interest: An odd number occurs that is, A_2 = {1, 3, 5}
Note that A_2 = {1, 3, 5} is a subset of S = {1, 2, 3, 4, 5, 6}. We write
$A_2 \subset S$.

EXAMPLE 2.3

Experiment E_3: Throw a fair die once.
Sample space S: S = {1, 2, 3, 4, 5, 6}
Event of interest: At most 3 appears, that is, A_3 = {1, 2, 3}
Note that A_3 = {1, 2, 3} is a subset of S = {1, 2, 3, 4, 5, 6}. We write
$A_3 \subset S$.

EXAMPLE 2.4

Experiment E_4: Throw a fair die once
Sample space S: S = {1, 2, 3, 4, 5, 6}
Event of Interest: At least 4 appears, that is A_4 = {4, 5, 6}
Note that A_4 = {4, 5, 6} is a subset of S = {1, 2, 3, 4, 5, 6}. We write
$A_4 \subset S$.

EXAMPLE 2.5
Find the sample space of experiment of tossing a coin repeatedly and of counting the number of tosses required until the first head appears.

It is easy to realize that all possible outcomes for this experiment are the terms of the sequence 1, 2, 3, . . . Thus the sample space S is given by
$$S = \{1, 2, 3, ...\}$$

EXAMPLE 2.6

Find the sample space of the experiment of measuring (in hours) the lifetime of a transistor.

Clearly, all possible outcomes are all positive real numbers, i.e. the lifetime can take all values on a positive number line. Thus the sample space S is given by

$$S = \{0 \leq x \leq \infty\}$$

It is useful to distinguish between two types of sample spaces: discrete and continuous. A sample space is discrete if it contains a finite or countable infinite set of outcomes. In this case the elements in the sample space can be listed or the list does not terminate as in example 2.5.
A sample space is continuous if it contains an interval (either finite or infinite)
of real numbers as in example 2.6.

EXAMPLE 2.7

All the following describe a continuous sample space EXCEPT:
(a) $S = \{(x, y): x \geq 0, \; y \geq 0\}$
(b) $S = \{(a, b): 1 \leq a \leq b, \; 1 \leq b \leq 6\}$
(c) $S = \{k: k \text{ is odd or } k = 2, 4, ...\}$
(d) $S = \{x: x = 1, 2, 3 \text{ or } x \geq 8\}$

Clearly, (c) i.e $S = \{k: k \text{ is odd or } k = 2,4, ...\}$ is not continuous sample space.

We can now use the various methods of combining sets (that is, events) and obtain the new sets (that is, events), which we introduced earlier.

(a) If A and B are events, A∪B is the event which occurs if and only if A or B (or both) occur.

(b) If A and B are events, A∩B is the event that occurs if and only if both A and B occur.

(c) If A is an event, A^c is the event that occurs if and only if A does not occur.

(d) If A_1, A_2, ... A_n is any finite collection of events, then
$A_1 \cup A_2 \cup A_3 ... \cup A_n = \cup_1^n A_i$, where A_i is the event which occurs if and only if at least
    one of the events A_i occurs.

(e) If A_1, A_2, ..., A is any finite collection of events, then $A_1 \cap A_2 \cap A_3$
.... $\cap A_n = \cap_1^n A_i$,
    where A_i is the event which occurs if and only if all the events A_i
occur.

(f) Suppose that S represents the sample space associated with some experiment E and we perform E twice. Then the Cartesian product S x S may be used to represent all outcomes of these two repetitions. That is, $(s_1, s_2) \in S \times S$ means that s_1 resulted when E was performed the first time and s_2 resulted when E was performed the second time.

2.3. Mutually Exclusive events

Definition (Mutually Exclusive Events)

Two sets A and B are disjoint or mutually exclusive if they contain no common element, that is $A \cap B = \emptyset$, that is, the intersection of A and B is the empty set. This means that A and B cannot occur together.

An athlete registered to do high jump and long jump track and field events during a sport meeting illustrates a simple example of mutually exclusive events. The athlete cannot perform both events at the same time. It has to be one event or the other.

One of the basic characteristics of the concept of "experiment" as discussed in the previous section is that we do not know which particular outcome will occur when the experiment is performed. Saying it differently, if A is an event associated with the experiment, then we cannot state with certainty that A will or will not occur. Hence, it becomes very important to try to associate a number with the event A which will measure, in some sense, how likely it is that the event A occurs. This task leads us to the theory of probability.

2.4. Concept of Probability

Chance and the assessment of risk play a part in everyone's life. As a loan officer you may want to evaluate the probability that a customer defaults; as an investment officer you are interested in finding the likelihood of an investment results in a loss.

Probability has a wide range of business applications. In addition to the calculation of risk in the banking and insurance industries, probability provides the basis of many of the sampling procedures used in quality control and market research. Most events in life are unpredictable, i.e. events in life are non-deterministic. When we begin to treat the world as stochastic, we can assess the chances of particular outcomes happening in a given situation. Probability measures the likelihood of the occurrence of an event. The probability of an event can be thought of as the relative frequency of the event; that is, the fraction of the time that an event can be expected to occur if the experiment is repeated over and over many times. The probability of an outcome can also be interpreted as a subjective probability, or degree of belief, that the outcome will occur. Different individuals will no doubt assign different probabilities to the same outcomes. Probabilities are numbers between 0 and 1. An event with probability 0 has no chance of occurring, and an event with probability 1 is certain to occur.

2.4.1. RELATIVE FREQUENCY DEFINITION OF PROBABILITY

The probability of an outcome is interpreted as the limiting value of the proportion of times the outcome occurs in n repetitions of the random experiment as n increases beyond all bonds. For example if we assign a probability of 0:4 to the outcome that there is a defective item in a production batch, we might interpret this assignment as implying that, if we analyze many items in the production batch, approximately 40% of them will be defective. This example illustrates a relative frequency interpretation of probability. Mathematically, if we assume that the random experiment is repeated n times, and event A occurs n(A) times, then the probability of event A, denoted P(A) is defined as

$$P(A) = \lim_{n \to \infty} \frac{n(A)}{n}$$

Where $\frac{n(A)}{n}$ is called the relative frequency of event A. Stated differently, we accept
the relative frequency ratio $\frac{n(A)}{n}$ as the probability of A if n is a large number.

Suppose that we decide to toss a fair die, and we are interested in finding the chance (or probability) that it will come up heads (H) or tails (T). One solution would be to toss it repeatedly, observing the outcome each time. Suppose the experiment is repeated 25 times and the following results are obtained.

RESULTS OF 25 TOSSES OF A FAIR COIN

HTHHT HTTHH HTHHH HTTTT HTTTH

We can count the number of heads (or tails) showing up. The results show 13 heads (H) and 12 tails (T); giving the number of heads $n_H = 13$, the number of tails $n_T = 12$ and $n_H + n_T = n$ = total number of tosses = 25.

The relative frequency of heads is, therefore, given as

$$f = \frac{number\ of\ Heads\ showing\ up}{Total\ number\ of\ tosses} = \frac{n_H}{n} = \frac{13}{25} = 0.52,$$

and the relative frequency of tails is given as

$$f_T = \frac{number\ of\ Tails\ showing\ up}{Total\ number\ of\ Tosses} = \frac{n_T}{n} = \frac{12}{25} = 0.48$$

Both figures, 0.52 and 0.48, become 0.5 when written to one decimal place. Of course tossing the coin 25 times would not be enough to average chance fluctuations. But over the long run, the relative frequency of Heads (or Tails) would settle down to a limiting value, which is probability. The answer 0.5 (or half) gives the indication that if a fair coin is tossed a large number of times, approximately the same amount of heads as tails is obtained (i.e. the relative frequency for both outcomes is always approximately half). We now say that the probability of a head (or tail) showing up is half and we can write,

Probability = Proportion, in the long run. Or more formally,

Probability of Heads: $P(H) = \lim_{n \to \infty} \frac{n_H}{n}$

Probability of Tails: $P(T) = \lim_{n \to \infty} \frac{n_T}{n}$,

Where $\lim_{n \to \infty} \frac{n_H}{n}$ is the limiting value of $\frac{n_H}{n}$ (a real number associated with the event H) as the number of tosses n approaches infinity, and

$\lim_{n \to \infty} \frac{n_T}{n}$ is the limiting value of $\frac{n_T}{n}$ (a real number associated with the event T) as the number of tosses n approaches infinity.

2.4.2. BASIC PROBABILITY FORMULA: CLASSICAL DEFINITION (EQUALLY LIKELIHOOD MODEL)

In an experiment for which the sample space is S and the outcomes are equally likely (that is each outcome has the same chance of occurring), the probability of an event A occurring is given by the formula

$$P(A) = \frac{\text{number of outcome in A}}{\text{total number of outcomes}}$$

$$= \frac{\text{number of ways that A can occur}}{\text{number of ways the sample space S can occur}}$$

$$= \frac{n(A)}{n(S)}$$

EXAMPLE 2.8

There are 5 balls in a box, 3 balls are red and 2 are black. The box is shaken, and you select a ball without looking. What is the probability that this ball is red?

The event that a ball selected is red consists of 3 outcomes out of 5 possible outcomes in all. According to the basic probability formula, the probability of selecting a red ball is 3 over 5 that is

$$P(red\ ball) = \frac{3}{5}$$

2.4.3. SUBJECTIVE PROBABILITY

Subjective probability is derived from an individual's personal judgment about whether a specific outcome is likely to occur. It contains no formal calculations and only reflects the subject's opinions and past experience. Subjective probability differs from person to person and they contain a high degree of personal bias.

EXAMPLE 2.9

Take a scenario where a person is asked to predict the percentage chance of whether a flipped coin will land with heads or tails. The person's initial response may be the mathematically true 50%. If 10 coin flips occur, all resulting in the coin landing tails up, the person may change his percentage chance to a number other than 50%. For example, saying the chance of it landing tails up is 75%. Even though the new prediction is mathematically inaccurate, the individual's persona experience of the previous 10 coin flips has created a situation in which he chooses to use subjective probability.

2.5. Elementary Properties of Probability

We consider now the general case of n repetitions of an experiment with r equally likely outcomes A_i ($i=1, 2, \ldots, r$). Suppose n_i out of the n repetitions result in outcome A_i ($i=1, 2, \ldots, r$). The relative frequency $\frac{n_i}{n}$ of any outcome A_i is positive, since both the numerator and denominator are positive; moreover, since the numerator cannot exceed the denominator, relative frequency cannot exceed the value 1. Thus

$0 \leq \dfrac{n_i}{n} \leq 1$. This is also true in the limit; that is

$$0 \leq \lim \dfrac{n_i}{n} = P\left(A_i\right) \leq 1.$$

Also, the frequencies of all possible outcomes sum to n:

$n_1 + n_2 + n_3 + \ldots\ldots + n_r = n$, so that the relative frequencies

$\dfrac{n_1}{n} + \dfrac{n_2}{n} + \ldots + \dfrac{n_r}{n} = \dfrac{n}{n} = 1$ sum to one. This also is true in the limit;

that is

$P\ (A_1)\ +\ P\ (A_2)\ +\ P\ (A_3)\ +\ \ldots\ldots.+\ P\ (A_r)\ =\ 1$.

2.5.1   Definition (Probability of an event)

Let A be an experiment. Let S be a sample space associated with A. With each event A we associate a real number, designated by P(A), and called the probability of A, satisfying the following properties.

(i) $0 \leq P\ (A) \leq 1$ i.e Probability is a number between 0 and 1 (inclusive)

(ii) $P\ (S) = 1$ i.e In performing an experiment, we assume that the result will be one of the simple outcomes. In other words, we assume that the event S will occur.

(iii)   If A and B are mutually exclusive events,
$$P(A \cup B) = P(A) + P(B)$$
If the sample space S is not finite then (iii) must be modified as follows:

(iv)   If $A_1, A_2, \ldots$, is an infinite sequence of mutually exclusive events in S $(A_1 \cap A_j = \emptyset)$ for $i \neq j$, then
$$P(A_1 \cup A_2 \cup \ldots) = P(\cup_{i=1}^{\infty} A_i)$$
$$= P(A_1) + P(A_2) + \cdots = \sum_{i=1}^{\infty} P(A_i)$$

This means that if $A_1, A_2, \ldots$, are events that cannot occur simultaneously, then the probability that one among them will occur is the sum of their probabilities. For example, the probability of either a 2 or a 6 in a roll of a die is $\dfrac{1}{6} + \dfrac{1}{6} = \dfrac{2}{6}$

2.5.2 Probability of ϕ (The Empty Set).

2.5.2.1 Theorem: If ϕ is the empty set, then $P(\emptyset) = 0$. (i.e. The probability of the impossible event is zero).

Proof:
Given that $\emptyset \cup S = S$, applying probability to both side gives
$$P(\emptyset \cup S) = P(S)$$
Applying property 2 and 3 gives

$$P(\emptyset) + P(S) = 1$$
since
$$P(S) = 1$$
This implies that
$$P(\emptyset) = 0$$

EXAMPLE 2.5

Consider a coin that is biased, such that both sides are marked Heads. What is the probability of obtaining tails in a toss of this coin?

The sample space $S = \{H, H\}$. The event $A = \{T\} = \phi$; this is an impossible event, and therefore, $P(A) = P(\phi) = 0$.

2.5.2.2 Corollary: If 'A' is an event that is always certain, then $P(A) = 1$.

EXAMPLE 2.6

Consider tossing the coin of Example 2.5 above. The sample space $S = \{H, H\}$. Since the outcome is always heads, the event $A = \{H\}$ is always certain and $P(A) = P(H) = 1$.

EXAMPLE 2.7

From a box containing only black marbles, a marble is chosen and its colour noted. Find the probability that the chosen marble is black.

SOLUTION
The required probability is 1. Can the student reason this out?

2.5.3 Probability of the Complementary Event

2.5.3.1 Theorem.   Let A be an event. If A^c is the complementary event of A, then
$$P(A^c) = 1 - P(A)$$

Proof.
Using the identity $A \cup A^c = S$, we have
$$P(A \cup A^c) = P(S)$$
applying property 2 and 3, we have
$$P(A) + P(A^c) = 1$$
This implies that
$$P(A^c) = 1 - P(A)$$

EXAMPLE 2.8

Let A be an event such that P(A) = 0.4. If A^c is the complement of A, then
$$P(A^c) \equiv 1 - P(A) = 1 - 0.4 = 0.6.$$

EXAMPLE 2.9

Let E be the experiment (tossing a fair coin once). The sample space associated with E, is S = {H, T}.

Let A be the event: head shows up i.e. A = {H}, $P(A) = P(H) = \dfrac{1}{2}$

$A^c = \{T\}$ $P(A^c) = 1 - P(A) = \dfrac{1}{2}$

EXAMPLE 2.10

Let E be the experiment (throwing a Ludo die once). The sample space S = {1, 2, 3, 4, 5, 6}. Let A be the event: an odd number shows up. i.e. A = {1} $\cup$ {3} $\cup$ {5} = {1} or {3} or {5} = {1, 3, 5}

$\therefore P(A) = P(1 \cup 3 \cup 5) = P(1) + P(3) + P(5) = \dfrac{1}{6} + \dfrac{1}{6} + \dfrac{1}{6} = \dfrac{3}{6} = \dfrac{1}{2}$

$A^c = (2) \cup (4) \cup (6) = 2 \ or \ 4 \ or \ 6$

$\therefore P(A^c) = P(2 \cup 4 \cup 6) = P(2) + P(4) + P(6) = 1- P(A) = 1-\frac{1}{2} = \frac{1}{2}$

2.5.4 Probability of the Union of Events ($A \cup B$)

2.5.4.1 Theorem: Let A and B be any two events, then
$P(A \cup B) = P(A) - P(B) - P(A \cap B).$

Proof

Method 1

Write:
$$A \cup B = A \cup (B \cap A^c) \dots \dots \dots \dots \dots \dots \dots \dots \dots . (1)$$
$$B = (A \cap B) \cup (B \cap A^c) \dots \dots \dots \dots \dots \dots \dots .. (2)$$

This implies that
From (1) we can write,
$$P(A \cup B) = P\{A \cup (B \cap A^c)\}$$

$$= P(A) + P(B \cap A^c) \dots \dots \dots \dots \dots \dots \dots \dots (3)$$

since $A \; and \; (B \cap A^c)$ are mutually exclusive.

Next
From (2), we can write
$$P(B) = P\{(A \cap B) \cup (B \cap A^c)\} = P(A \cap B) + P(B \cap A^c)$$
since $(A \cap B) \; and \; (B \cap A^c)$ are mutually exclusive.
This implies that
$$P(B \cap A^c) = P(B) - P(A \cap B) \dots \dots \dots \dots \dots \dots .. (4)$$
Now we put equation 4 into equation 3 and we have
$$P(A \cup B) = P(A) + P(B) - P(A \cap B) \; as \; claimed$$

Method 2
We will use Union {Intersection Counting Formula}. Now given that

$$n(A \cup B) = n(A) + n(B) - n(A \cap B)$$
We have
$$P(A \cup B) = \frac{n(A \cup B)}{n(S)}$$
$$= \frac{n(A) + n(B) - n(A \cap B)}{n(S)}$$

$$= \frac{n(A)}{n(S)} + \frac{n(B)}{n(S)} - \frac{n(A \cap B)}{n(S)}$$
$$= P(A) + P(B) - P(A \cap B)$$

where S is the universal set.

2.5.4.1.1 Corollary: Let A and B be any two events, if A and B are mutually exclusive, then $A \cap B = \emptyset$ and $P(A \cap B) = 0$, giving
$$P(A \cup B) = P(A) + P(B)$$

EXAMPLE 2.11

In a certain college, the men engage in various sports in the following proportions: 60% of all men play football (F), 50% of all men play basketball (B), and 30% of all men play both football and basketball. If a man is selected at random for an interview, what is the probability that he will: (a) play football or basketball? (b) play neither sport?

SOLUTION:
60% of men play football (F) implies $P(F) = 0.60$
50% of men play basketball implies $P(B) = 0.50$
30% of men play both football and basketball implies $P(F \cap B) = 0.30$

(a) Probability that the selected men will play football or basketball:
   $P(F \cup B) = P(F) + P(B) - P(F \cap B) = 0.60 + 0.50 - 0.30 = 0.80$

(b) Probability that the selected man will play neither sport: is the probability of the complement of (F$\cup$B), i.e. $P(F \cup B)^c = 1 - P(F \cup B)$
$$= 1 - 0.80 = 0.2$$
Alternatively: $(F \cup B)^c = F^c \cap B^c$ (*De Morgan's Law*):
$$\therefore P(F \cup B)^c = P(F^c \cap B^c) = P(F^c).P(B^c)$$
$$= 0.4 \times 0.5 = 0.2$$

EXAMPLE 2.12
Let A and B be two events such that $P(A) = 0.3$, $P(B) = 0.6$ and $P(A \cap B) = 0.1$.

(a) Compute P(A∪B).   (b) If A and B are mutually exclusive, find P(A∪B).

SOLUTION:

(a) $P(A∪B) = P(A) + P(B) – P(A∩B)$
$$= 0.3 + 0.6 – 0.1$$
$$P(A∪B) = 0.8$$

(b) If A and B are mutually exclusive, then
$A ∩ B = ∅$ and $P(A ∩ B) = 0$
∴ $P(A∪B) = P(A) + P(B)$
$$= 0.3 + 0.6$$
$$P(A∪B) = 0.9$$

EXAMPLE 2.13

Let A and B be two events such that $P(A) = 0.4$, $P(A∩B) = 0.3$ and $P(A∪B) = 0.6$ compute (a) P(B)   (b) $P(A^c ∪ B^c)$

SOLUTION:

(a) $P(A∪B) = P(A) + P(B) – P(A∩B)$
$$0.6 = 0.4 + P(B) – 0.3$$
∴ $P(B) = 0.5$
(b) $P(A^c ∪ B^c) = P(A ∩ B)^c = 1 – P(A ∩ B) = 1 – 0.3 = 0.7$

EXAMPLE 2.14

Let A and B be two mutually exclusive events such that,
$P(A) = 0.5$ and $P(B) = 0.1$. Compute $P(A ∪ B)$.

SOLUTION:

$P(A∪B) = P(A) + P(B)$
$$= 0.5 + 0.1 = 0.6$$
Can the student explain why in Example above P(A∪B) = 0.6?

2.5.5 Theorem: If A, B and C are any three events, then
$$P(A ∪ B ∪ C) = P(A) + P(B) + P(C) – P(A ∩ B) – P(A ∩ C)$$
$$– P(B ∩ C) + P(A ∩ B ∩ C)$$

Proof

Hint: Let $A \cup B = D$ and use property 3 to calculate $P(D \cup C)$

NOTE: An extension to the above Theorem is given as follows:-
Let, $A_1, A_2, \ldots, A_k$ be any k events. Then,

$$P(A_1 \cup A_2 \cup \ldots \cup A_k) = \sum_{i=1}^{k} P(A_i) - \sum_{i<j-2}^{k} P(A_i \cap A_j) + \sum_{i<j<r-3}^{k} P(A_i \cap A_j \cap A_r) + \ldots + (-1)^{k-1} P(A_1 \cap A_2 \cap \ldots \cap A_k).$$

EXAMPLE 2.15

Suppose that A, B and C are events such that
$$P(A) = P(B) = P(C) = \frac{1}{4}, \quad P(A \cap B) = P(C \cap B) = 0, \text{ and } P(A \cap C) = \frac{1}{8}.$$

Calculate the probability that at least one of the events A, B or C occurs.

SOLUTION:

$$P(A \cup B \cup C) = P(A) + P(B) + P(C) - P(A \cap B) - P(B \cap C) - P(A \cap C) + P(A \cap B \cap C)$$

We are given that $P(A \cap B) = 0$, implying that $A \cap B = \phi$.
Also $P(C \cap B) = 0$ implying that $C \cap B = \emptyset$.

$\therefore$ We can write

$A \cap B \cap C = (A \cap B) \cap C = \phi \cap B = \phi$ $\quad \ldots \ldots \ldots (2.5.5.1)$
Also $A \cap B \cap C = A \cap (B \cap C) = A \cap \phi = \emptyset$ $\quad \ldots \ldots (2.5.5.2)$
Equations 2.5.5.1 and 2.5.5.2 imply that $\quad P(A \cap B \cap C) = P(\phi) = 0$

Hence; the probability that at least one of the events A, B or C occurs, i.e. P(AUBUC) is given by

$$P(A \cup B \cup C) = P(A) + P(B) + P(C) - P(A \cap B) - P(A \cap C)$$
$$- P(B \cap C) + P(A \cap B \cap C)$$
$$= \frac{1}{4} + \frac{1}{4} + \frac{1}{4} + -0 - \frac{1}{8} - 0 + 0 = \frac{5}{8}$$

$$P(A \cup B \cup C) = \frac{5}{8}$$

2.5.6 Theorem: If A and B are two events such that A ⊆ B, then P(A) ≤ P(B).

Proof
$$B = A \cup (B \cap A^c) \Rightarrow$$
$$P(B) = P\{A \cup (B \cap A^c)\} \Rightarrow$$
$$P(B) = P(A) + P(B \cap A^c)$$

Since *A and* $(B \cap A^c)$ are mutually exclusive.

Hence $P(B) \geq P(A)$

Since $P(B \cap A^c) \geq 0$

2.5.7. Definition: Two events A and B are said to be equally likely if $P(A) = P(B)$. Event A is said to be more likely than B if $P(A) > P(B)$

2.5.8. Theorem. (Finite Sample Theorem)
If all the outcomes of a random experiment consist of *n* equally likely simple events, then the probability of an event E made up of *k* simple events is

$$P(E) = \frac{k}{n}$$

Proof
Let S be the sample space and *p* the probability of a simple event. Then since all the simple events are mutually exclusive

$$P(S) = \underbrace{p + p + \cdots + p}_{n \, times} \Rightarrow$$

$$1 = np$$

from which $p = \dfrac{1}{n}$

Next

$$P(E) = \underbrace{p + p + \cdots + p}_{k \, times} \Rightarrow$$

$$P(E) = kp$$

Therefore $P(E) = k\left(\dfrac{1}{n}\right)$

2.6. Conditional Probability

The simplest way to illustrate the idea of conditional probability is by means of an example.

EXAMPLE 2.16

Suppose that a packet contains 6 marbles; 2 being white and 4 yellow, and are marked 1 to 6 respectively as shown below: (W = white marble, Y = yellow marble).

```
1   2   3   4  5  6
W   W   Y   Y  Y  Y
```

Furthermore, let the marbles be classified according to their conditions (O = old marble, N = new marble), as shown in the sample space S_1 below.

$$S_1 = \begin{Bmatrix} 1 & 2 & 3 & 4 & 5 & 6 \\ W & W & Y & Y & Y & Y \\ N & 0 & N & N & O & N \end{Bmatrix}$$

Let a number (representing a marble) be chosen at random, and let A be the event: a white marble is selected. Then $P(A) = \frac{2}{6}$ (since there are only two white marbles in the sample space S_1). Also let B be the event: a new marble was selected, the event

$P(A \cap B) = \frac{1}{6}$ (Since there is only one marble that is both white and new). Note that, both the probabilities $P(A)$ and $P(A \cap B)$ are computed with respect to the sample space S_1.

There is now an important question to be asked. Suppose that a marble is chosen at random and is observed to be white, what is the probability that the chosen marble is new, i.e. what is the value of $P(B/A)$? In this situation, we have a reduced sample space

$$S_2 = \begin{Bmatrix} \overset{1}{W} & \overset{2}{W} \\ N & O \end{Bmatrix}$$

and the probability $P(B/A)$ is computed with respect to this reduced sample space S_2 (not with respect to the original sample space S_1). Of the two white marbles shown in the sample space S_2, only one of them is new; hence $P(B/A) = \frac{1}{2}$.

Previously, we computed $P(A \cap B) = \frac{1}{6}$

and $P(A) = \frac{2}{6}$, giving the ratio $\frac{P(A \cap B)}{P(B)} = \frac{\frac{1}{6}}{\frac{2}{6}} = \frac{1}{2}$ which is the same value computed for $P(B/A)$.

Hence, we can now write. $P(B/A) = \frac{P(A \cap B)}{P(A)}$

called the conditional probability of event B occurring given that event A has already occurred.

EXAMPLE 2.17

In a family of 3 children, suppose it is known that A (fewer than two girls) has occurred. What is the probability that H (all of the same sex) has occurred? That is, if we imagine many repetitions of this experiment and consider just those cases in which A has occurred, how often will H occur i.e. P(H/A)?

SOLUTION:

The sample space S for a family of 3 children is given as
S = {GGG, GGB, GBG, GBB, BGB, BBG, BGG, BBB)
Where, B = Boy and G = Girl.
P(A) = P(fewer than two girls) = P(0 or 1 girl).

With respect to S, $P(A) = \frac{4}{8} = \frac{1}{2}$. Similarly, $P(H \cap A) = \frac{1}{8}$.

The reduced sample space, A = (GBB, BGB, BBG, BBB). With respect to A,

$$P(H/A) \;=\; P(BBB) \;=\; \frac{1}{4}. \quad \text{But} \quad \frac{P(H \cap A)}{P(A)} = \frac{\frac{1}{8}}{\frac{4}{8}} = \frac{1}{4} = P(H/A)$$

EXAMPLE 2.18

Let us consider the difference between choosing an item at random from a lot with or without replacement. Suppose that we choose two marbles from a box containing 3 black and 2 white marbles in Figure 2.3 below:–
(a) with replacement (b) without replacement.

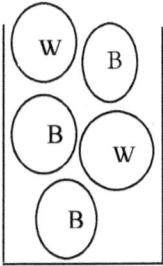

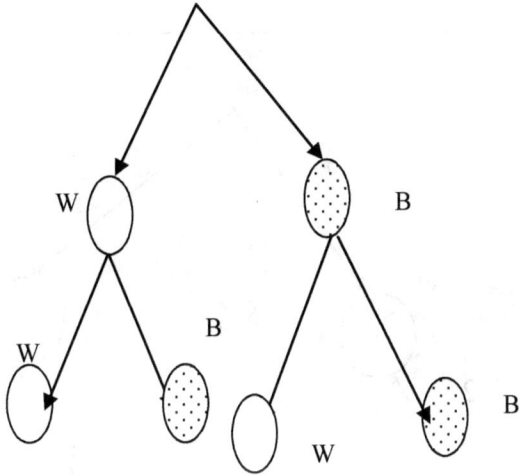

Figure 2.3

Figure 2.4

The selection can be explained in the various ways shown in Figure 2.4 above: There are four possible outcomes:

First, white, then white or (ii) white, then black (iii) black, then white or (iv) black then black

If the chosen marble is replaced, $P(W) = \dfrac{2}{5}$ and $P(B) = \dfrac{3}{5}$. For each time we choose a marble from the box there are always 2 white and 3 black marbles among the total of 5 marbles. The probabilities for the first and second selections are shown in the tree diagram Figure 2.5 below:

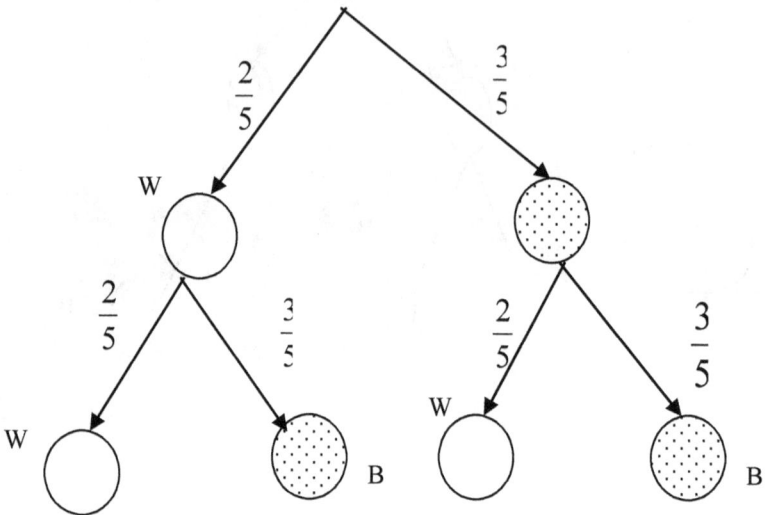

Figure 2.5

However, if we choose a marble without replacing it, the results are not quite immediate. The probabilities for the first and second selections are shown in the tree diagram, Figure 2.6 below:

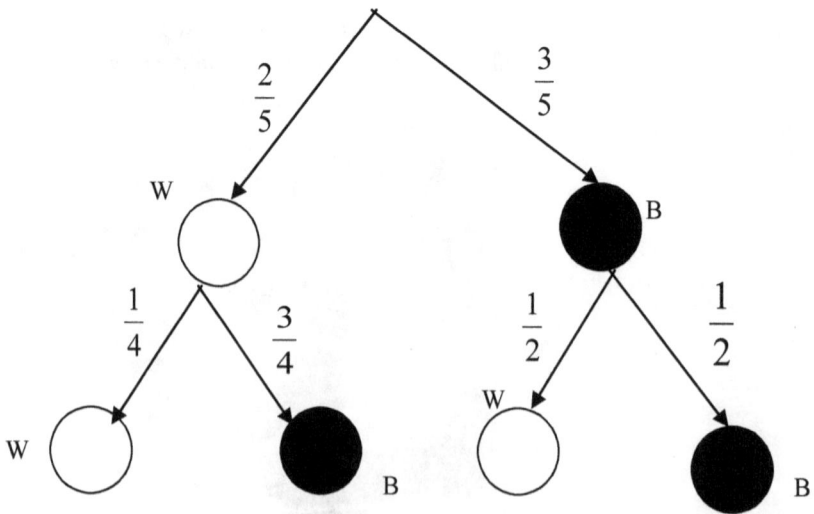

Figure 2.6

It is seen from Figures 2.5 and 2.6 above, that the probabilities of the first selection (with or without replacement) are the same, but what about the second? In order to compute this probability we should know the composition of the lot at the time the second marble is chosen. That is, we should know whether a particular event did or did not occur in the first selection. This example indicates the need to introduce the concept of conditional probability.

Let A and B be two events associated with an experiment E. We denote by P(B/A) the conditional probability of the event B, given that A has occurred.

We define the following two events:
$$A = \{\text{the first marble is black}\}$$
$$B = \{\text{the second marble is black}\}$$

In Figure 2.6 above (selection without replacement). $P(A) = \dfrac{3}{5}$

but $P(B/A) = \dfrac{2}{4} = \dfrac{1}{2}$. For if A has occurred, then on the second drawing there are only 4 marbles left, 2 of which are black.

EXAMPLE 2.19

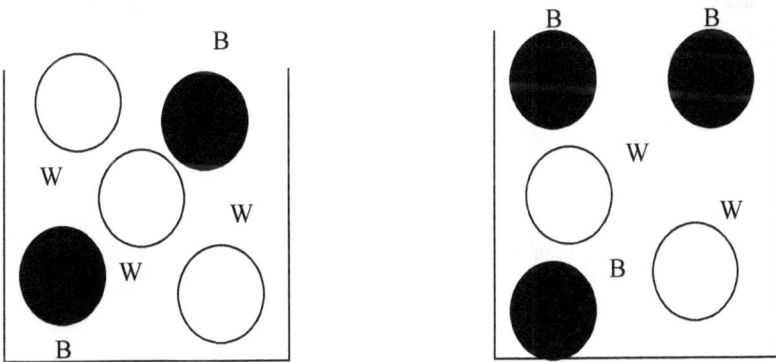

Box (1)
Figure 2.7 (i)

Box (2)
Figure 2.7 (ii)

Consider both boxes (1 and 2) shown in Figure 2.7 above. Box 1 consists of 5 marbles; 3 of which are white and two black, whilst box 2 contains 3 black and 2 white marbles. One chooses at random, a marble from box 1 and places it in box 2 without observing the marble's colour. Next, one chooses a marble from Box 2. One is interested in the colour of the marble in the second selection. The probabilities for the first and second selections are shown in the tree diagram, Figure 2.8 below: could the reader reason this out?

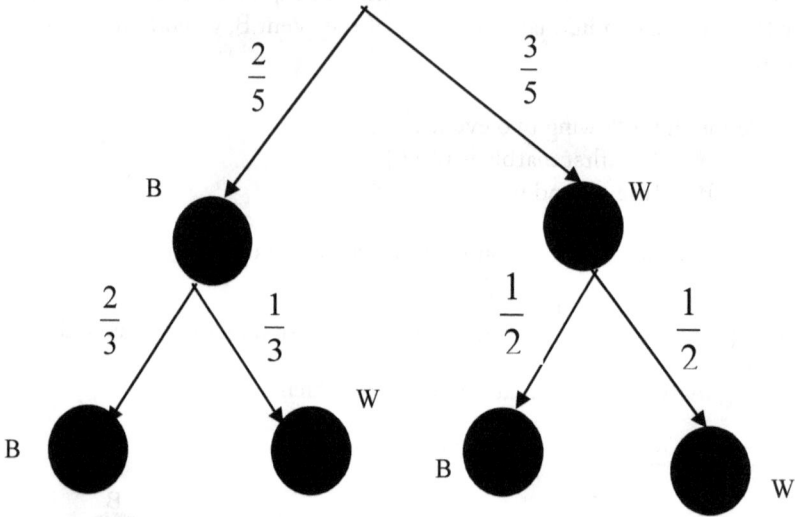

Figure 2.8

2.6.1 Definition (Conditional probability)

Suppose S is the sample space for an experiment with events A and B. Then for experiments with equally likely outcomes, the conditional probability of A given B is the fraction of outcomes in B that are also in A. Since only outcomes in B are taken into consideration, in effect, B becomes the sample space, so we have

$$P(A/B) = \frac{n(events\ that\ A\ occcurs\ given\ that\ B\ occurs)}{n(event\ B)}$$

However, the event that A occurs given that B occurs is nothing but $A \cap B$. Hence

$$P(A/B) = \frac{n(A \cap B)}{n(B)} = \frac{P(A \cap B)}{P(B)}, P(B) \geq 0$$

Let A and B be two events. The conditional probability that the event B occurs, given that A has occurred: $P(B/A) = \dfrac{P(A \cap B)}{P(A)}$

Provided that P(A) > 0.   ......2.6.1.1

The most important consequence of the definition 2.6.1 of conditional probability is obtained by writing it in the following form:

$$P(A \cap B) \; = \; P(A).P(B/A)$$
Or equivalently $P(A \cap B) \; = \; P(B).P(A/B)$ 2.6.1.2

This is sometimes known as the Multiplication Theorem of probability. We may apply this theorem to compute the probability of the simultaneous occurrence of two events A and B. As before, in Example 2.18 of section 2.6 above, we define the events A and B as follows:

> A = (the first marble is black)
> B = (the second marble is black)

We require $P(A \cap B)$ which we may compute (see Figure 2.8 of section 2.6 above), according to the above equation, 2.6.1.2,

$P(A \cap B) = P(B/A).P(A)$. But $P(B/A) = \dfrac{1}{2}$,

While $P(A) = \dfrac{3}{5}$. Hence $\quad P(A \cap B) = \dfrac{1}{2} \times \dfrac{3}{5} = \dfrac{3}{10}$

NOTE: The above multiplication Theorem, 2.6.1.2, may be generalized to more than two events in the following way:

$$P(A_1 \cap A_2 \cap ....\cap A_n)$$
$$= P(A_1)P(A_2/A_1)P(A_3/A_1, A_2) ... P(A_n /A_1,,,, A_{n-1})$$

2.7. Partitions

A collection of several events is defined as mutually exclusive if there is no overlap whatsoever, i.e. if no outcome belongs to more than one event. The mutually exclusive events cover the whole of the sample space S.

EXAMPLE 2.20

Let E be the experiment of tossing a Ludo die once. The sample S associated with E is given as S = {1, 2, 3, 4, 5, 6}.

Consider the following events A_1, A_2, A_3 and B_1, B_2 given below:
A_1= (at most 3 shows up)= (1,2,3),
A_2= (a four shows up)= (4)
A_3= (at least 5 shows up)= (5,6)
B_1= (at most 3 shows up)= (1,2,3)
B_2= (at least 3 shows up)= (3,4,5,6).
A_1, A_2 and A_3 are mutually exclusive and their union $A_1 \cup A_2 \cup A_3$ cover the entire sample space.
The events B_1 and B_2 are not mutually exclusive since there is an overlap in the outcome '3', even though their union, $B_1 \cup B_2 = S$, covers the entire sample space S. Thus, the events A_1, A_2 and A_3 would represent a partition of the sample space S, whereas B_1 and B_2 would not.

2.7.1 Definition (Partition of a sample space)

A partition of a sample space S is a collection of mutually exclusive events,
A_1, A_2, A_3, ...A_k, whose union is the whole of S. That is,

(i) $A_i \cap A_j = \emptyset$, *for all* $i \neq j$
(ii) $\cup_{i=1}^{k} = S$
(iii) $P(A_i) \geq 0$ for all i.

In words: when the experiment E is performed one and only one of the events A_i occurs.
The idea of partitions is quite useful in describing events associated with a sample space S. In Figure 2.9 below; (a) represents a sample space S associated with an experiment E, (b) represents an event D associated with the above sample space S, and (C) shows events A_1, A_2, ... A_7 as a partition of S; subdividing D into 5 non-overlapping regions, such that we can now write:

$$D = (D \cap A_1) \cup (D \cap A_2 \cup (D \cap A_3) \cup (D \cap A_4) \cup (D \cap A_5)$$

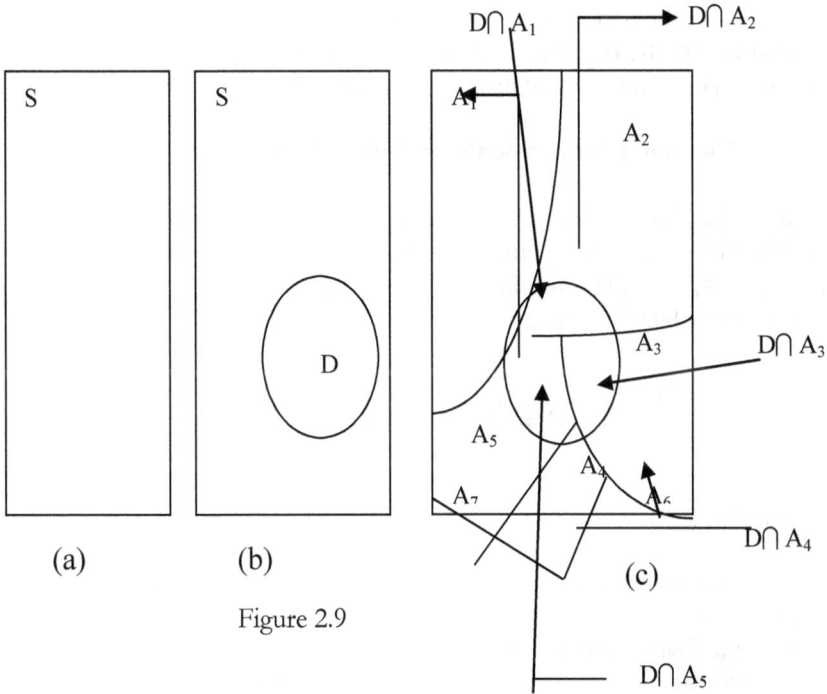

Figure 2.9

We may apply the addition property of mutually exclusive events (2.7.1) and write

$$P(D) = P(D \cap A_1) + P(D \cap A_2) + P(D \cap A_3)) + P(D \cap A_4) + P(D \cap A_5)$$

However, each term $P(D \cap A_i)$ *may be expressed as* $P(A_i)P(D / A_i)$ and we obtain the Theorem of total probability:

$$P(D) = P(A_1)P(D/A_1) + P(A_2)P(D/A_2) + P(A_3)P(D/A_3) + P(A_4)P(D/A_4) + P(A_5)P(D / A_5)$$

The above multiplication theorem may be generalized in the following way:

$$P(D) = P(A_1)P(D/A_1) + P(A_2)P(D/A_2) + P(A_3)P(D/A_3) + \cdots$$

Of course, some of the sets $D \cap A_i$ may be empty, but this does not invalidate the above decomposition of D. The important point is that all the events $D \cap A_1, D \cap A_2, \ldots D \cap A_k$ are pairwise mutually exclusive. Hence we apply the addition property for mutually exclusive events.

2.7.2. Theorem (Total Probability Rule (TPR)

Suppose that the sample space of a random experiment can be partitioned into a set of "k" mutually exclusive and exhaustive events $A_1, A_2, \ldots, A_k$. Let D be another event in the sample space with a strictly positive probability of taking place. Then

$$P(D) = \sum_{i=1}^{k} P(D/A_i) * P(A_i)$$

EXAMPLE 2.21

We revisit Figure 2.8 of section 2.6 where we defined the following events:
  A = {the first marble is black}
  B = {the second marble is black}, the selection is without replacement.

We may now compute P(B), using the following routes:-

Either (i) the first marble is white and the second is black
or   (ii) the first marble is black and the second is also black
      Hence P(B) $= P(A) P(B/A) + P(A^c)P(B / A^c)$
      Where $P(A^c) = 1 - P(A)$.

Hence Figure 2.6 gives
$$P(B) = \frac{2}{5} \bullet \frac{3}{4} + \frac{3}{5} \bullet \frac{1}{2} = \frac{12}{20} = \frac{3}{5}$$

This result may be a bit startling, particularly if we recall that in Figure 2.5 of Example 2.18 above, we found that $P(B) = \dfrac{3}{5}$ when we chose the marble with replacement.

2.8. Bayes Theorem

In this section, we introduce an important branch of Statistics called Bayesian analysis, which is based on an imaginative use of definition 2.6.1 above. We shall deduce a formula called Baye's Theorem; also called the formula for the probability of causes.

Certain causes, say $C_1, C_2, C_3, \dots C_k$ (e.g. class of Education, types of diseases, types of machines used to manufacture items in a factory, types of Jars containing different types of sweets etc.) have prior probabilities, denoted by $P(C_k)$. That is, we have knowledge of the proportions of each category (or group) that make up the population under consideration.

Suppose, for example, that we decide to focus on the population "class of Education"; certain proportions of this population may have no formal education. Some may have only primary education, some have secondary education, whilst others may have tertiary education. The various classes or groups may, consequently, be employed (or unemployed) depending on their educational backgrounds (which are probably the main causes of employment). In other words, these causes produce an effect E (unemployment), not with certainty but with conditional probabilities $P(E/C_k)$. We eventually calculate $P(E/C_k)$, the posteriori probability of a cause, once the effect has been observed. In other words:

<u>If we are given</u>   <u>We can Deduce</u>
(i) Probability of Cause $P(C_k)$ Probability of a cause having
observed the effect: $P(E/C_k)$)
(ii) Probability of the effect produced
by a given cause $P(E/C_k)$

2.8.1 Theorem: Let $C_1, C_2, C_3, \dots C_k$ be a partition of the sample space S and let E be an event associated with S. Applying the definition of conditional probability, we may write:

$$P(C/E) = \frac{P(C \cap E)}{P(E)} = \frac{P(C)P(E/C)}{P(C)P(E/C) + P(C^c)P(E/C^c)}$$

where C^c is the complement of C.
More generally, we can write

$$P(C_k/E) = \frac{P(C_k)P(E \ / \ C_k)}{\sum_{k=1}^{n} P(C_k)P(E \ / \ C_k)}$$

This is known as Bayes' Theorem.

EXAMPLE 2.22

A company employs 100 persons: 75 men and 25 women. The finance department provides data for 12% of the men and 20% of the women. If a name is chosen at random from the finance department, what is the probability that it refers to (i) a man? (ii) a woman?

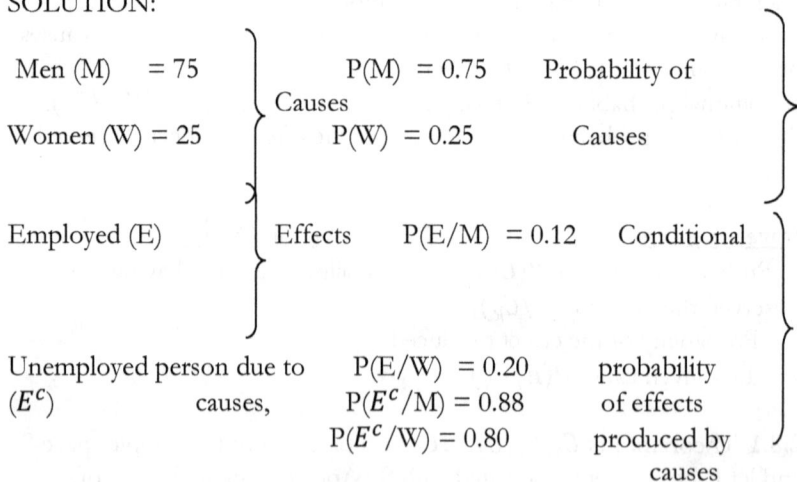

The above probabilities are illustrated in the tree diagram of figure 2.10 below:

SOLUTION:

Men (M)   = 75   P(M) = 0.75   Probability of
                   Causes
Women (W) = 25   P(W) = 0.25   Causes

Employed (E)   Effects   P(E/M) = 0.12   Conditional

Unemployed person due to   P(E/W) = 0.20   probability
(E^c) causes, $P(E^c/M)$ = 0.88   of effects
                              $P(E^c/W)$ = 0.80   produced by
                                                      causes

The above probabilities are illustrated in the tree diagram of figure 2.10 below:-

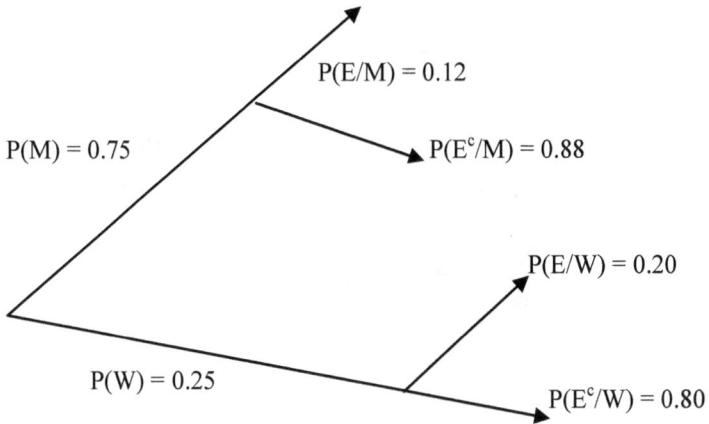

Figure 2.10

P(M/E)   Posterior
P(W/E)   Probabilities of causes,
P(M/E^c)   once effect has
P(W/E^c)   been observed.

We require *Posterior* probabilities: P(M/E) and P(W/E) respectively.

$$P(M/E) = \frac{P(M).P(E/M)}{P(E)} = \frac{P(M).P(E/M)}{P(M)P(E/M) + P(W)P(E/W)}$$

$$= \frac{0.75 \; x \; 0.12}{(0.75 \; x \; 0.12) + (0.25 \; x \; 0.20)}$$

$$= \frac{0.09}{0.14} = \frac{9}{14}$$

(b)

$$P(W/E) = \frac{P(W).P(E/W)}{P(E)} = \frac{P(W).P(E/W)}{P(W)P(E/W) + P(M)P(E/M)}$$

$$= \frac{0.05}{0.14} = \frac{5}{14}$$

<u>Note</u>: Could the reader compute *posterior* probabilities $P(M / E^c)$ and $P(W / E^c)$?

EXAMPLE 2.23

In a certain country, it rains 40% of the days and shines 60% of the days. A barometer manufacturer, in testing his instrument in the lab, has found that it sometimes errs: on rainy days it erroneously predicts "shine" 10% of the time, and on shiny days it erroneously predicts "rain" 30% of the time.
In predicting tomorrow's weather before looking at the barometer, the (prior) chance of rain is 40%. After looking at the barometer and seeing it predicts "rain", what is the (posterior chance that it will: (a) rain? (b) shine?

SOLUTION:

Rain (R_B) ⎫ Effects
        ⎬ due to
Shine (S_B) ⎭ Causes,
effects

$P(R_B/S) = 0.90$ Conditional
$P(S_B/S) = 0.10$ probabilities of

$P(R_B^c/S) = 0.30$ produced by
$P(S_B^c/S) = 0.70$ causes

The above probabilities are illustrated in the tree diagram of figure 2.11 below:

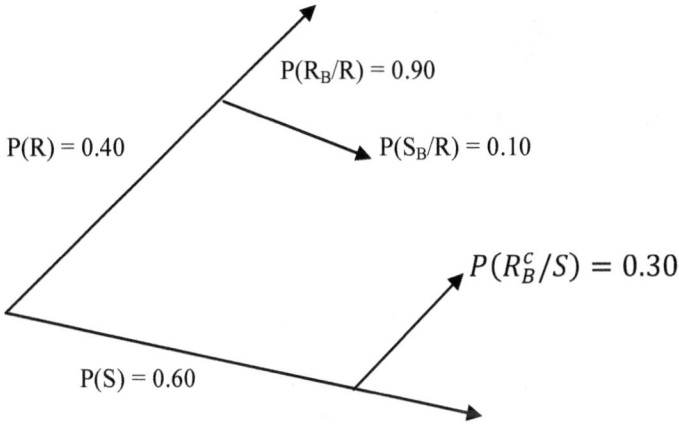

Figure 2.11

$P(R/R_B)$
$P(R/S_B)$ posterior probabilities of causes
$P(S/R_B^c)$ once effect has been observed.
$P(S/S_B^c)$

We require posterior probabilities:

(a) $P(R/R_B)$ (b) $P(S/R_B^c)$

(a) $\quad P(R/R_B) = \dfrac{P(R) . P(R_B/R)}{P(R_B)} = \dfrac{P(R) . P(R_B/R)}{P(R) . P(R_B/R) + P(S) P(R_B/S)}$

$\qquad = \dfrac{0.40 \ x \ 0.90}{(0.40 \ x \ 0.90) + (0.60 \ x \ 0.30)}$

$\qquad = \dfrac{0.36}{0.54} = \dfrac{36}{54} = 0.67$

(b) $P(S/R_B^c)=\dfrac{P(S).P(R_B^c/S)}{P(R_B^c)}=\dfrac{0.60\times0.3}{0.54}=0.33$

2.8.1.1 Assignment 2.1: Could the reader compute *posterior* probabilities $P(R/S_B)$ and $P(S/S_B^c)$?

2.9. Independent Events

We have considered events that could not occur simultaneously (i.e. mutually exclusive events). For mutually exclusive events, the occurrence of one event precludes the occurrence of the other. However, there are situations in which knowing that some event did occur has no bearing whatsoever on the occurrence or non-occurrence of the other. An example illustrates:

Two events, A and B that are related in such a way that the occurrence of either one has no bearing on the occurrence of other are said to be independent. For instance, if you toss a coin repeatedly, the result of one toss does not affect the result of any other toss. If you roll a die repeatedly, the results are independent of one another.

Two events A and B are said to be independent if knowledge of the occurrence of A in no way influences the probability of the occurrence of B and vice versa.

EXAMPLE 2.24

Suppose a fair die is tossed twice.
Define the events:
A = {the first die shows an odd number}
B = {the second die shows a 1 or 2}
It will be clear in the illustrations below that events A and B are totally unrelated. Knowing the result B yields no information about the occurrence of A.

The probabilities for the events A, B and A∩B are computed below:

$P(A) = P$(odd number on first die)

$$P(A) = \begin{cases} (1,1), & (1,2), & (1,3), & (1,4), & (1,5), & (1,6) \\ (3,1), & (3,2), & (3,3), & (3,4), & (3,5) & (3,6) \\ (5,1), & (5,2), & (5,3), & (5,4), & (5,5), & (5,6) \end{cases} = \frac{18}{36} = \frac{1}{2}$$

P(B) = P(1 or 2 on second die).

$$P(B) = \begin{cases} (1,1), & (1,2), & (2,1), & (2,2), & (3,1), & (3,2) \\ (4,1), & (4,2), & (5,1), & (5,2), & (6,1), & (6,2) \end{cases} = \frac{12}{36} = \frac{1}{3}$$

P(A∩B) = P(odd number on first die and (1 or 2 on second))

$$P(A \cap B) = \{(1,1), \quad (1,2), \quad (3,1), \quad (3,2), \quad (5,1) \quad (5,2)\} = \frac{6}{36} = \frac{1}{6}$$

Hence, $P(A/B) = \dfrac{P(A \cap B)}{P(B)} = \dfrac{\frac{1}{6}}{\frac{1}{3}} = \dfrac{1}{2}$

We note that, the unconditional probability $P(A) = \dfrac{1}{2}$ is equal to the

conditional probability $P(A/B) = \dfrac{1}{2}$.

Similarly, $P(B/A) = \dfrac{P(B \cap A)}{P(A)} = \dfrac{P(A \cap B)}{P(A)} = \dfrac{\frac{1}{6}}{\frac{1}{2}} = \dfrac{1}{3} = P(B)$

We are therefore tempted to conjecture that P(A/B) = P(A) and
$P(B/A) = P(B)$ and
may write,

$$P(A \cap B) = P(B) . P(A/B) = P(B) . P(A)$$
and $\quad P(A \cap B) = P(A) . P(B/A) = P(A) . P(B)$

Provided that neither P(A) or P(B) equals zero, the unconditional
probabilities are equal to the conditional probabilities if and only if
$P(A \cap B) = P(A) . P(B)$. [If either P(A) or P(B) equals zero, this
definition is still valid.

2.9.1 Definition (Independent Events)

Two events A and B are statistically independent if and only if

$P(A/B) = P(A)$ and $P(B/A) = P(B)$.

Hence $P(A \cap B) = P(A).P(B)$ for two independent events.

EXAMPLE 2.25

(a) Suppose A, B and C are three independent events: A is independent of B and also

independent of C, but B and C are mutually exclusive. If P(A) = 0.1, P(B) = 0.2 and

P(C) = 0.3. Compute P(C/(A∪B)).

(b) Suppose that A and B are independent events with P(A) = 0.6 and P(B) = 0.2.

What is

(i) $P(A/B)$
(ii) $P(A \cap B)$
(iii) $P(A \cup B)$?

(c)   A fair coin is tossed three times. Compute the probability that heads show up in all three tosses.

(d)   A box contains 4 bad and 6 good apples. Two are drawn out together. One of them is tested and found to be good. What is the probability that the other one is also good? (a) with replacement (b) without replacement

SOLUTION:

(a) $P(C/(A \cup B)) = \frac{P(C \cap (A \cup B))}{P(A \cup B)}$, by definition of conditional probability.

But $P(A \cup B) = P(A) + P(B) - P(A \cap B)$

Since A and B are independent: $P(A \cap B) = P(A).P(B)$

$\therefore P(A \cup B) = P(A) + P(B) - P(A)\,P(B)$

$= 0.1 + 0.2 - (0.1 \times 0.2) = 0.28$

$C \cap (A \cup B) = (C \cap A) \cup (C \cap B)$

But B and C are mutually exclusive,

$\therefore C \cap B = \phi$

$\therefore C \cap (A \cup B) = (C \cap A) \cup \phi = C \cap A$

$P(C \cap (A \cup B)) = P(C \cap A) = P(C).P(A);$
Since A and C are independent $= 0.3 \times 0.1 = 0.03$
$\therefore P(C/(A \cup B)) = \dfrac{0.03}{0.28} = \dfrac{3}{28}$

(b)(i) If A and B are independent, $P(A \cap B) = P(A).P(B)$,
hence, $P(A/B) = P(A)$ [see section 2.9 above].

By definition of conditional probability

$$P(A/B) = \frac{P(A \cap B)}{P(B)} = \frac{P(A).P(B)}{P(B)} = P(A)$$

$\therefore P(A/B) = P(A) = 0.6$
(ii) $P(A \cap B) = P(A).P(B)$ since A and B are independent
$= 0.6 \times 0.2 = 0.12$
(iii) $P(A \cup B) = P(A) + P(B) - P(A \cap B)$, since A and B are
independent
$= P(A) + P(B) - P(A).P(B)$
$= 0.6 + 0.2 - 2 (0.6 \times 0.2) = 0.68$

(c) Probability that heads show up in all three tosses of a fair coin:
$=$ P[Head on first toss and Head second toss and Head third toss]

Define the events: A $=$ {Head on first toss}; $P(A) = \dfrac{1}{2}$

B $=$ {Head on second toss} ; $P(B) = \dfrac{1}{2}$

C $=$ {Head on third toss}; $P(C) = \dfrac{1}{2}$

Then P{Head on all three tosses} $= P\{A \cap B \cap C\}$
(Since the tosses are independent of each other)
$$P(A \cap B \cap C) = P(A).P(B).P(C) = \frac{1}{2} \times \frac{1}{2} \times \frac{1}{2} = \frac{1}{8}$$
(d) The tree diagrams for the two successive selections are shown in
figure 2.12
below:
Denote G $=$ (Good apple)
B $=$ (Bad apple)

(i)   With replacement

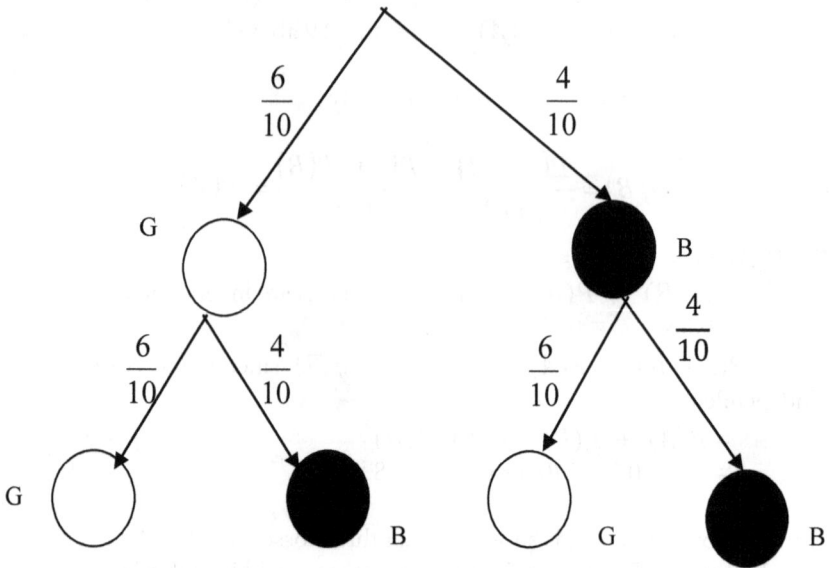

Figure 2.12

P (first apple is good and second apple is good)

$= P(G \cap G)$ = [*selections are independent*]

$= P(G).P(G)$ $\quad = \dfrac{6}{10} \cdot \dfrac{6}{10} = \dfrac{36}{100} = 0.36$

(ii)Without replacement

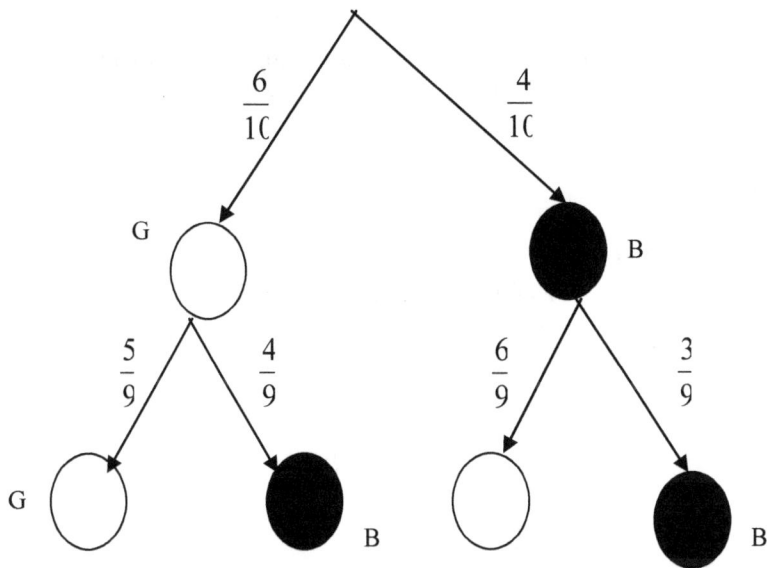

Figure 2.13

(first apple is good and second apple is good)
= P(G∩G) (selections are independent) = $P(G).P(G) = \frac{6}{10} \times \frac{5}{9} = \frac{30}{90} = \frac{1}{3}$

2.9.2 Mutual Independence

We say that three events A, B, and C are jointly or mutually independent if, and, only
if the following conditions hold:

$$P(A \cap B) = P(A) * P(B)$$
$$P(A \cap C) = P(A) * P(C)$$
$$P(B \cap C) = P(B) * P(C)$$

And $\qquad P(A \cap B \cap C) = P(A) * P(B) * P(C)$

Example 2.2.6

A ball is drawn at random from an open box containing four balls numbered 1, 2, 3, 4. Let A={1,2}, B = {1,3}, C = {1,4} : Examine whether or not A, B, C are independent.

Solution

For events A, B, C to be independent then
$$P(A \cap B) = P(A) * P(B)$$
$$P(A \cap C) = P(A) * P(C)$$
$$P(B \cap C) = P(B) * P(C)$$

and
$$P(A \cap B \cap C) = P(A) * P(B) * P(C)$$

Now

$A \cap B = \{1\}$ so that $P(A \cap B) = \frac{1}{4}$

$A \cap C = \{1\}$ so that $P(A \cap C) = \frac{1}{4}$

$B \cap C = \{1\}$ so that $P(B \cap C) = \frac{1}{4}$

Also $P(A) = \frac{2}{4}, P(B) = \frac{2}{4}$

and so $P(A) * P(B) = \frac{1}{4} = P(A \cap B)$

Similarly, $P(A) = \frac{2}{4}, P(C) = \frac{2}{4}$

$$P(A) * P(C) = \frac{1}{4} = P(A \cap C)$$

$$P(B) = \frac{2}{4}, P(C) = \frac{2}{4}$$

$$P(B) * P(C) = \frac{1}{4} = P(B \cap C)$$

Furthermore, $A \cap B \cap C = \{1\} \Rightarrow P(A \cap B \cap C) = \frac{1}{4}$

However,
$$P(A) * P(B) * P(C) = \frac{1}{2} * \frac{1}{2} * \frac{1}{2} = \frac{1}{8} \neq P(A \cap B \cap C)$$
Hence A, B, C are pairwise independent but not jointly or mutually independent.

2.9.3. Theorem. If A and B are independent events, then
a: A^c and B^c are independent
b: A^c and B are independent
   c: A and B^c are independent

Proof:
a:
$$P(A^c \cap B^c) = 1 - P\{(A \cup B)^c\}$$
since $(A \cup B)^c = A^c \cap B^c$

$$P(A^c \cap B^c) = 1 - \{P(A) + P(B) - P(A \cap B)\}$$
$$= 1 - \{P(A) + P(B) - P(A) * P(B)\}$$
since A and B are independent.
$$P(A^c \cap B^c) = 1 - P(A) - P(B) + P(A) * P(B)$$
$$= 1 - P(A) - P(B)\{1 - P(A)\}$$
$$= \left(1 - P(A)\right)\left(1 - P(B)\right)$$
$$= P(A^c)P(B^c)$$
This implies A^c and B^c are independent.

b: To prove (b), you need to realize that $A^c \cap B = (A \cup B) - A$. Can you see that? You
   may use Venn diagram.
$$P(A^c \cap B) = P\{(A \cup B) - A\} = P(A \cup B) - P(A)$$
$$= P(A) + P(B) - P(A \cap B) - P(A)$$
$$= P(B) - P(A \cap B)$$
$$= P(B) - P(A) * P(B)$$
since A and B are independent.
$$= P(B)[1 - P(A)]$$
$$= P(B)P(A^c)$$

This implies A^c and B are independent. The independence of A and B^c can similarly be established by noting that $A \cap B^c = (A \cup B) - B$.

2.9.4. A comparison of Mutually Exclusive and Independent Events

Most times students find it difficult to distinguish `mutually exclusive' events from `independent' events. Two events A and B are mutually exclusive if $A \cap B = \emptyset$; that
is they cannot occur together. This is not necessarily the same as saying that A and
B are independent; in fact it is quite possible for dependent events to be mutually exclusive.

For instance considering a group of people consists of 10 single and 30 married men with 25 single and 5 married women. Let event A denote a single man and B, a single woman. What is the probability of selecting a single person?

For the above,

$$P(single\ man) = P(A) = \frac{10}{70}$$
$$P(single\ woman) = P(B) = \frac{25}{70}$$
$$P(single\ person) = P(A) + P(B) = \frac{10}{70} + \frac{25}{70}$$
$$= \frac{35}{70}$$

In this case a single person can either be a man or a woman. $P(A \cap B) = 0$. Both
events are mutually exclusive, but they are not independent, since $P(A/B) = 0$ (if
only a single person is selected and it is a woman, then it cannot be a man) so
that $P(A/B) \neq P(A)$.

DIAGNOSTIC TEST

D 2.1 (a) A, B, C are three mutually independent events, such that P(A) = 0.5, P(B) =
0.6 and P(C) = 0.7. What is the conditional probability that event A occurs
given that at least one of A, B and C occur?

(b) A and B are two events such that P(B) = 0.4, P(A∪B) = 0.6. Compute $P(A/B^c)$.

SOLUTION (a):

$$P(A/(A \cup B \cup C)) = \frac{P(A \cap (A \cup B \cup C))}{P(A \cup B \cup C)} = \frac{P(A)}{P(A \cup B \cup C)}$$

$P(A \cup B \cup C) = 1 - P(A^c \cap B^c \cap C^c)$ (since, A, B, C are independent)

$$= 1 - P(A^c) P(B^c) P(C^c)$$
$$= 1 - (0.5 \times 0.4 \times 0.3) = 0.94$$

$\therefore P(A/A \cup B \cup C)$ $= \dfrac{0.5}{0.94} = 0.5319$

Alternatively:

$$P(A \cup B \cup C) = P(A) + P(B) + P(C) - P(A)P(B) - P(A)P(C)$$
$$- P(B)P(C) + P(A)P(B)P(C) = 0.94$$

SOLUTION (b):

$$P(A/B^c) = \frac{P(A \cap B^c)}{P(B^c)}$$
$$= \frac{P(A \cup B) - P(B)}{P(B^c)}$$
$$= \frac{0.6 - 0.4}{0.6} = \frac{1}{3}$$

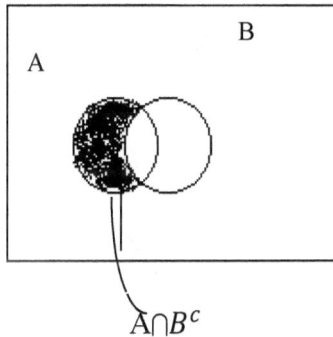

$A \cap B^c$

Figure 2.14

D2.2 Two baskets are taken on an outing:
    The first contains 2 ham and 5 cheese sandwiches; the second contains 6 ham
    and 3 cheese sandwiches. In the confusion of darkness, a basket is picked at
    random.

(a)   If a sandwich is drawn from the basket, what is the probability that it is a ham sandwich?

(b) If a second sandwich is drawn from the same basket, what is the conditional probability that this second sandwich will be (i) ham (ii) cheese, if the first is ham.

SOLUTION:
The probabilities for the various selections are shown in the branches of the tree diagram below. Probability of selecting a basket: $P(B_1) = P(B_2) = \dfrac{1}{2}$

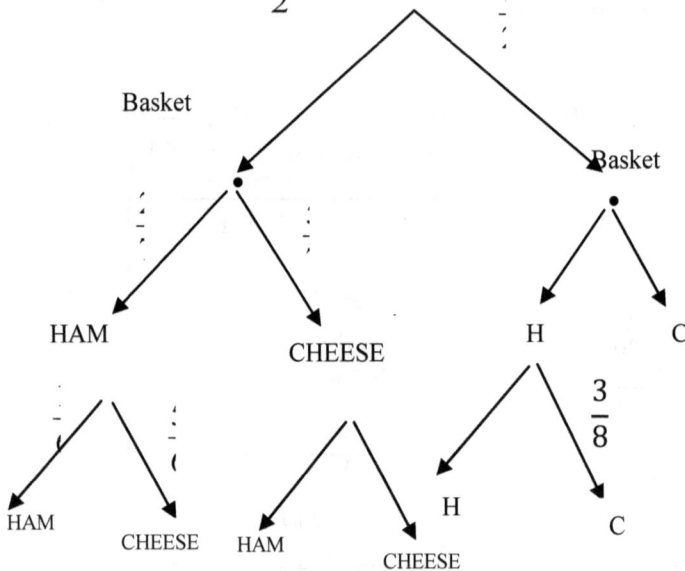

Figure 2.15

$$P(Ham) = \left(\frac{1}{2} \times \frac{2}{7}\right) + \left(\frac{1}{2} \times \frac{6}{9}\right) = \frac{1}{2}\left(\frac{20}{21}\right) = \frac{10}{21}$$

(a) (i) $$P(Ham/Ham) = \frac{1}{2}\left(\frac{2}{7} \times \frac{1}{6}\right) + \frac{1}{2}\left(\frac{6}{9} \times \frac{5}{8}\right) = \frac{117}{252}$$

(ii) $$P(Cheese/Ham) = \frac{1}{2}\left(\frac{5}{7} \times \frac{2}{6}\right) + \frac{1}{2}\left(\frac{3}{9} \times \frac{6}{8}\right) = \frac{123}{504}$$

Could the student compute the following probabilities?
(c) (i) P(Cheese) (ii) P(Cheese/Cheese) (iii) P(Ham/Cheese)?
D2.3 A bag contains 2 blue, 4 red and 6 white sockets. One chooses at random two sockets from the bag. What is the probability that they are of the same colour?

SOLUTION:
The probabilities for the various stages of selection are shown in the tree diagram of Figure 2.15 below:

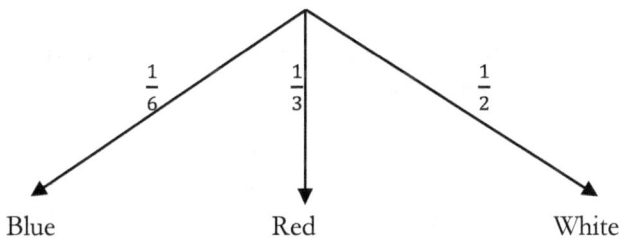

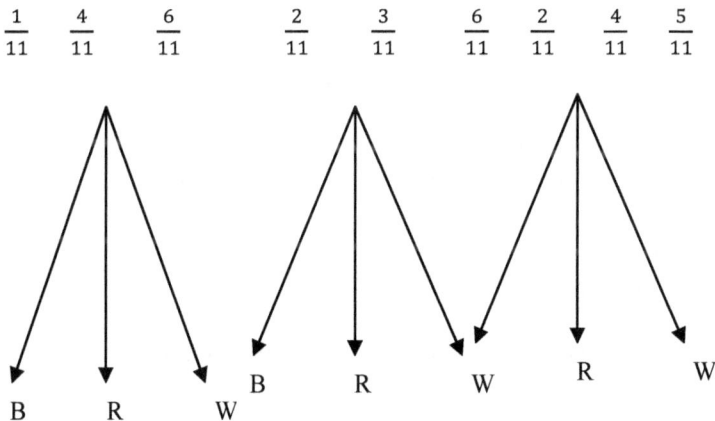

Figure 2.16

Required probability is:

$$\left(\frac{1}{6} \times \frac{1}{11}\right) + \left(\frac{1}{3} \times \frac{3}{11}\right) + \left(\frac{1}{2} \times \frac{5}{11}\right) = \frac{22}{66} = \frac{1}{3}$$

EXERCISE 2

2.1.   Consider an experiment of drawing two cards at random from a bag containing four cards marked with the integers 1 through 4.

(a)   Find the sample space S_1 of the experiment if the card is not replaced. The first number indicates the first number drawn.

(b) Find the sample space S_2 of the experiment if the first card is replaced before the second is drawn.

2.2.   In a group of 400 college seniors, there are 170 science majors and 230 non-science majors. Of the science majors, 55 are women, and there are 120 women non-science majors. One student is selected at random from the group. Find the probability that the selected student is:
(a) A science major
(b) A woman
    (c) Neither a woman nor a science major

2.3.   (a) A and B are two events such that
$P(A) = 0.4, \ P(A \cap B) = 0.3$ and $P(A \cup B) = 0.6$

Compute:
(i) $P(A \cap B^c)$ (ii) $P(B)$ (iii) $P(A^c \cup B^c)$

(b) A and B are two events such that P(A) = 0.5 and P(B) = 0.2.

Compute:
(i) P(A∪B)   (ii) $P(A^c \cup B^c)$

(c) A and B are two events such that
$PA) = 0.4, P(B) = 0.6, P(A \cap B^c) = 0.75$

Compute: $P(A \cup B)$.

2.4. (a) A and B are two events such that P(A) = 0.3, P(B) = 0.6 and $P(A \cap B) = 0.1$

Compute:
(i) $P(A \cup B)$ (ii) $P((A \cap B) \cup (A \cap B))$
(iii) $P((A^c \cap B) \cup (A^c \cap B^c))$

(b) A and B are two independent events such that P(A) = 0.6, P(B) 0.3

Compute: $P(A/A \cup B)$.

(c) An electronic assembly consists of two subsystems called A and B. From previous testing procedures, the following probabilities are assumed to be known:

P(A fails)   = 0.20
P(B fails alone) = 0.15
P(A and B fail) = 0.15

Evaluate the following probabilities:
P(A/B has failed).
P(A fails alone).

2.5. Define the events A, B, C as follows:
A: It might rain today
B: It might be very sunny today
C: It is very windy today.

If $P(A) = 0.50, P(B) = 0.40, P(C) = 0.20,$
$P(A \cap B) = 0.24, P(B \cap C) = 0.10, P(A \cap C) = 0.10$ and
$P(A \cap B \cap C) = 0.06$

Compute:
(i) $(A \cap B^c)$ (ii) $P(A^c \cap B^c)$ (iii) $P(C / A)$ (iv) $P(C / B$
(v) $P(C/(A \cap B))$

2.6.   Consider ten integers, five of which are positive and five negative. One chooses at random (without replacement) two integers and computes their product. What is the probability that the product of the chosen integers is

(i) Positive?   (ii) Negative?

2.7.   Twenty items, 12 of which are defective and 8 non-defective, are inspected one after the other. If these items are chosen at random (without replacement), what is the probability that:

(i) The first two items inspected are defective?
(ii) The first two items inspected are non-defective?
(iii) Among the first two items inspected there is one defective and one non-defective?

2.8.   During a manufacturing process, the probability of manufacturing a defective item
        is 0.2. If three items are manufactured in succession, what is the probability that
(i) None is defective?
(ii) Only one is defective?

2.9.   In a class consisting of 25 pupils, 13 are girls and the rest are boys. Among the girls
        9 are 16 years old and the rest are 15 years old; whilst among the boys, 8 are 16 years old and the rest are 15 years old. If one chooses at random a pupil from the class, compute the probability that the chosen pupil is

(i)   a boy   (ii)   a girl who is 16 years old
(iii) 15 year old   (iv)   a boy who is 15 years old.

2.10   A box contains 3 black and 7 white marbles. One chooses at random (without replacement) three marbles from the box. Compute the probability that:

(i)   one chooses three white marbles.
(ii)   one chooses exactly two black marbles.

2.11. Consider two boxes (1 and 2). Box 1 contains m black marbles and n yellow marbles. Box two contains p black and q yellow marbles. A marble is chosen at random from box 1 and put into box 2 (without observing the colour). Then a marble is chosen at random from box 2. What is the probability that this marble is black?

2.12. Consider two boxes (Box 1 and Box 2). Box 1 contains 4 white and 2 black marbles whilst Box 2 contains 4 black and 2 white marbles. A box is chosen at random. If a marble is chosen from the box, what is the probability that it is
(i) White? (ii) Black?
are
2.13. Consider three boxes (1, 2 and 3), each containing two black and one white marbles respectively. A marble is transferred from Box 1 to Box 2 followed by a marble transferred from Box 2 to Box 3 (without observing the colour). Finally, a marble is chosen from box 3. What is the probability that this marble is black?

2.14. Suppose that a test has been discovered for a disease that afflicts 2% of the population. In clinical trials, the test was found to have the following error rates:

(i)   8% of the people free of the disease had a positive reaction ("false alarm").
(ii)   1% of the people having the disease had a negative reaction ("missed alarm").

A testing programme is being proposed for the whole population; anyone with a positive reaction will be suspected of having the disease, and will be brought to the hospital for further observation.

(a)   Of all the people with positive reaction, what proportion actually will have the disease?
(b)   Of all the people with negative reactions, what proportion will be free of the disease?
(c)   Of the whole population, what proportion will be misdiagnosed one
        way or the other?

2.15. A vacuum tube may come from any one of three manufacturers A, B, C, with probabilities P(A) = 0.25, P(B) = 0.50 and P(C) = 0.25. The probabilities that the tube will function properly during a specified period of time equal 0.1, 0.2 and 0.4 respectively, for the three manufacturers

(a) Compute the probability that a randomly chosen tube will function properly for the specified period of time.

(b) Assume that a vacuum tube is chosen at random and is:

(i) found to function properly. What, then, is the probability that it came from manufacturer A?

(ii) found not to function properly. What, then is the probability that it came from manufacturer B?

2.16.   (a) Given that $P(B) = 0.4$, $P(A/B) = 0.65$ and $P(A/B^c) = 0.35$. Calculate the following probabilities:

(i) $P(A \cap B)$ (ii) $P(A)$ (iii) $P(A \cup B)$

(b) Given that $P(A) = 0.5$, $P(B/A) = 0.4$, $P(A \cup B) = 0.9$. Evaluate

(i) $P(A \cap B)$ (ii) $P(B)$ (iii) $P(B/A)$

2.17.   Let A and B denote two disjoint events such that $P(A) = \frac{1}{2}$ and $P(B) = \frac{1}{3}$. If A^c is the complement of A and B^c is the complement of B, compute the conditional probability $P(A^c/A \cup B^c)$.

CHAPTER 3
COUNTING PROCESSES

3.1. Equally Likely Outcomes

In this chapter, we shall deal mainly with experiments for which the sample space S consists of a finite number of elements: that is, S = {a₁, a₂, a₃, ... aₙ}, where *n* is a finite number. Suppose A is an event consisting of any of the outcomes $\{a_i\}$, ($i = 1, 2, ..., n$) of the sample space S, for example A = {a₁} or A = {a₂} or A = {a₃}, and so on. Each of these events, A is called an elementary event and we can compute the probability P(A) = $P\{a_i\}$, ($i = 1, 2, ..., n$) associated with each event A. These probabilities $\{a_i\}$, will satisfy the following conditions:

(a) $P(a_i) \geq 0, \quad i = 1, 2, ..., n$
(b) $P(a_1) + P(a_2) + \cdots + P(a_n) = 1$

In order to compute the above probabilities, suppose we assume that all outcomes (aᵢ), ($i = 1, 2, ..., n$) are equally likely (i.e. each outcome has the same chance of occurring as the other). If the sum of the probabilities
$P(a_1) + P(a_2) + \cdots + P(a_n)$ should have the value 1, it follows that
$P(a_i) = \frac{1}{n}$ ($i = 1, 2, ..., n$). For the condition, $P(a_1) + P(a_2) + \cdots + P(a_n)$ becomes $nP(a_i)$ for all i. From this it follows that, for any event A consisting of *k* outcomes, i.e. A = (a₁, a₂, a₃, . . ., aₖ), associated with a sample space S = (a₁, a₂, a₃,. . . aₙ) consisting of *n* equally likely outcomes, we have

$$P(A_1) = \frac{k}{n}$$

Figure 3.1 below gives a specific illustration where *k* = 6 and *n* = 15.

A = (., ., ., ., ., .);   k = 6 outcomes
S = (., ., ., ., ., ., ., ., ., ., ., .,.,.,.);
n = 15 equally likely outcomes

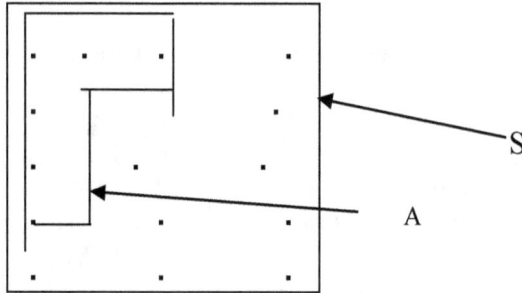

Figure 3.1

$$P(A) = \frac{6}{15}$$

This method of evaluating P(A) is stated as follows:

$$P(A)$$
$$= \frac{number\ of\ points\ (or\ outcomes) associated\ with\ the\ event\ A}{Total\ number\ of\ points\ (or\ outcomes) in\ the\ sample\ space\ S.}$$

3.1.1.   Theorem. (Finite Sample Theorem)

If all the outcomes of a random experiment consist of n equally likely simple events, then the probability of an event E made up of k simple events is

$$P(E) = \frac{k}{n}$$

Proof

Let S be the sample space and p the probability of a simple event. Then since all the simple events are mutually exclusive

$$P(S) = \underbrace{p + p + \cdots + p}_{n\ times} \Rightarrow$$
$$1 = np$$

from which $p = \frac{1}{n}$

Next $P(E) = \underbrace{p + p + \cdots + p}_{k\ times} \Rightarrow$

$$P(E) = kp$$

Therefore $P(E) = k\left(\frac{1}{n}\right)$

We shall mention below a few examples of equally likely outcomes:
(a) The experiment E_1: Tossing a Ludo die once.
The sample space S = (1, 2, 3, 4, 5, 6)
$P(A) = P(A_i), = P(1) = P(2) = P(3) = P(4) = P(5) =$
$P(6) = \dfrac{1}{6}$

(b) The experiment E_2: Tossing a fair coin once.
The sample space S = (H, T): H = Head, T = Tail

$P(A) = P(A_i) = P(H) = P(T) = \dfrac{1}{2}$

(c) A box consists of n marbles of which r are white and the rest $(n - r)$ black.
Each of these marbles are equally likely to be selected.

Define the events: A_1 = select a white marble
A_2 = select a black marble. Then
$$P\{A_1\} = \frac{r}{n} \text{ and } P\{A_2\} = \frac{n - r}{n}$$

(d) A box consists of n items of which p are defective and the rest $(n - p)$ nondefective.

Define the events:
A_1 = select a defective item
A_2 = select a nondefective item.

Then $P(A_1) = \dfrac{P}{n}$ and $P(A_2) = \dfrac{n-P}{n}$

EXAMPLE 3.1

(a) A book contains 30 pages numbered from 1. If the book is opened at random, what
  is the probability that the page number has
(i)   a digit 2 or 5?

(ii)   Two digits?

(b) What is the probability that the page opened has a page number divisible by 3?
(c) What is the probability that the sum of the digits of the page selected is divisible by 3 given that it is a two-digit page number?

SOLUTION:
 (a) the sample space S consists of 30 equally likely outcomes (i.e. the pages of the
    book).

$$S = \left\{ \begin{array}{l} 1, \quad 2, \ 3, \ 4, \ 5, \ 6, \quad 7, \ 8, \ 9, \ 10 \\ 11, 12, \ 13, 14, 15, \ 16, \ 17, 18, 19, \ 20 \\ 21, 22, 23, 24, 25, 26, 27, 28, 29, 30 \end{array} \right\} = 30 \; equally \; likely \; outcomes$$

(i)   Event A_1 = {page number has digit 2 or 5}

$$A_1 = \left\{ \begin{array}{l} 2, \ 5, \ 12, \ 15, \ 20, 21, \ 22 \\ 23, \ 24, \ 25, \ 26, \ 27, \ 28, 29 \end{array} \right\}$$

Thus 14 out of 30 equaly likely outcomes.

$$P(A_1)\frac{14}{30} = \frac{7}{15}$$

(ii)   Event A_2 = {the opened page has two digits}

$$A_2 = \left\{ \begin{array}{lllllll} 10, & 11, & 12, & 13, & 14, & 15, & 16 \\ 17, & 18, & 19, & 20, & 21, & 22, & 23 \\ 24, & 25, & 26, & 27, & 28, & 29, & 30 \end{array} \right\}$$

Thus 21 out of 30 equally likely outcomes

$$P(A_2) = \frac{21}{30} = \frac{7}{10}$$

(b) Event A_3 = {Opened page number is divisible by 3}.

$$A_3 = \left\{ \begin{array}{l} 3, 6, 9, 12, 15 \\ 18, 21, 24, 27, 30 \end{array} \right\}$$

Thus 10 out of 30 are equally likely outcomes.

$$P(A_3) = \frac{10}{30} = \frac{1}{3}$$

(c) Event A_4 = {sum of digits on selected page is divisible by 3 given that it is a 2 digit number}.

$P(A_4)$ = P {sum divisible by 3/two digit number}

There are two ways of computing this conditional probability: (i) with respect to the reduced sample space A_2 (ii) with respect to the original sample space S.

I. With respect to the reduced sample space A_2, there are only 7 outcomes out of 21 i.e. (12, 15, 18, 21, 24, 27, 30) which are divisible by 3.

Hence $P(A_4) = \dfrac{7}{21}$ is the required conditional probability.

II. With respect to the original sample space S:

$$\text{P(divisible by 3 and 2 digits)} = \left\{ \begin{array}{l} 12, 15, 18, 21, \\ 24, 27, 30 \end{array} \right\} = \frac{7}{30}$$

$$\text{P(2 digits)} = P(A_2) = \frac{21}{30}$$

$\therefore$ P(A_4) = P(divisibly by 3/2 digits) = $\dfrac{P(divisible\ by\ 3\ and\ 2\ digits)}{P(2 digits)} = \dfrac{\frac{7}{30}}{\frac{21}{30}} =$

$\dfrac{7}{21}$

which is the same as in (i) above.

EXAMPLE 3.2

A boy tossed a fair coin three times. Give the possible outcomes obtained and their corresponding probabilities.

SOLUTION:

In three tosses of a fair coin, the possible outcomes are expressed as follows:- (Heads shows up first toss, heads second toss, and heads third toss)
i.e. $(H \cap H \cap H)$
or
Heads first toss, heads second toss, and tails third toss
i.e. $(H \cap H \cap T)$
or Heads first toss, tails second toss, and heads third toss
i.e. $(H \cap T \cap H)$ and so on. The eight possible combinations for the three tosses are summarized in Table 3.1 below. For brevity, we write $H \cap H \cap H$ as H H H, etc.

Table 3.1

Possibility	Combination of Outcomes
1	$H \cap H \cap H = HHH$
2	$H \cap H \cap T = HHT$
3	$H \cap T \cap H = HTH$
4	$T \cap H \cap H = THH$
5	$T \cap H \cap T = THT$
6	$T \cap T \cap H = TTH$
7	$H \cap T \cap T = HTT$
8	$T \cap T \cap T = TTT$

Each of the 8 combinations shown above is regarded as a point within the sample space S.
That is,
S = (H H H, H H T, H T H, T H H, T H T, T T H, H T T, T T T)

The various combinations are the results of the Cartesian product S_1 x S_1 x S_1, where S_1 = (H, T) is the sample space associated with a single toss of a coin.

The 8 outcomes are equally likely, that is, each has probability $\frac{1}{8}$ of occurring: Let us define the following events:

A_1 = (H H H), A_2 = (H H T), A_3 = (H T, H),
A_4 = (T H H), A_5 = (T T H), A_6 = (H T T),
A_7 = (T H T), A_8 = (T T T).

Then, $P(A_1) = P(A_2) = P(A_3) = P(A_4) = P(A_5) = P(A_6) = P(A_7) = P(A_8)$
$= \frac{1}{8}$

Suppose we now look at any of the events, say $A_1 = \{H\ H\ H\} = \{H \cap H \cap H\}$.

$H \cap H \cap H$ are the results of three independent tosses of a coin;

$\therefore P(H \cap H \cap H) = P(H).P(H).P(H)$

$$= \frac{1}{2} \cdot \frac{1}{2} \cdot \frac{1}{2} = \frac{1}{8},$$

Since the probability of obtaining a Head in a single toss of a coin is $\frac{1}{2}$,

similar reasoning follows for the other events A_2 through A_8.

EXAMPLE 3 .3

Consider the sample space of Example 3.2 above, and compute probabilities for the following events:

(i) A_1: At least one Head shows up.
(ii) A_2: At least 2 Tails show up
(iii) A_3: At most two Heads show up
(iv) A_4: More Heads than Tails show up
(v) A_5: No Head shows up

SOLUTION :
(i)   At least one head means, one or more heads.

$A_1 = \{THT, HTT, TTH, THH, HTH, HHT, HHH\}$

$$P(A_1) = \frac{7}{8}$$

(ii)   At least 2 tails means, two or more tails

$A_2 = (TTH, HTT, THT, TTT)$

$$P(A_2) = \frac{4}{8} = \frac{1}{2}$$

(iii)   At most two heads means, not more than two heads (0, 1 or 2 Heads)

$A_3 = (THT, HTT, TTH, THH, HTH, HHT, TTT)$

$$P(A_3) = \frac{7}{8}$$

(iv)   More Heads than Tails

$$A_4 = \{HHT, HTH, THH, HHH\}$$

$$P(A_2) = \frac{4}{8} = \frac{1}{2}$$

(v)   No head shows up.

$A_5 = (TTT), \quad P(A_5) = \frac{1}{8}$

NOTE: The probability of the event A_1: at least one head shows up, can also be obtained as $P(A_1) = 1 - (\text{no head}) = 1 - \frac{1}{8} = \frac{7}{8}$.

EXAMPLE 3.4

Two fair dice are tossed once. What is the probability of obtaining a total score of (i) at least 11 (ii) Ten on the two dice?

SOLUTION:

The sample space S associated with the toss of a single die is given as $S = \{1, 2, 3, 4, 5, 6\}$. Each of these outcomes are equally likely,

$$P(1) = P(2) = P(3) = P(4) = P(5) = P(6) = \frac{1}{6}$$

The sample space of the experiment of tossing two dice can be regarded as the Cartesian product S X S = {1, 2, 3, 4, 5, 6} x {1, 2, 3, 4, 5, 6} (see for example assignment 1.5 (iii) of chapter 1).

The sample space S associated with S x S is shown below:

$$S = \begin{cases} (1,1) & (2,1) & (3,1) & (4,1) & (5,1) & (6,1) \\ (1,2) & (2,2) & (3,2) & (4,2) & (5,2) & (6,2) \\ (1,3) & (2,3) & (3,3) & (4,3) & (5,3) & (6,3) \\ (1,4) & (2,4) & (3,4) & (4,4) & (5,4) & (6,4) \\ (1,5) & (2,5) & (3,5) & (4,5) & (5,5) & (6,5) \\ (1,6) & (2,6) & (3,6) & (4,6) & (5,6) & (6,6) \end{cases} = 36 \; outcomes$$

The 36 outcomes are equally likely each with probability $\dfrac{1}{36}$.

The event: A: total of at least 11 means, total of 11 or more. Thus

$A = \{(5,6),\ (6,5),\ (6,6)\}$ and $P(A) = P(5,6) + P(6,5) + P(6,6)$.

Hence $P(A) = \dfrac{1}{36} + \dfrac{1}{36} + \dfrac{1}{36} = \dfrac{3}{36} = \dfrac{1}{12}$

We note again that,

$$P(5,6) = P(5 \cap 6) = P(5) \times P(6) = \frac{1}{6} \times \frac{1}{6} = \frac{1}{36}.$$

Similarly for $P(6,5)$ and $P(6,6)$.

(ii) The event A: total of 10 on two dice

$$A = \{(4,6)\ (5,5),(6,4)\}$$
$$P(A) = P(4 \cap 6) + P(5 \cap 5) + P(6 \cap 4)$$
$$P(4,6) = P(4) \times P(6) \quad \text{[Independent events]}$$

$$= \frac{1}{6} \times \frac{1}{6} = \frac{1}{36}$$

Similarly $P(5 \cap 5) = P(6 \cap 4) = \dfrac{1}{36}$

$$\therefore P(A) \ = \ \frac{3}{36} = \frac{1}{12}$$

Many problems in probability theory require that we count the number of ways that a particular event can occur. For this we study the topics of permutations and combinations. We will consider permutations in this section and combinations in the next sections. Before discussing permutations, it is useful to introduce a general counting technique that will enable us to solve a variety of counting problems, including the problem of counting the number of possible permutations of "n" objects.

3.2. The multiplication principle

Many problems in probability theory require that we count the number of ways that a particular event can occur. For this we study the topics of permutations and combinations.

Suppose there are "n" routes from city A to city B, and "m" routes from B to a third city C. If we decide to go from A to C via B, then for each route that we choose from A to B, we have "m" choices from B to C. Therefore, altogether we have "nm" choices to go from A to C via B.

Theorem 3.1.
Suppose a task is to be carried out in a sequence of "r" stages. There are n_1 ways to carry out the first stage. For each of the n_1 ways, there are n_2 ways to carry out the second stage; for each of the n_2 ways, there are n_3 ways to carry out the third stage, and so forth. The total number of ways the task can be accomplished is given thus:

$$N = n_1 * n_2 * \ldots n_r$$

Example. How many outcomes are there if we throw five dice?

Solution. Let $S_i, 1 \leq i \leq 5$, be the set of all possible outcomes of the ith die.

Then $S_i = \{1,2,3,4,5,6\}$.

The number of possible outcomes if five dice are thrown = the number of ways we can first, choose an element of S_1, then an element of $S_2,\ldots,$ and finally an element of S_5.

Thus we get 6^5 possible outcomes.

Example. You are eating at Emiles's restaurant and the waiter informs you that you have (a) two choices for appetizers: soup or juice; (b) three for the main course: a meat, fish, or vegetable dish; and (c) two for dessert: ice cream or cake. How many possible choices do you have for your complete meal?

Solution. We illustrate the possible meals by a tree diagram shown in figure 3.2.

Your menu is described in three stages; at each stage the number of possible choices does not depend on what is chosen in the previous stages: Two choices at the first stage, three at the second, and two at the third. From the tree diagram in figure 3.2 we see that the total number of choices is the product of the number of choices at each stage. In this example we have $2\times3\times2 = 12$ possible menus.

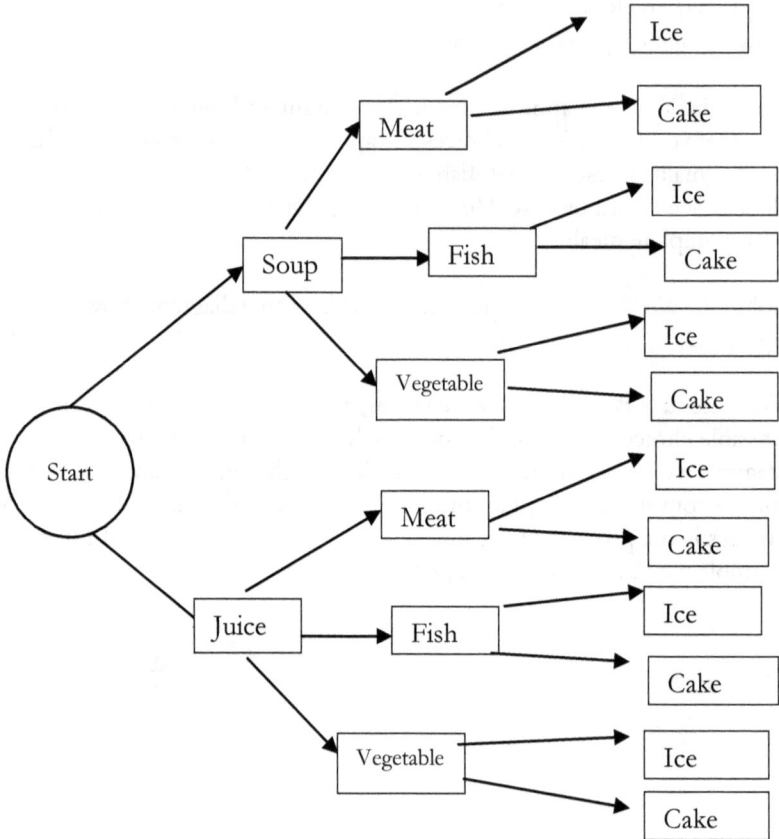

Figure 3.2

3.2.1. The General basic Principle of Counting

If k experiments that are to be performed are such that:

$$the\ first\ experiment\ \longrightarrow\ n_1\ outcomes$$
$$the\ second\ experiment\ \longrightarrow\ n_2\ outcomes$$
$$the\ third\ experiment\ \longrightarrow\ n_3\ outcomes$$
$$.$$

the kth experiment $\longrightarrow$ n_k outcomes

then there is a total of $n_1 * n_2 * ... n_k$ possible outcomes of the k experiments.

3.2. Permutations

If we have one event that can happen in a number of different ways and another event that can also happen in a number of different ways, the total number of ways in which the two events can happen is the product of the number of the separate ways. Suppose we have four distinct objects, A, B ,C, D in a given order, the sequence or arrangement of these objects is called a permutation. Arranging the 4 objects is equivalent to putting them into a box with 4 partitions, in some specified order. The first partition may be filled in any 4 ways, the second in any one of (4 – 1) = 3 ways, the third in any one of (3 – 1) = 2 ways, and the last partition in exactly one way as illustrated in figure 3.3 below:

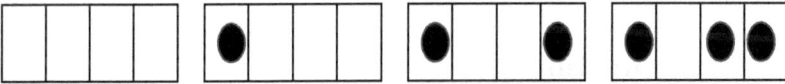

Figure 3.3

Hence, applying the above multiplication principle, we see that the box may be filled in 4 x 3 x 2 = 24 ways. Similarly, if we have three distinct objects, A, B, C, we see that six different permutations of these three objects are possible.

A B C, B A C, C A B, A C B, B C A, C B A

The rules of the preceding section are often applied when repeated selections are made from one and the same set, and the order in which the selections are made is of no significance. Frequently, we are interested in all possible orders or arrangements of a group of objects. For example, we might want to know how many different arrangements are possible for sitting 6 people around a table, or we might ask how many different orders are possible for drawing 2 lottery tickets from a total of 20. The different arrangements are called permutations.

Definition. A permutation is an ordered arrangement of all or part of a set of objects.

Example. In a club of 24 members from which one is to be elected president and another is to be elected vice-president, there are altogether 24×23 = 552 ways in which the selection can be made. If a third member of the club is to be elected treasurer, and another one is to be elected secretary, the total number of ways in which all four officers can be selected is 24×23×22×21 = 255024. After each choice there is one less to choose from the next section.

The number of different permutations of n objects is always equal to $n(n-1)(n-2)\ldots1$, i.e. $n!$. This symbol is called factorial n. Factorial n, or $n!$ means the product of all the integers from n to 1; ie. $n!=n(n-1)(n-2)(n-3)(n-4)\ldots1$. We indicate the number of permutations of n objects taken n at a time by $_nP_n$. Thus

$$nPn = n! = n(n-1)(n-2)\ldots.1$$

Definition (*n* factorial)

If n is a positive integer, we define $n! = n(n-1)(n-2)\ldots1$ and call it n factorial. We also define $0! = 1$.
For n = 3 the number of permutations is 3! = 3 x 2 x 1 = 6.
Example. Evaluate the following:

a: $\dfrac{9!}{6!}$ 　　　　　b: $\dfrac{10!}{3!4!}$

Solution

a. $\dfrac{9!}{6!} = \dfrac{9*8*7*6*5*4*3*2*1}{6*5*4*3*2*1} = 9*8*7 = 504$

b. $\dfrac{10!}{3!4!} = \dfrac{10*9*8*7*6*5*4*3*2*1}{3*2*4*3*2*1} = \dfrac{10*9*8*7*6*5}{3*2} = 25200$

Consider now a different situation, where we have four objects and we decide to choose 2 of them, and arrange (or permute) the chosen 2 objects. We may again consider filling a box with 4 partitions: This time we simply stop after the second partition has been filled. The first partition may be filled in 4 ways, the second in 3 ways. Thus, the entire process may be accomplished, using the multiplication principle, in 4 x 3

= 12 ways. The number of different permutations of n objects taken r at a time is given by

$n(n-1)(n-2)\ldots.(n-r+1)$ ways using the factorial notation introduced above, we may write $nP_r = \dfrac{n!}{(n-r)!}, 0 \leq r \leq n.$

For our four objects, taken 2 at a time, we have

$$4P_2 = \frac{4!}{(4-2)!} = \frac{4*3*2*1}{2!} \text{ different orders or permutations.}$$

3.2.2 Examples of Permutations when some objects are alike

Suppose we have n objects, and that these objects can be divided into k sets so that the objects within each set are alike. We let $r_1, r_2 \ldots r_k$ represent the number of objects within each of the respective sets, with $n = r_1 + r_2 + \ldots + r_k$. Then the number of permutations of the n objects will be given by

$$nP_{r_1, r_2 r_3, \ldots, r_k} = \frac{n!}{r_1! r_2! \ldots \ldots \ldots r_k!} \quad ,$$

where $nP_{r_1, r_2 r_3, \ldots, r_k}$ is the number of permutations of n objects of which r_1 are alike, r_2, are alike and so on.

For example, if we have 6 objects whose composition are 2As, 2Bs and 2Cs that is

A A B B C C,

then the number of permutations of these 6 objects would be

$$_6P_{2,2,2} = \frac{6!}{2!\,2!\,2!} = 18$$

3.3. Combinations

The number of different ways in which we can select r objects from a set of n objects with the order of arrangements of the r objects ignored is called the number of combinations of the n objects taken r at a time. The number of combinations of n objects taken r at a time is given by:

$$nC_r = \frac{n!}{r!\,(n-r)!} = \binom{n}{r}$$

This number arises in many contexts in Mathematics.

If we have 3 objects, A, B, C, then we see that taking any of these objects two at a time, we have 6 permutations: AB, BA, BC, CB, AC, CA. Ignoring the order, AB and BA are one and same combination, BC and CB another same combination, and AC and CA a third same combination. Thus, in this instance, we have 3 combinations. With $n = 3$ and $r = 2$, we have by substitution in the formula,

$$3C_2 = \frac{3!}{2!\,(3-2)!} = 3 \; combinations$$

For the present purpose, $\binom{n}{r}$ is defined only if n is a positive integer and if r is an integer $0 \leq r \leq n$. However, we can define $\binom{n}{r}$ quite generally for any real number n and for any non-negative integer r as follows:

$$\binom{n}{r} = \frac{n(n-1)(n-2) \dots (n-r+1)}{r!}$$

The numbers $\binom{n}{r}$ are often called Binomial coefficients, for they appear as coefficients in the expansion pertinent to the Binomial Theorem often met with in Pure Mathematics:

$$(a+b)^n = \sum_{k=0}^{n} \binom{n}{r} a^k b^{n-k}$$

DIAGNOSTIC TEST

D.3.1 A certain chemical substance is made by using 5 separate liquids. It is proposed to pour one liquid into a tank, and then to add the other liquids in turn. All possible combinations must be tested to see which gives the best yield. How many tests must be performed?

SOLUTION: $5! = 5 \times 4 \times 3 \times 2 \times 1 = 120$ tests.

D.3.2 (a) Show that

$$\binom{n}{r} = \binom{n}{n-r} = \binom{n-1}{r-1} + \binom{n-1}{r}$$

 (b) Hence if $\binom{99}{5} = a$ and $\binom{99}{4} = b$, express $\binom{100}{95}$ in terms of a and b.

 [Hint: Do not evaluate the above expressions to solve this problem].

Solution:

(a)   By definition
$$\binom{n}{r} = \frac{n!}{r!\,(n-r)!} = \frac{n!}{(n-r)!\,r!} = \binom{n}{n-r}$$

$$\binom{n-1}{r-1} + \binom{n-1}{r} = \frac{(n-1)!}{(r-1)!\,(n-r)!} + \frac{(n-1)!}{r!\,(n-1-r)!}$$
$$= \frac{(n-1)!}{(r-1)!\,(n-r-1)!}\left(\frac{1}{n-r} + \frac{1}{r}\right)$$
$$= \frac{(n-1)!}{(r-1)!\,(n-r-1)!}\left(\frac{n}{r(n-r)}\right)$$
$$= \frac{n!}{r!\,(n-r)!}$$
$$= \binom{n}{r}$$

Hence from (a) we have $\binom{100}{95} = \binom{100}{5} = \binom{99}{4} + \binom{99}{5}$

$= a + b$

Ans. $(a + b)$

D.3.3 A fair die is tossed once. What is the probability that
(i)   At most 2 shows up?
(ii)   An odd number shows up?
(iii)   At least a 2 shows up?

SOLUTION:
$$S = [1,\ 2,\ 3,\ 4,\ 5,\ 6]$$

Event A: (i) At most 2 show up.
$$A = \{1, 2\}$$
$$P(A) = P(1\ or\ 2) = P(1) + P(2) = \frac{1}{6} + \frac{1}{6} = \frac{1}{3}$$

(ii)   A: An odd number shows up.
$$A = \{1, 3, 5\}$$
$$P(A) = P(1\ or\ 3\ or\ 5) = P(1) + P(3) + P(5) = \frac{1}{6} + \frac{1}{6} + \frac{1}{6} = \frac{1}{2}$$

(iii)   At least 2 shows up.
$$A = \{2, 3, 4, 5, 6\}$$

$P(A) = P(2 \ or \ 3 \ or \ 4 \ or \ 5 \ or \ 6) = P(2) + P(3) + P(4) + P(5) + P(6) = \frac{5}{6}.$

EXERCISE 3

3.1 A symmetric die is tossed twice. Determine the probabilities of the following events:

(a)   Sum of 7
(b)   Sum of at least 8
(c)   Second die gives more points than the first.
(d)   One of the tosses gives at least 5 points
(e)   The sum is divisible by 4
(f)   Both numbers are even
(g)   The numbers on the dice are equal.
(h)   The numbers on the dice differ by at least 4.

3.2 A girl tossed a fair coin twice.
(a)   Give the set of all possible outcomes and their corresponding probabilities.
(b)   Determine the probabilities of the following events:

(i) No heads show up.
(ii)   At least one head shows up.
(iii)   At least 1 tail shows
(iv)   Equal number of heads and tails show up.

3.3 A product is assembled in three stages. At the first stage there are 5 assembly lines, at the second stage there are 4 assembly lines, and at the third stage there are 6 assembly lines. In how many different ways may the product be routed through the assembly process?

3.4 An inspector visits 6 different machines during the day. In order to prevent operators from knowing when he will inspect, he varies the order of his visits. In how many ways may this be done?

3.5 How many outcomes are there in the sample space of each of the following experiments, using the most detailed classification?

(a)   selection of 4 persons from a group of 10, without regard to order?

(b)   selection of 4 persons from 10, to serve as chairman, vice chairman, secretary, treasurer?

3.6 There are 12 ways in which a manufactured item can have a minor defect and 10 ways in which it can have a major defect. In how many ways can (a) 1 minor and 1 major defect occur? (b) 2 minor and 2 major defects occur?

3.7 A complex mechanism may fail at 15 stages. If it fails at 3 stages, in how many ways can this happen?

3.8 Use the Binomial Theorem: $(x+y)^n = \sum_{k=0}^{n} \binom{n}{k} x^k y^{n-k}$,

$n = 1, 2, ...$ to show that $\sum_{k=0}^{n} \binom{n}{k} = 2^n$, $n = 1, 2, ...$

3.9 Compute the following:

(i) $\binom{12}{4}$ (ii) $\binom{100}{97}$ (iii) $\binom{n}{1}$

3.10 Assuming n to be a positive integer and r an integer, $0 \le r \le n$, show that

(i) $\binom{n}{r} = \binom{n}{n-r}$ (ii) $\binom{n}{r} = \binom{n-1}{r-1} + \binom{n-1}{r}$

3.11 How many outcomes are there in the following experiments?

(a)   The selection of 4 persons from a group of 10, without regard to order.

(b)   The selection of 4 persons from a group of 10, with regard to order.

3.12. A bag contains two red, three green and four black balls of identical size except
for colour. Three balls are drawn at random without replacement.

a: Show that the probability is $\frac{55}{84}$ that two balls have the same colour and the third
      is different.

b: What is the probability that at least one is red?

CHAPTER 4
DISCRETE RANDOM VARIABLES

4.1. Random Variables (One Dimensional)

We have, up to this point, discussed a random experiment E, the set of all possible outcomes defining the sample space associated with experiment E, certain subsets of the outcomes defining events, and an assignment of numbers to each event, which satisfy the axioms of probability. We have observed that some experiments yield sample spaces whose elements are numbers, (for example, the throwing of a die), whilst others yield sample spaces whose elements are non-numerical (for example, the toss of a coin). We need one more discussion to extend the basic ideas of probability. The discussion, which we will now start, leads to what is called a random variable.

4.1.1. The Meaning of a Random Variable

Suppose that some function, say X, acts on each outcome or sample point of a sample space S, to yield a numerical value. This mapping can be regarded as a simple rule of measurement or counting that associates the outcomes with numbers. For instance, consider the experiment of tossing a fair coin twice. The sample space S, consisting of all possible outcomes of the experiment can be obtained. Suppose that for each sample point we apply the rule: let the function X count the number of heads, H, occurring. Thus for a single toss of the fair coin with $S = \{H,\ T\}$, we can write

X (Head) $= X(H)$: number of heads in this sample point $= 1$

X (Tail) $= X(T)$: number of heads in this sample point $= 0$

Similarly, for two tosses of the same coin, the sample space $S = \{HH,\ HT,\ TH,\ TT\}$, and we can write,

X (Heads) $= X(HH)$: number of heads in this sample point $= 2$

X (Head and Tail) $= X(HT)$: number of heads in this sample point $= 1$

X (Tail and Head) $= X(TH)$: number of heads in this sample point $= 1$

X (Tails) $= X(TT)$: number of heads in this sample point $= 0$

If on the other hand, we were interested in counting the number of tails occurring in each sample point, we could write, for the single toss of the coin;

$X(H) = 0$ and $X(T) = 1$ whilst for the two tosses, we have

$X(HH) = 0$, $X(HT) = 1$, $X(TH) = 1$, $X(TT) = 2$

Other experiments for which we can use this rule of measurement are

(i)   Manufactured items: X is the count of the number of defective items they contain

(ii) Choosing a person at random from a group: X is a measure of the person's height, and so on

Such a rule for assigning numbers to sample points is termed a random variable. Thus, in the experiment of tossing a fair coin twice, the random variable X assumes the values 0, 1, 2 corresponding to the sample points no head, 1 head and 2 heads respectively. To specify the value of the random variable for a particular sample point s, we will use the notation $X(s)$.

4.1.2.   Definition (Random Variable)

Let E be an experiment and S a sample space associated with the experiment.

A function X that assigns to each sample point $s \in S$, a real number $X(s)$, is called a random variable. Therefore *A random variable is a function that assigns a real number to each outcome of an experiment.*

A random variable is denoted by an upper letter such as X. after an experiment is conducted the measured value of the random variable is denoted by a lower case such as $x = 20$ millimeters.

The following correspondence rules (mappings) define random variables,

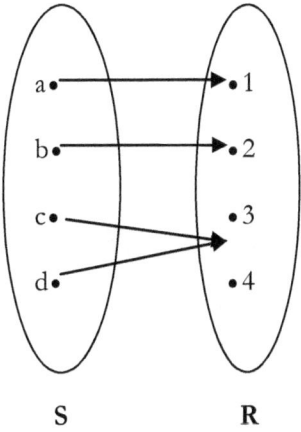

Figure 4.1

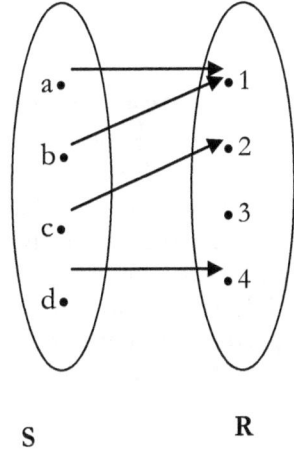

Figure 4.2

Whilst the following correspondence rules (mappings) do not define random variables

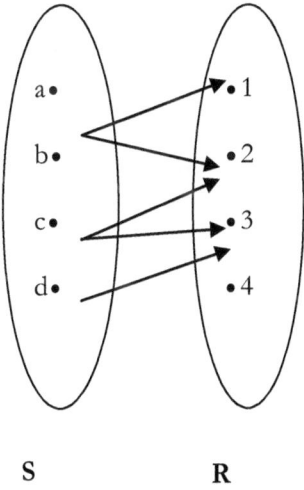

Figure 4.3

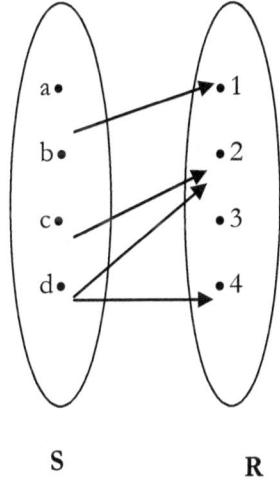

Figure 4.4

The sample space S is termed the domain of the random variable or function, and the collection of all numbers that are values of X is termed the range of the random variable. Thus the range is a certain subset of the set of all real numbers.

Notice, in particular, that two or more different sample points in S might give the same value for X (two different people might have the same height), but two different values (or numbers) in the range R, cannot be assigned to the same sample in S (one person cannot have two different heights). See figures 4.3 to 4.4 respectively.

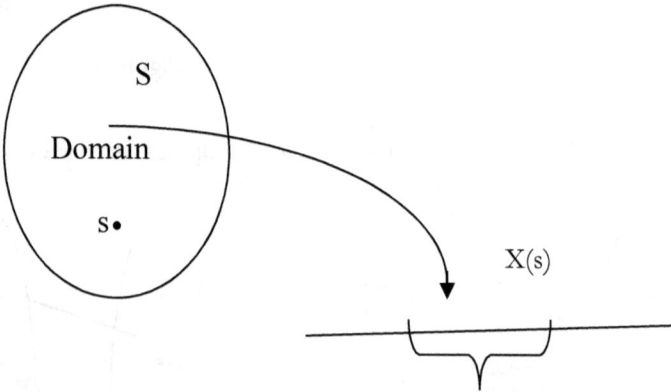

Figure 4.5

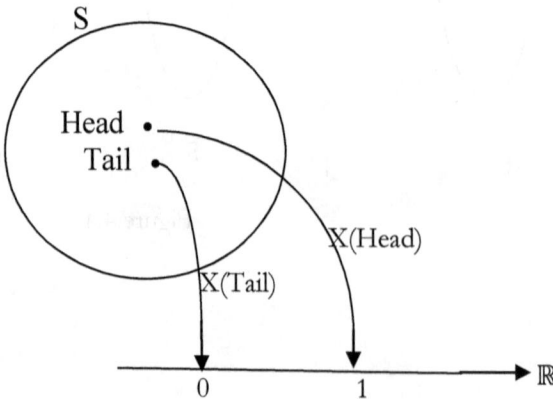

Figure 4.6

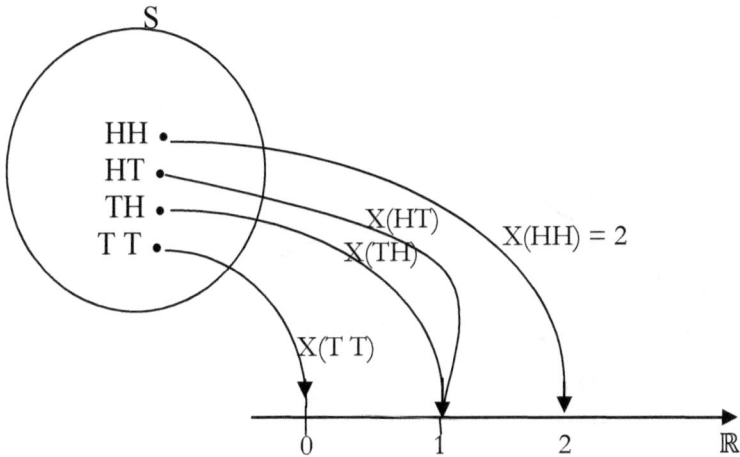

(HH) = 2

(HT)= X (TH) = 1   Real variable X associated with coin tossing

(TT) =0   Figure 4.7

4.1.3.   Events Defined by Random Variables

In the previous section, we discussed the rule of measurement, whereby, the random variable X assigns a number x to a sample point, s, within the sample space S, associated with an experiment E. Let A be an event consisting of one or more points within the sample space S, that is, $A = \{s : X(s) = x\}$. The event A will have probability $p = P(A) = P\{X(s) = x\}$. This probability p is a set - function – a rule that assigns a number between 0 and 1 to the set of points in S. Its domain is the set of events of a random experiment, and its range is contained in the closed interval $0 \leq p \leq 1$. We might interpret p as the probability that the random variable X takes on the value x. We will again consider the example of tossing a fair coin twice. The outcomes are illustrated in Figure 4.8 below:

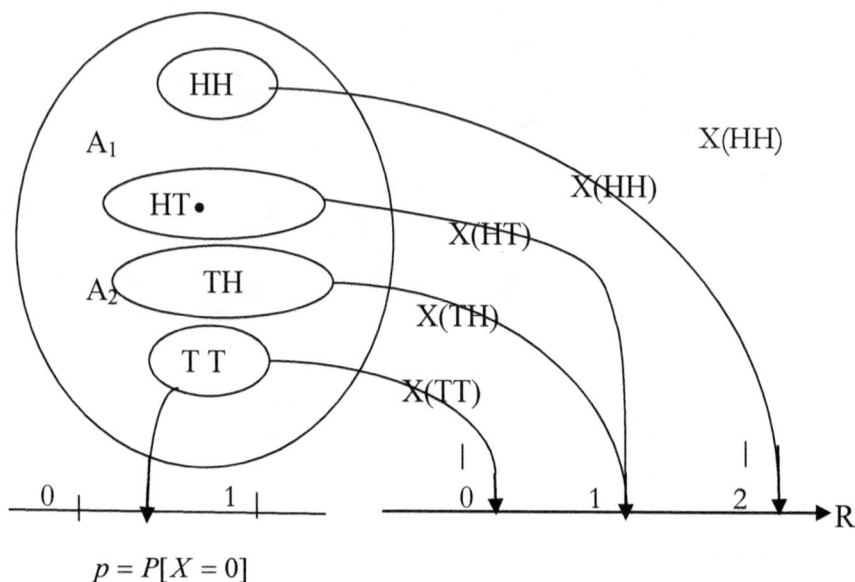

Figure 4.8

The sample space S consists of 4 equally likely outcomes:
$S = \{HH, HT, TH, TT\}$ with individual probabilities

$$P(HH) = P(HT) = P(TH) = P(TT) = \frac{1}{4}, \text{ and}$$

$$\sum P(\cdot \cdot) = P(HH) + P(HT) + P(TH) + P(TT) = 1.$$

Define the random variable X by
$X(HH) = 2$, $X(HT) = X(TH) = 1$ and $X(TT) = 0$.
Thus in figure 4.8 above,

$$A_1 = \{HH\} \text{ and } P(A_1) = P\{HH\} = P\{X(HH) = 2\} = \frac{1}{4}$$

$$A_2 = \{HT, TH\} \text{ and } P(A_2) = P\{HT, TH\}$$
$$= P\{[X(HT) = 1] \cup [X(TH) = 1]\}$$
$$= P\{[X(HT) = 1] + [X(TH) = 1]\}$$

$$= \frac{1}{4} + \frac{1}{4} = \frac{1}{2}$$

$A_3 = \{TT\}$ and $P(A_3) = P\{TT\} = P\{X(TT) = 0\} = \frac{1}{4}$

For simplicity, we can write $P[X(s)] = x = P[X = s]$, hence,

$$P[X(HH) = 2] = P[X = 2] = \frac{1}{4}$$

$$P\{[X(HT) = 1] \cup [X(TH) = 1]\} = P\{X = 1\} = \frac{1}{2}$$

$$P\{X(TT) = 0\} = P\{X = 0\} = \frac{1}{4}$$

Thus, by means of the random variable X, we have assigned probabilities to the values 0, 1, 2 along the real line, although the original probabilities were only defined for subsets of the set S. We say that the random variable X, has induced probabilities on the real line. Note, however, that the possible values of the random variable X, that is $X = \{0, 1, 2\}$ are not equally likely. To find what the probabilities are, we must examine the original sample space S.

4.2. Discrete Random Variables (One Dimensional Case)

DEFINITION 3. *Let X be a random variable. If the number of possible values of X (that is the range of X) is finite or countably infinite, we call X a discrete random variable. In this case the possible values of X may be listed as x_1, x_2, ,...,x_n. In the finite case the list terminates; in the infinite case the list continues indefinitely.*

In the previous section, we discussed the following ideas:

(i)   The random variable , X, considered as a function

(ii)   The specific value x, which is assigned to the random variable X, and

(iii)   The probability $p(x) = P(X = x)$ associated with the specific value x, assigned to the random variable X.

EXAMPLE 4.1. Consider the experiment of tossing a coin twice, and let X denote the random variable that measures the number of times heads comes up.

a)   List all of the possible outcomes of this experiment and the value that the random variable X assigns to each outcome

b)   List the outcomes that make up the event $[X = 1]$

SOLUTION

(a)   The possible outcomes for the experiment can be represented as HH, HT, TH, and TT. Since X assigns to each output the number of times heads comes up, it assigns 2 to HH, 1 to HT and to TH and 0 to TT, as indicated in table 1. This is a clear example of a discrete random variable since the possible values of X are 0, 1, 2. In this case the list terminates.

(b)   The event $[X = 1]$ is the set of outcomes for which $X = 1$; that is for which exactly one head comes up. Thus, $[X = 1]$ denotes the event $\{HT, TH\}$

We will again consider the experiment of tossing a fair coin twice. The random variable X assumes the values 0, 1, 2 respectively for the number of heads occurring in the experiment. In this case, we write $X = 0, 1, 2$ respectively for the number of heads occurring in the experiment, with corresponding probabilities

$$P(0) = P[X = 0] = \frac{1}{4}, P(1) = P[X = 1] = \frac{1}{2}.$$
$$P(2) = P[X = 2] = \frac{1}{4},$$

That is, each specific value x of the random variable X , is associated with probability

$P(x) = P(X = x)$. In figure 4.9 below we obtain the probability distribution of the random variable X .

Original Sample Space (S)	Random Variable (X)	Range space (Specific values assigned to X)	Probability Distribution $p(x)$

HH • ———————— X

$p(2) = P(X = 2) = \frac{1}{4}$

HT • ————————

$p(1) = P(X = 1) = \frac{1}{2}$

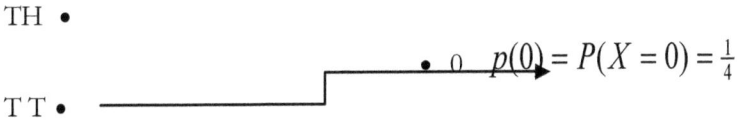

TH •

TT •
$p(0) = P(X = 0) = \frac{1}{4}$

Figure 4.9

4.1.4. Definition (Discrete Probability Distribution)

A discrete random variable takes on various values with probabilities specified by its probability distribution.

Suppose that, X is a discrete random variable. Then R_X , the range space of X , consists of countably finite values of $x_1, x_2, x_3, \cdots, x_n$ (list terminates) or countably infinite values of $x_1, x_2, x_3, \cdots$ (list continues indefinitely). With each possible outcome x_i we associate a number $p(x_i) = P(X = x_i)$, called the probability of x_i. The numbers $p(x_i)$, $i = 1, 2, \cdots\cdots$ must satisfy the following conditions:

(a) $p(x_i) \geq 0$ for all i

(b)   (i) $\sum_{i=1}^{n} p(x_i) = 1$ (countably finite)

(ii) $\sum_{i=1}^{\infty} p(x_i) = 1$ (countably infinite)

The function $p(x)$ mentioned above, is called the probability mass function of the random variable X . The collection of pairs $(x_i, \ p(x_i))$, $i = 1, 2, \cdots$ is sometimes called the probability distribution of X . This probability distribution of X may be presented equally well in any of the three usual forms for a function:

(1) Table form as shown in Tables 4.1 and 4.2 below:

Specific Values Random Variable	X	x_1	x_2	x_3	...
Probability	$P(X = x_i) = p(x_i)$	$p(x_1)$	$p(x_2)$	$p(x_3)$	...

Table 4.1

x

X	0	1	2
$p(x)$	$\dfrac{1}{4}$	$\dfrac{1}{2}$	$\dfrac{1}{4}$

Table 4.2

(2) Graph form, as shown in Figures 4.10 and 4.11 below.

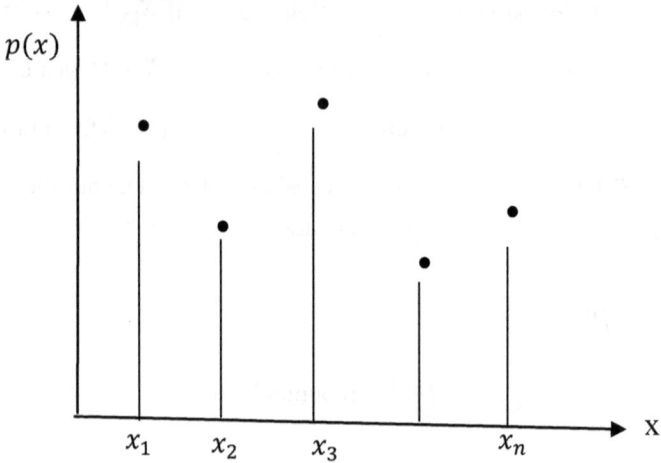

Figure 4.10

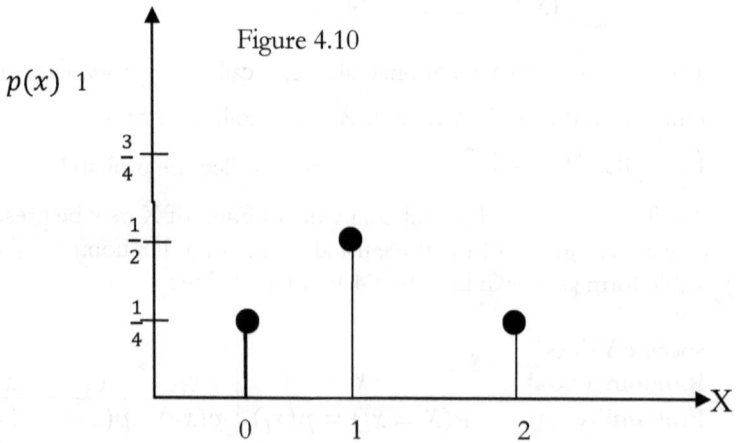

Figure 4.11

Probability distribution: two tosses of a fair coin

NOTE: Each of the probabilities in Figure 4.10 above, is greater than or equal to zero, and satisfies the condition $0 \le p(x) \le 1$. Also, the sum of the probabilities in the table equals 1, i.e.

$\sum_x p(x) = \dfrac{1}{4} + \dfrac{1}{2} + \dfrac{1}{4} = 1$. This is always true for any probability distribution of a discrete random variable.

(3)   A formula: a typical discrete probability distribution in a formula form is shown as follows:

$P(X = x) = \dfrac{1}{10}(x - 1), (x = 1, 2, 3, 4, 5)$

Verify that $\sum_{i=1}^{4} P(X = x_i) = 1$

You now see that the probability distribution of a random variable X is a description of the probabilities associated with the possible values of X. For a discrete random variable, the distribution is often specified by just a list of the possible values along with the probability of each. In some cases, it is convenient to express the probability in terms of formula.

EXAMPLE 4.2

A fair die is thrown twice, find the probability distribution of
(a)   The sum of the dots on both dice; $Y = X_1 + X_2$
(b)   The difference (absolute value) between the numbers, $D =| X_1 - X_2 |$

SOLUTION:
The sample space of 36 equally likely outcomes

Table 4.3: Outcomes from first die

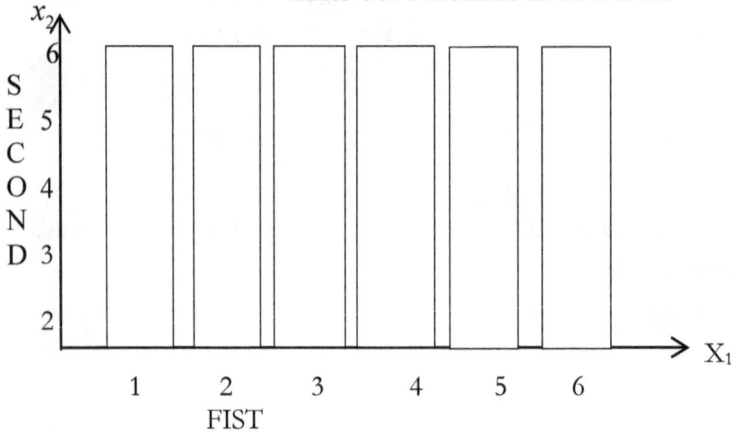

FIST

(a)   The random variable $Y = X_1 + X_2$, takes the values

$2, 3, 4, \cdots, 12$ with associated probabilities $p(2) = \dfrac{1}{36}$, $p(3) = \dfrac{2}{36}$,

$p(4) = \dfrac{3}{36}$, $p(5) = \dfrac{4}{36}$, $p(6) = \dfrac{5}{36}$ $p(7) = \dfrac{6}{36}$, $p(8) = \dfrac{5}{36}$,

$p(9) = \dfrac{4}{36}$, $p(10) = \dfrac{3}{36}$, $p(11) = \dfrac{2}{36}$, $p(12) = \dfrac{1}{36}$

Thus, the probability distribution of Y is given in the table below:

Y	2	3	4	5	6	7	8	9	10	11	12
$p(y_i)$	$\dfrac{1}{36}$	$\dfrac{2}{36}$	$\dfrac{3}{36}$	$\dfrac{4}{36}$	$\dfrac{5}{36}$	$\dfrac{6}{36}$	$\dfrac{5}{36}$	$\dfrac{4}{36}$	$\dfrac{3}{36}$	$\dfrac{2}{36}$	$\dfrac{1}{36}$

Table 4.4

(b)   The random variable $D = |X_1 - X_2|$ take values , 0, 1, 2, 3, 4, 5 with associated probabilities $p(0) = \dfrac{6}{36}$, $p(1) = \dfrac{10}{36}$, $p(2) = \dfrac{8}{36}$, $p(3) = \dfrac{6}{36}$, $p(4) = \dfrac{4}{36}$, $p(5) = \dfrac{2}{36}$

Thus, the probability distribution of D is given below:

D	0	1	2	3	4	5
$p(d_i)$	$\dfrac{6}{36}$	$\dfrac{10}{36}$	$\dfrac{8}{36}$	$\dfrac{6}{36}$	$\dfrac{4}{36}$	$\dfrac{2}{36}$

Table 4.5

EXAMPLE 4.3

A population consists of five elements A, B, C, D and E, and the values of the random variable X are shown in the table below. A random sample of 3 elements is selected, without replacement, from the population. If Z denotes the median of the sample, construct the probability distribution of Z .

	A	B	C	D	E
x	1	5	7	9	11

Table 4.6

SOLUTION:

There are $C_3^5 = \begin{pmatrix} 5 \\ 3 \end{pmatrix} = 10$ equally likely samples. See table 4.7.

Sample	Median Z	Sample	Median Z
1, 5, 7	5	1, 9, 11	9
1, 5, 9	5	5, 7, 9	7
1, 5, 11	5	5, 7, 11	7
1, 7, 9	7	5, 9, 11	9
1, 7, 11	7	7, 9, 11	9

Table 4.7

Z	5	7	9
$P(Z = z)$	$\dfrac{3}{10}$	$\dfrac{4}{10}$	$\dfrac{3}{10}$

Table 4.8

EXAMPLE 4.4. A random variable X has probability mass function (pmf)

$$P(X = x) = \left(\frac{1}{2}\right)^x, x = 0,1,2,\dots$$

Show that this pmf is legitimate and find the probability that X is even.

SOLUTION.

$$P(X = x) = \left(\frac{1}{2}\right)^x, x = 0,1,2,\dots$$

Clearly $P(x) \geq 0$, for every x
Next,

$$\sum_{x=0}^{\infty} P(x) = \sum_{x=0}^{\infty} \left(\frac{1}{2}\right)^x = \left(\frac{1}{2}\right)^0 + \left(\frac{1}{2}\right)^1 + \cdots$$
$$= 1$$

which is the sum to infinity of a geometric progression with first term 1 and common ratio $\frac{1}{2}$. Hence $P(x)$ is a legitimate pmf.

$$P(X = \text{even}) = P(X = 2,4,6,8,10,\dots)$$
$$= P(x = 2) + P(x = 4) + P(x = 6) + \cdots$$
$$= \left(\frac{1}{2}\right)^2 + \left(\frac{1}{2}\right)^4 + \left(\frac{1}{2}\right)^6 + \cdots$$
$$= \frac{\left(\frac{1}{2}\right)^2}{1 - \left(\frac{1}{2}\right)^2}$$

$$= \frac{1}{3}$$

4.2.2 The Expected Value (or Mean)

We noted before (section 2.4 chapter 2) the close relationship between the relative frequency distribution and the probability distribution, when we considered the repeated tossing of a fair coin. If the number of tosses is repeated without limit, the relative frequency distribution will settle down to the probability distribution. That is, the limiting relative frequency becomes the probability.

From the relative frequency distribution, it is possible to calculate the mean $\overline{X}$ and variance S^2 of the sample. It is natural to calculate analogous values from the probability distribution and call them the mean (or expected value $E(X)$) and variance σ^2 of the probability distribution $p(x)$, or of the random variable X itself. Thus we define:

4.2.2.1 Definition (Expected Value of a Discrete Random Variable)

Let X be a discrete random variable with possible values x_1, x_2, x_3, ...

Let $p(x_i) = P(X = x_i), i = 1,2,3, ... n,$ be the probability associated with x_i, $i = 1, 2, \cdots n$ respectively. Then the expected value of X, denoted by $E(X) = \mu,$ is defined as

$$E(X) = \sum_{i=1}^{\infty} x_i p(x_i) :$$ if X assumes infinite number of

values, or

$$E(X) = \sum_{i=1}^{n} x_i p(x_i) :$$ if X assumes finite number of values

The expected value $E(X)$ is said to exist if the series $\sum_{i=1}^{\infty} x_i p(x_i)$

converges absolutely, i.e. if $\sum_{i=1}^{\infty} | x_i | p(x_i) < \infty$. This number is

referred to as the mean of X. This sum, however, may be infinite and may not converge(diverge) as an infinite sum. Indeed, if it does not converge absolutely, the mean $E(X)$ is said not to exist, and one writes $E\mid X\mid = \infty$.

4.2.2.2 Properties of Expected value

(i)   If $X = k$, where k is a constant, then
$$E(k) = k$$
Proof:

$$E(x) = \sum_{i=1}^{\infty} x_i P(X = x_i)$$

$$= \sum_{i=1}^{\infty} kP(X = x_i)$$

$$= k \sum_{i=1}^{\infty} P(X = x_i)$$

Hence
$$E(x) = k$$

Since for legitimacy,

$$\sum_{i} P(X = x_i) = 1$$

(i.e. the expected value of a constant, is equal to the constant value).

(ii)   If k is a constant and X is a random variable, then
$E(kX) = kE(X)$
(i.e. the expected value of the product of a constant and a random variable, is equal to the constant multiplied by the expected value of the random variable.

(iii)   If a and b are any constants then
$$E(aX + b) = aE(X) + b$$

Proof:

$$E(aX + b) = \sum_{i=1}^{\infty} (ax + b)P(x_i) = \sum_{i=1}^{\infty} ax_i P(x_i) + \sum_{i=1}^{\infty} bP(x_i)$$

$$= a\sum_{i=1}^{\infty} x_i P(x_i) + b\sum_{i=1}^{\infty} P(x_i)$$

$$aE(X) + b$$

$$since \sum_{i=1}^{\infty} P(x_i) = 1$$

This property illustrates the fact that if X and Y are two random variables, and Y can be written as a function of X, then one can compute the expected value of Y using the distribution function of X.

A generalization of the property states that if $a_0, a_1, a_2, \ldots, a_n$ are constants and if $E(X), E(X^2), \ldots, E(X^n)$, exists, then

$$E(a_0 + a_1 X + a_2 X^2 + \cdots + a_n X^n) = a_0 + a_1 E(X) + a_2 E(X^2) + \cdots + a_n E(X^n)$$

(iv)   Let X and Y be two random variables, a and b are constants, then $E(aX + bY) = a\,E(X) + b\,E(Y)$

{i.e. the expected value of the sum of two random variables equals the sum of the expected values of the random variables. This result can be generalised to more than two random variables}.

If X_i $(i = 1, 2, \cdots n)$ is a random variable and a_i $(i = 1, 2, \cdots n)$ is a constant, then

$$E(a_1 X_1 + a_2 X_2 + \cdots a_n X_n) = E\left(\sum_{i=1}^{n} a_i X_i\right)$$

$$= E(a_1 X_1) + E(a_2 X_2) + \cdots E(a_n X_n)$$

$$= \sum_{i=1}^{n} a_i E(X_i)$$

If $a_i = 1$ $\forall i = 1, 2, \cdots n$, then we have

$$E(X_1 + X_2 + \cdots X_n) = E\left(\sum_{i=1}^{n} X_i\right)$$

$$= \sum_{i=1}^{n} E(X_i)$$

4.2.3. The Variance of a Discrete Random Variable

The variance of a distribution is commonly used as a measure of its dispersion or spread, and is usually denoted by σ^2. In referring to a random variable X having the given distribution, the notation Var(X) or V(X) is used. The term variance refers to the fact that observations on X are unpredictable and will vary from one trial to another and the variance describes the extent of the variability in the value of X that can be observed.

4.2.3.1 Definition (Variance and Standard Deviation of a Discrete Random Variable)

Let X be a random variable. We define the variance of X, denoted by $V(X)$ or σ_X^2 as follows:

$$V(X) = E[(X - \mu)^2] = \sum (X - \mu)^2 p(x)$$
$$= \sum x^2 p(x) - \mu^2$$

The positive square root of $V(X)$ is called the standard deviation of X and is denoted by σ_X.

Let us at this point show that

$$\sigma^2 = \sum (X - \mu)^2 p(x) = \sum x^2 p(x) - \mu^2 = E(X^2) - [E(X)]^2$$

Proof :

$$\sigma^2 = \sum (X - \mu)^2 p(x) = \sum (X^2 - 2\mu X + \mu^2) p(x)$$

$$= \sum \left(X^2 p(x) - 2\mu X\, p(x) + \mu^2 p(x) \right)$$

$$= \sum X^2 p(x) - 2\mu \sum X\, p(x) + \mu^2 \sum p(x)$$

$$= \sum x^2 p(x) - 2\mu \sum x\, p(x) + \mu^2 \sum p(x)$$

$$= \sum x^2 p(x) - 2\mu^2 + \mu^2$$

[Since $\sum p(x) = 1$, and μ is a constant]

$$= \sum x^2 p(x) - \mu^2$$

$$= \sum x^2 p(x) - [E(X)]^2$$

$$= E(X^2) - [E(X)]^2$$

4.2.3.2 Properties of the Variance

(i)   If k is a constant, $V(k) = 0$ (that is, the variance of a constant, equals zero)

(ii)   If k is a constant and X is a random variable, $V(KX) = K^2 V(X)$

(iii)   If a and b are constants then $Var(aX + b) = a^2 Var(X)$

Proof:

$$Var(aX + b) = E\{(aX + b - E(aX + b)\}^2$$
$$= E\{aX + b - aE(X) - b\}^2$$
$$= E\{aX - aE(X)\}^2$$
$$= Ea^2\{X - E(X)\}^2$$
$$= a^2 E\{X - E(X)\}^2$$
$$= a^2 Var(X)$$

REMARK. *Expected value is a long range average in the sense that if repeated trials are made of an experiment with a random variable X, the sample average of the observed values of x will get closer and closer to the expected value* E(X) *i.e.* μ, *as the number of repetitions gets larger.*
The layout for computing expected value and variance of X is as follows:

x_i	$P(x_i)$	$x_iP(x_i)$	$(x_i-\mu)^2$	$p_i(x_i-\mu)^2$
x_1	$P(x_1)$	$x_1P(x_1)$	$(x_1-\mu)^2$	$p_1(x1-\mu)^2$
x_2	$P(x_2)$	$x_2P(x_2)$	$(x_2-\mu)^2$	$p_2(x2-\mu)^2$
.	.	.	.	.
.	.	.	.	.
.	.	.	.	.
x_n	$P(x_n)$	$x_nP(x_n)$	$(x_n-\mu)^2$	$p_n(x_n-\mu)^2$
Total		$E\ (X) =$ $\mu =$ $\sum_{i=1}^{n} x_iP(x_i)$		$Var(x) = \sum_{i=1}^{n} P_i(x_i-\mu)^2$

EXAMPLE 4.5

Let $Y = 2X + 8$, then
$$E(Y) = E(2X + 8) = 2E(X) + E(8)$$
$$= 2E(X) + 8$$

$$V(2X + 8) = 2^2 V(X)$$
$$= 4V(X)$$

Using properties 4.2.3.2 (i), (ii) and (iii) above.

EXAMPLE 4.6

Suppose a fair die is tossed once. Let X be the discrete random variable assuming the following values:

$X = 1$ if the die shows a 1 on the face

$X = 2$ if the die shows 2 or 3

$X = 3$ if the die shows 4, 5 or 6

Compute the probability distribution of X, and hence find the mean of the distribution.

SOLUTION:

In the single toss of a die, the sample space $S = \{1, 2, 3, 4, 5, 6\}$.

Define the events:

A_1: 1 shows up, $A_1 = \{1\}$

A_2: 2 or 3 shows up, $A_2 = \{2, 3\}$

A_3: 4, 5 or 6 shows up, $A_3 = \{4, 5, 6\}$

$$P(A_1) = \frac{1}{6}, \qquad P(A_2) = \frac{2}{6}, \qquad P(A_3) = \frac{3}{6}$$

Hence, the probability distribution of X is presented in the table below:

Table 4.9

X	1	2	3
$P(X = x)$	$\dfrac{1}{6}$	$\dfrac{2}{6}$	$\dfrac{3}{6}$

Mean of X: $\displaystyle E(X) = \sum_{i=1}^{3} x_i\, p(x_i) = \left(1 \times \frac{1}{6}\right) + \left(2 \times \frac{2}{6}\right) + \left(3 \times \frac{3}{6}\right)$

$$= \frac{1}{6} + \frac{4}{6} + \frac{9}{6} = \frac{14}{6} = \frac{7}{3}$$

EXAMPLE 4.7

The discrete random variable X has the following probability distribution:

X	0	1	4
$P(X = x)$	0.2	0.4	0.4

Table 4.10

Compute   (i)   The mean.

(ii)   The variance, and hence the standard deviation of X.

SOLUTION:

(i)
$$E(X) = \sum_{i=1}^{3} x_i P(x_i) = (0 \times 0.2) + (1 \times 0.4) + (4 \times 0.4)$$
$$= 0 + 0.4 + 1.6$$
$$= 2.0$$

(ii)
$$V(X) = E(X^2) - [E(X)]^2$$

$$E(X^2) = \sum_{i=1}^{3} x_i^2 P(x_i) = (0^2 \times 0.2) + (1^2 \times 0.4) + (4^2 \times 0.4)$$
$$= 0 + 0.4 + (16 \times 0.4)$$
$$= 6.8$$

$$E(X) = 2.0 \quad \Rightarrow \quad [E(X)]^2 = 4$$

Hence
$$V(X) = E(X^2) - [E(X)]^2$$
$$= 6.8 - 4.0$$
$$= 2.8$$

The standard deviation of X is given by
$$\sigma_X = \sqrt{V(X)} = \sqrt{2.8} \cong 1.673$$

EXAMPLE 4.8

A red die has its faces marked 1, 1, 2, 2, 3, 4 whilst a green die has its faces marked 1, 1, 1, 2, 2, 3. Both dice are thrown simultaneously. Let the random variable X denote the number of dots that show up on the red die, and Y the number of dots on the green die. Compute the mean and variance of the random variable $Z = X + Y$.

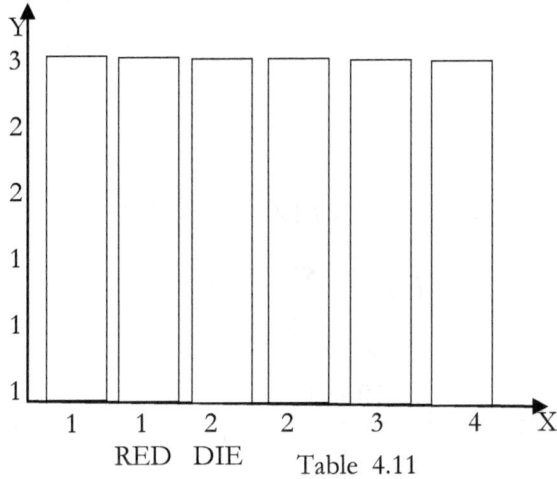

RED DIE   Table 4.11

$Z = X + Y$ has values 2, 3, 4. 5, 6, 7, with probability distribution given in the table below:

Table 4.12

Z	2	3	4	5	6	7
$P(Z = z)$	$\dfrac{6}{36}$	$\dfrac{10}{36}$	$\dfrac{9}{36}$	$\dfrac{7}{36}$	$\dfrac{3}{36}$	$\dfrac{1}{36}$

Mean of Z : $E(Z) = \displaystyle\sum_{i=1}^{6} z_i \, p(z_i)$

$$= \left(2 \times \frac{6}{36}\right) + \left(3 \times \frac{10}{36}\right) + \left(4 \times \frac{9}{36}\right) + \left(5 \times \frac{7}{36}\right) + \left(6 \times \frac{3}{36}\right) + \left(7 \times \frac{1}{36}\right)$$

$$= \frac{12}{36} + \frac{30}{36} + \frac{36}{36} + \frac{35}{36} + \frac{18}{36} + \frac{7}{36}$$

$$= \frac{138}{36} = \frac{23}{6} \cong 3.83$$

EXAMPLE 4.9

Compute the variance of the random variable $Z = X + Y$ defined in example 4.8.

Solution

Variance of Z: $V(Z) = E(Z^2) - [E(Z)]^2$

$$E(Z^2) = \sum_{i=1}^{6} z_i^2 \, p(z_i)$$

$$= \left(2^2 \times \frac{6}{36}\right) + \left(3^2 \times \frac{10}{36}\right) + \left(4^2 \times \frac{9}{36}\right) + \left(5^2 \times \frac{7}{36}\right) + \left(6^2 \times \frac{3}{36}\right) + \left(7^2 \times \frac{1}{36}\right)$$

$$= \left(4 \times \frac{6}{36}\right) + \left(9 \times \frac{10}{36}\right) + \left(16 \times \frac{9}{36}\right) + \left(25 \times \frac{7}{36}\right) + \left(36 \times \frac{3}{36}\right) + \left(49 \times \frac{1}{36}\right)$$

$$= \frac{24}{36} + \frac{90}{36} + \frac{144}{36} + \frac{175}{36} + \frac{108}{36} + \frac{49}{36}$$

$$= \frac{590}{36}$$

But $E(Z) = \dfrac{23}{6}$ $\Rightarrow$ $[E(Z)]^2 = \dfrac{529}{36}$

Hence $V(Z) = E(Z)^2) - [E(Z)]^2 = \dfrac{590}{36} - \dfrac{529}{36} = \dfrac{61}{36}$

4.3. Function of a Discrete Random Variable

In section 4.1.3 above, we discussed the rule of measurement, whereby, a discrete random variable, X, assigns a number, x, to a sample point, s, within a sample space S associated with an experiment, E. That is, the random variable, X, defined as a function of the sample points. This random variable takes on the values $X(s)$ with probabilities specified by its probability distribution.

We are now interested in determining a new random variable that is a function, not directly of points within the sample space S, but of some other random variable whose values and probability distribution are known. In other words, our new random variable say Y can be determined directly by another random variable, say X, mentioned above. This means that there can be some other function, say H, that can be used to determine the values of Y directly, from the values of X, i.e. $Y = H(X)$, and for each point, s, within the sample space S, the value of Y is obtained as $Y(s) = H[X(s)]$, as shown in Figure 4.12 below.

Range X   Range H

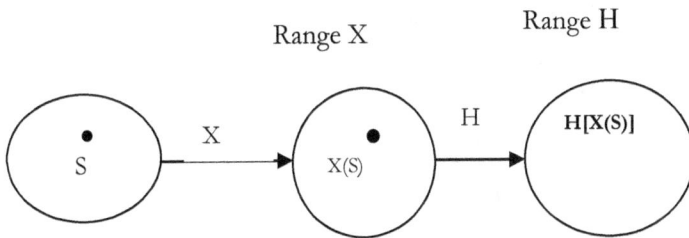

Figure 4.12

Since X is a discrete random variable, it follows immediately that Y is also a discrete random variable. The probability distribution of Y can be determined from the distribution of X.

Generally, suppose that the possible values of X are enumerated as x_1, x_2, $\cdots$, x_n, $\cdots$, and that $Y = H(X)$. Then certainly, the possible values of Y may be listed as $y_1 = H(x_1)$, $y_2 = H(x_2)$, ...,

$y_n = H(x_n)$, ... (Some of the above Y values may be the same, but this does not detract from the fact that these values may be enumerated). If $p(x_i) = P[X = x_i]$, are the respective probabilities assigned to the values of X, and H is a function such that to each value of y there corresponds exactly one value of X, then the probability distribution of Y is obtained as follows:

Possible values of Y: $y_i = H(x_i)$, $i = 1, 2, 3, \cdots n$

Probabilities of Y : $p(y_i) = P[Y = y_i] = p(x_i)$,

that is, deducing the probabilities of X from the earlier probabilities in the original sample space. This is illustrated in the example below:

EXAMPLE 4.10

Let us consider Example 4.4 above: A fair die is tossed once. Let X be the random variable assuming the following values:

$X = 1$ if the die shows 1

$X = 2$ if the die shows 2 or 3, and

$X = 3$ if the die shows 4, 5 or 6. The probability distribution of X is given in the Table below:

X	1	2	3
$P[X = x_i]$	$\dfrac{1}{6}$	$\dfrac{2}{6}$	$\dfrac{3}{6}$

Table 4.13

The mean of X, i.e. $E(X) = \sum_{i=1}^{3} x_i \, p(x_i) = \dfrac{7}{3}$

We may now define a new random variable $Y = H(X) = 2X + 1$ and wish to obtain:

(i)   The probability distribution of Y

(ii)   The mean and variance of Y.

SOLUTION:

(i)   The values of Y corresponding to the given values of X are given in

Table 4.14 below:

Value of X	$p(x)$	Value of $Y = H(X)$	$P(Y = y)$
1	$\dfrac{1}{6}$	$Y(1) = (2 \times 1) + 1 = 3$	$\dfrac{1}{6}$
2	$\dfrac{2}{6}$	$Y(2) = (2 \times 2) + 1 = 5$	$\dfrac{2}{6}$
3	$\dfrac{3}{6}$	$Y(3) = (2 \times 3) + 1 = 7$	$\dfrac{3}{6}$

Table 4.14

The probability distribution of Y is given as shown in the Table 4.15 below:

Y	3	5	7
$P[Y = y_i]$	$\dfrac{1}{6}$	$\dfrac{2}{6}$	$\dfrac{3}{6}$

Table 4.15

(ii) The expected value of Y may be calculated either from the probability function of Y given in Table 4.15 i.e.

$E(Y) = \sum_{i=1}^{3} y_i\, P(Y = y_i)$ or alternatively from the probability

function of X according to $E(Y) = E\{H(X)\} = \sum_{x} H(x)p(x)$

Case 1

$$E(Y) = \sum_{i=1}^{3} y_i p(y_i) = \left(3 \times \frac{1}{6}\right) + \left(5 \times \frac{2}{6}\right) + \left(7 \times \frac{3}{6}\right)$$

$$= \frac{3}{6} + \frac{10}{6} + \frac{21}{6} = \frac{34}{6} = \frac{17}{3}$$

$$E(Y^2) = \sum_{i=1}^{3} y_i^2\, p(y_i) = \left(3^2 \times \frac{1}{6}\right) + \left(5^2 \times \frac{2}{6}\right) + \left(7^2 \times \frac{3}{6}\right)$$

$$= \frac{9}{6} + \frac{50}{6} + \frac{147}{6} = \frac{206}{6} = \frac{103}{3}$$

$$V(Y) = E(Y^2) - [E(Y)]^2 = \left(\frac{103}{3}\right) - \left(\frac{17}{3}\right)^2$$

$$= \frac{103}{3} - \frac{289}{9} = \frac{309 - 289}{9} = \frac{20}{9} \cong 2.222$$

<u>Case 2</u>

From Example 4.4, $E(X) = \frac{7}{3}$

$E(Y) = E\{H(X)\}$

$$= \sum_x H(x)p(x) = \{(2 \times 1) + 1\} \times \frac{1}{6}$$

$$+ \{(2 \times 2) + 1\} \times \frac{2}{6} + \{(2 \times 3) + 1\} \times \frac{3}{6}$$

$$= \left(3 \times \frac{1}{6}\right) + \left(5 \times \frac{2}{6}\right) + \left(7 \times \frac{3}{6}\right)$$

$$= \frac{3}{6} + \frac{10}{6} + \frac{21}{6} = \frac{34}{6}$$

$$V(Y) = E\left\{(H(X))^2\right\} - \{E[H(X)]\}^2$$

$$E\{(H(X))\}^2 = \Sigma_x [H(X)]^2 p(x) = \left(3^2 \times \frac{1}{6}\right) + \left(5^2 \times \frac{2}{6}\right) + (7^2 \times \frac{3}{6})$$

$$= \frac{9}{6} + \frac{50}{6} + \frac{147}{6} = \frac{206}{6}$$

$$[E\{H(X)\}]^2 = \left(\frac{34}{6}\right)^2 = \frac{1156}{36}$$

$$\therefore V(Y) = E[H(X)]^2 - [E(H(X))]^2 = \frac{1236}{36} - \frac{1156}{36} = \frac{80}{36} \approx 2.222$$

Alternatively:

$$E(Y) = E\{H(X)\} = E\{2X + 1\} = 2E(X) + 1$$

$$= 2\left(\frac{7}{3}\right) + 1$$

$$= \frac{14}{3} + \frac{3}{3} = \frac{17}{3}$$

Now $E(X^2) = \left(1^2 \times \frac{1}{6}\right) + \left(2^2 \times \frac{2}{6}\right) + \left(3^2 \times \frac{3}{6}\right)$

$$= \frac{1}{6} + \frac{8}{6} + \frac{27}{6} = \frac{36}{6} = 6$$

$$V(X) = E(X^2) - [E(X)]^2 = 6 - \left(\frac{7}{3}\right)^2$$

$$= \frac{6(9) - 49}{9} = \frac{54 - 49}{9} = \frac{5}{9}$$

Now $V(Y) = V\{H(X)\} = V\{2X + 1\} = 2^2 V(X)$

$$= 4V(X)$$

$$= 4 \times \frac{5}{9} = \frac{20}{9} \approx 2.222$$

Sometimes the function H does not have the properties demonstrated in Example 4.8 above, that is, one and only one value of X does not correspond to one and only one value of Y. It may happen that several values of X lead to the same value of Y, as in the following examples:

EXAMPLE 4.11

The random variable X has the following probability distribution

X	-2	0	2
$P(X = x)$	$\frac{1}{2}$	$\frac{1}{4}$	$\frac{1}{4}$

Table 4.16

Compute the mean and variance of X.

$$E(X) = \sum_{i=1}^{3} x_i p(x_i) = (-2) \times \frac{1}{2} + (0) \times \frac{1}{4} + (2) \times \frac{1}{4}$$

$$= -1 + 0 + \frac{1}{2} = -\frac{1}{2}$$

$$E(X^2) = \sum_{i=1}^{3} x_i^2 p(x_i) = (-2)^2 \times \frac{1}{2} + (0)^2 \times \frac{1}{4} + (2)^2 \times \frac{1}{4}$$

$$= 2 + 0 + 1 = 3$$

$$V(X) = E(X^2) - [E(X)]^2 = 3 - \left(-\frac{1}{2}\right)^2 = 3 - \frac{1}{4}$$

$$= \frac{11}{4} = 2\frac{3}{4}$$

Define the random variable $Y = X^2$, then the probability distribution of Y is given as:

x	$p(x)$	$y = x^2$	$p(y)$
-2	$\frac{1}{2}$	4	$\frac{1}{2}$
0	$\frac{1}{4}$	0	$\frac{1}{4}$
2	$\frac{1}{4}$	4	$\frac{1}{4}$

Table 4.17

Y	0	4
$P(Y = y)$	$\frac{1}{4}$	$\frac{3}{4}$

Table 4.18

Note that in the table above, $P(Y = 4)$ is obtained from

$P(X = -2) = \dfrac{1}{2}$ and $P(X = 2) = \dfrac{1}{4}$ since the square of $X = -2$ and $X = 2$ both yield $Y = X^2 = 4$.

The general procedure for situations described in the above example is as follows: Let $x_1, x_2, \cdots x_n, \cdots$ represent the X - values having the property $y_i = H(x_i)$ for all i. Then

$$p(y_i) = P[Y = y_i] = p(x_1) + p(x_2) + \cdots + p(x_n) + \cdots$$

In other words, to evaluate the probability of the event $\{Y = y_i\}$, we find the equivalent event in terms of X (in the range space R_X) and then add all the corresponding probabilities, as shown in Figure 4.13.

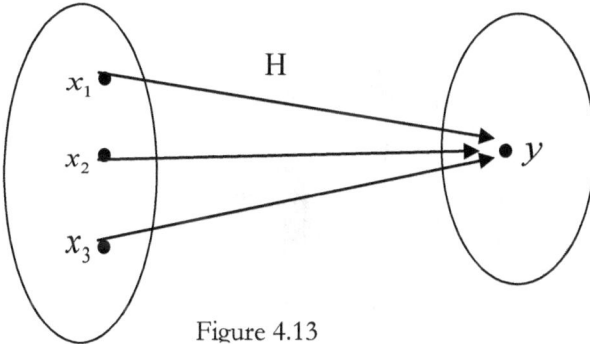

Figure 4.13

If $Y = H(X)$, then $p(y_i) = p[x_1] + p[x_2] + p[x_3]$

Case 1

$$E(Y) = \sum_{i=1}^{2} y_i p(y_i) = \left(0 \times \dfrac{1}{4}\right) + \left(4 \times \dfrac{3}{4}\right) = 0 + 3 = 3$$

$$V(Y) = E(Y^2) - [E(Y)]^2$$

$$E(Y^2) = \sum_{i=1}^{2} y_i^2 p(y_i) = \left(0^2 \times \dfrac{1}{4}\right) + \left(4^2 \times \dfrac{3}{4}\right) = 0 + 12 = 12$$

$$\therefore V(Y) = 12 - 9 = 3$$

Case 2

$$E(Y) = E[H(X)] = \sum_x H(x)p(x)$$

$$= \left\{(-2)^2 \times \frac{1}{2}\right\} + \left\{0^2 \times \frac{1}{4}\right\} + \left\{2^2 \times \frac{1}{4}\right\}$$

$$= \left\{4 \times \frac{1}{2}\right\} + \left\{0 \times \frac{1}{4}\right\} + \left\{4 \times \frac{1}{4}\right\}$$

$$= 2 + 0 + 1 = 3$$

$$V(Y) = E[H(X)^2] - [E(H(X))]^2$$

$$E[H(X)^2] = \sum_x H(x)^2 \, p(x)$$

$$= [(-2)^2]^2 \times \frac{1}{2} + [(0)^2]^2 \times \frac{1}{4} + [2^2]^2 \times \frac{1}{4}$$

$$= 16 \times \frac{1}{2} + 0 + 16 \times \frac{1}{4}$$

$$= 8 + 0 + 4 = 12$$

$$\therefore \ V(Y) = E[H(X)^2] - [E(H(X))]^2 = 12 - 3^2 = 12 - 9 = 3$$

EXAMPLE 4.12

The faces of a die are numbered 1, 2, 2, 3, 3 and 3 respectively. Let X and Y denote respectively the results obtained on the face of the die when the die is tossed twice. Obtain

(i)   The probability distribution of the random variable
        $Z = X - Y$

(ii)   Compute the mean and variance of Z

SOLUTION:

The sample space for the two tosses consists of 36 equally likely outcomes, shown below:

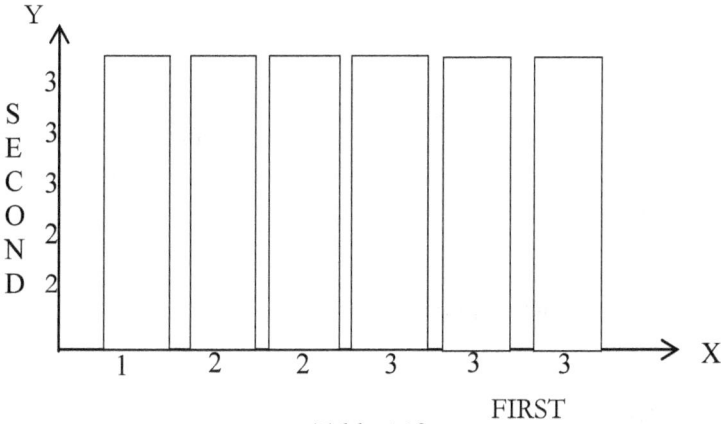

Table 4.19

The probability distribution of $Z = X - Y$ is given below:

Z	-2	-1	0	1	2
$P(Z = z)$	$\dfrac{3}{36}$	$\dfrac{8}{36}$	$\dfrac{14}{36}$	$\dfrac{8}{36}$	$\dfrac{3}{36}$

Table 4.20

$$E(Z) = \sum_i z_i P(z_i) = -2\left(\frac{3}{36}\right) - 1\left(\frac{8}{36}\right) + 0\left(\frac{14}{36}\right) + 1\left(\frac{8}{36}\right) + 2\left(\frac{3}{36}\right)$$

$$= 0$$

$$E(Z^2) = \sum_i z_i^2 P(z_i) = (-2)^2\left(\frac{3}{36}\right) + (-1)^2\left(\frac{8}{36}\right) + 0^2\left(\frac{14}{36}\right) + 1^2\left(\frac{8}{36}\right) + 2^2\left(\frac{3}{36}\right)$$

$$= \frac{12}{36} + \frac{8}{36} + \frac{0}{36} + \frac{8}{36} + \frac{12}{36}$$

$$= \frac{40}{36} = \frac{10}{9} = 1\frac{1}{9}$$

$$V(Z) = E(Z^2) - [E(Z)]^2 = \frac{10}{9} - 0 = \frac{10}{9}$$

NOTE:

The distribution of $Z = X - Y$ is symmetric about the point 0. We are therefore not surprised that $E(Z)$ turned out to be 0. We shall be dealing with situations of similar kind in later chapters.

4.3.1 Cumulative Distribution Function

Let s be a point within a sample space S, associated with an experiment E. If X is a random variable and x is a number, we have defined the event $[X \le x] = [s : X(s) \le x]$. For a given random variable, this event and its probability depend on the value used for x. Thus $P[X \le x]$ is a function whose value depends on x. In other words, it is a distribution function (or cumulative distribution function) of the random variable, and is denoted by $F(x)$ or $F_x(x)$.

$F(x) = P[X \le x]$, $-\infty < x < \infty$. Its domain is the set of real numbers, and its range is a set of numbers between 0 and 1 (since every value of $F(x)$ is a probability).

4.3.1.1 Definition (Cumulative Distribution Function)

Let X be a discrete random variable. We define $F(x)$ to be the cumulative distribution function of the random variable X (abbreviated as cdf) where

$$F(x) = P[X \le x] = P[X = x_j \ \text{for some } x_j \le x] = \sum_{x_j \le x} P[X = x_j] = \sum_{x_j \le x} p(x_j),$$

where the sum is taken over all indices j satisfying $x_j \le x$, and $p(x) = P[X = x]$. This relation is illustrated in Figure 4.14 below:

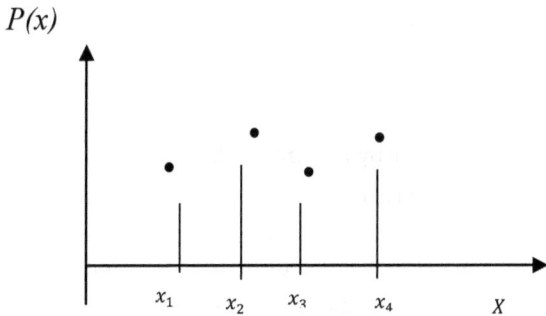

P(x)

Figure 4.14

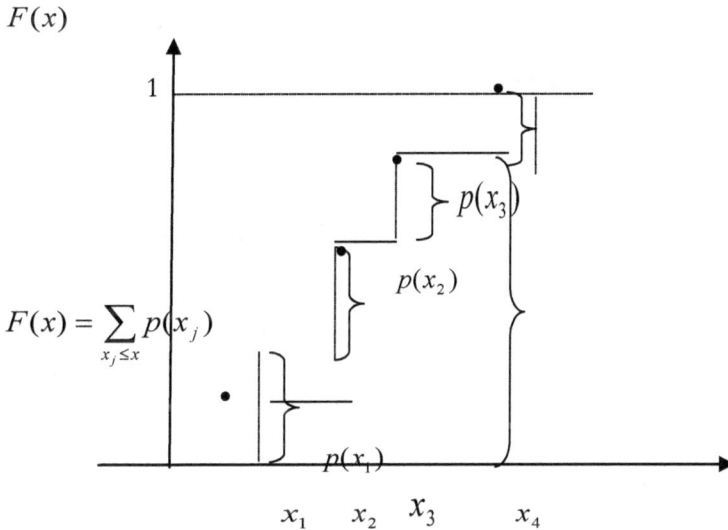

F(x)

$$F(x) = \sum_{x_j \le x} p(x_j)$$

Figure 4.15

Relation between the distribution function $F(x)$ and the probability density function $p(x)$ for a discrete random variable.

Since $p(x) = P[X = x]$ and $F(x) = \sum_{x_j \le x} p(x_j)$, we observe that $p(x)$ completely determines probabilities for events of the form:
$[X \le a]$, $[X < a]$, $[X > a]$, $[X \ge a]$, $[X = a]$,
$[a < X < b]$, $[a \le X < b]$, $[a < X \le b]$, $[a \le X \le b]$

We also notice from $F(x) = \sum_{x_j \le x} p(x_j)$ that the probabilities for these events are defined completely either by the probability distribution function $F(x)$ or by the probability mass function $p(x)$, when X is a discrete random variable. Hence, in practical application, we work with either $F(x)$ or $p(x)$, whichever is the easier one to use. Several properties of $F(x)$ follow directly from definition 4.3.1.1 above.

(a) $0 \le F(x) \le 1$ for $-\infty < x < \infty$ this follows because $F(x)$ is a probability.

(b) If $x_1 \le x_2$ then $F(x_1) \le F(x_2)$. This implies that if X increases, then $F(x)$ must also increase (or at least not decrease). $F(x)$ is said to be a non-decreasing function of X.

(c) $\lim_{x \to +\infty} F(x) = F(\infty) = 1$ $\lim_{x \to -\infty} F(x) = F(-\infty) = 0$.

Intuitively these formulae are clear since it would seem that, as
$x \to +\infty$, $F(x) = P[X \le x] \to P[X < +\infty] = P[s] = 1$ and as
$x \to -\infty$, $F(x) = P[X \le x] \to P[X \le -\infty] = P[\phi] = 0$, since
$-\infty < X(s) < \infty$ for every $s \in S$.

4.3.1.2 Determining Probabilities from Cumulative Distribution Function

We observed in the previous section that $F(x)$ gives probabilities of the form $P[X \le x]$. But we also want to compute other probabilities, such as $P[a < X \le b]$ or $P[X > a]$. We accomplish this as follows:

(a)   Since $[X \le a] \cup [X > a] = S$ we have
$P[X \le a] + P[X > a] = P(S) = 1$
$$\therefore \quad P[X > a] = 1 - P[X \le a] = 1 - F(a)$$

(b)   Since $[X \leq a] \cup [a < X \leq b] = [X \leq b]$, we have

$P[X \leq a] + P[a < X \leq b] = P[X \leq b]$ Thus

$P[a < X \leq b] = P[X \leq b] - P[X \leq a]$

$$= F(b) - F(a)$$

For example, $P\left[X \leq \frac{1}{4}\right] = F\left(\frac{1}{4}\right)$, $P\left[0 < X \leq \frac{1}{4}\right] = F\left(\frac{1}{4}\right) - F(0)$,

$P\left[X > \frac{1}{4}\right] = 1 - F\left(\frac{1}{4}\right)$

EXAMPLE   4.13

The cumulative distribution function $F_X(x)$ for the random variable X is given as,

$$F_X(x) = \begin{cases} 0 & if & x < 1 \\ 0.3 & if & 1 \leq x < 2 \\ 0.8 & if & 2 \leq x < 3 \\ 1 & if & x \geq 3 \end{cases}$$

(a)   Deduce the probability distribution for X.
(b)   Compute $P(X \leq 2)$ and $P[2 < x \leq 3]$.

SOLUTION:
(a)   The probability distribution for X is given in the table below.

X	1	2	3
$P(X)$	0.3	0.5	0.2

Table 4.21

(b) $P[X \leq 2] = P[x = 1] + P[x = 2] = 0.3 + 0.5 = 0.8$ and

$P[2 < x \leq 3] = P[x = 3] = 0.2$

Equivalently, we can read this from the distribution table given above.

EXAMPLE 4.12

The cumulative distribution function for the random variable X is given as

$$F_X(x) = \begin{cases} 0 & \text{if } x < 3 \\ \frac{1}{4} & \text{if } 3 \leq x < 5 \\ 1 & \text{if } x \geq 5 \end{cases}$$

Compute the mean and standard deviation of X.

SOLUTION :

The probability distribution for X is given in the table below:

Table 4.21

X	3	5
P(x)	$\frac{1}{4}$	$\frac{3}{4}$

Mean: $E(X) = \sum xP(x) = \left(3 \times \frac{1}{4}\right) + \left(5 \times \frac{3}{4}\right)$

$$\frac{3}{4} + \frac{15}{4} = \frac{18}{4} = \frac{9}{2}$$

Variance: $V(x) = E(x^2) - [E(x)]^2$

$$E(x^2) = \sum x^2 P(x) = \left(3^2 \times \frac{1}{4}\right) + \left(5^2 \times \frac{3}{4}\right) = \frac{9}{4} + \frac{75}{4} = \frac{84}{4}$$

$$[E(x)]^2 = \left(\frac{9}{2}\right)^2 = \frac{81}{4}$$

$$\therefore \quad V(x) = E(x^2) - [E(x)]^2 = \frac{84}{4} - \frac{81}{4} = \frac{3}{4}$$

Standard deviation of $X = \sqrt{V(X)} = \sqrt{0.75} \cong 0.866$

EXAMPLE 4.13

The random variable X has probability distribution given in the table below. Derive the cumulative distribution function $F_X(x)$.

Table 4.22

X	0	1	4
P(X)	0.2	0.4	0.4

SOLUTION:

The cumulative distribution function $F_X(x)$ for X is given as:

$$F_X(x) = \begin{cases} 0 & if \quad x < 0 \\ 0.2 & if \quad 0 \le x < 1 \\ 0.6 & if \quad 1 \le x < 4 \\ 1 & if \quad x \ge 4 \end{cases}$$

4.4. Bivariate (or Joint) Discrete Distributions

In our study of random variables we have, so far, considered only a single random variable, X, defined on a given sample space S. Many other applications, however, may require the use of two or more random variables, say (X,Y) or (X,Y,Z) etc. defined on the sample space S. In this section, we will consider the problem of two random variables and their associated distribution and density functions. Let X and Y be two random variables defined on some given sample space S, (for example X = colour and Y = size of a randomly selected marble from a box). Each random variable is a function whose domain consists of points s, the sample space S, and whose range is some real number in the x-y plane, with the values $X(s) = x$ and $Y(s) = y$ measured along the X and Y co-ordinate axis respectively. Unlike the one dimensional random variable which we discussed in the previous section (that is, a single point, s, in the domain yielding a single point $X(s) = x$ in the range space which is a section of the real line or the whole real line), the two dimensional random variables (X, Y) yield, for each point, s, in the domain, a point ($x = X(s)$, $y = Y(s)$) in the range space which is a plane. The values (x, y) can be treated as the coordinates of some point in the plane. The pair (X, Y) of random variables, can be considered as a function that to each point, s, in the sample space assigns a point (x, y), in the plane (see for example Figures 4.16 and 4.1.7 below).

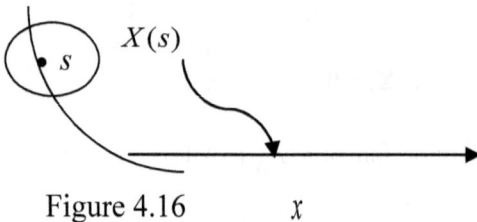

Figure 4.16 x

One dimensional case; X is a function from S to the real line.

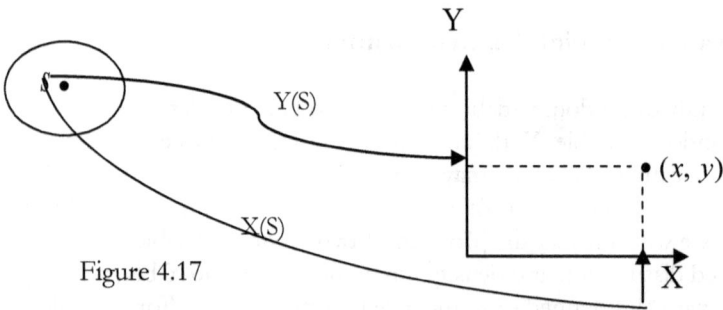

Figure 4.17

Two dimensional case: (x, y) as a function from S to the x-y plane

Thus, for fixed values x_1 and y_1, the event $A = \{X = x_1, \text{ and } Y = y_1\}$ (i.e. the event for which the results $X = x_1$ and $Y = y_1$ are both observed) is the set of all sample points, s, having X-value x_1, and Y-value y_1, $\{s : X(s) = x_1 \text{ and } Y(s) = y_1\}$. The corresponding region

in the x-y plane, therefore, will consist only of the single point with co-ordinates
$(x_1, \ y_1)$. (See Figure 4.18 below) :

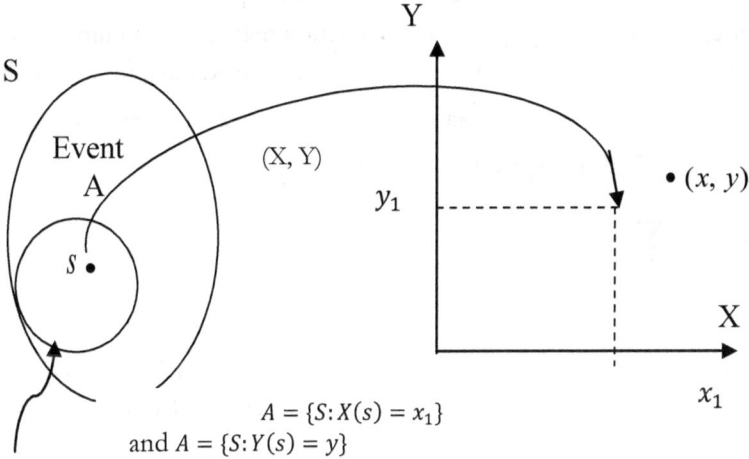

$$A = \{S:X(s) = x_1\}$$
and $A = \{S:Y(s) = y\}$

Figure 4.18: correspondence between points in the plane and event
A, a subset of S.

4.4.1. Definition (Joint Random Variables)

Let E be an experiment and S a sample space associated with E . If X and Y are two functions assigning the real values $x = X(s)$ and $y = Y(s)$ respectively, to each outcome $s \in S$, we call $(X, \ Y)$ a two dimensional random variable (sometimes called Bivariate or Joint random variables).

4.4.2 The Joint Probability Mass Function

A pair of random variables $(X, \ Y)$ are said to have a discrete joint distribution if X and Y individually are discrete random variables. Let us now assume that X and Y are integer –valued random variables (i.e. the range space consists of a set of integers). Whenever (x, y) is a pair of integers, define

$$p(x, \ y) = P[X = x \text{ and } Y = y].$$

The fact that X and Y are integer valued random variables implies that the sum $\sum \sum p(x, y)$ over all integer pairs $(x, \ y)$ is 1. If the pair

(x, y) is not a point in the joint range of (X, Y), then $p(x, y) = 0$. Clearly, $p(x, y)$ is a function that assigns a number to each pair of integers (x, y). This function is termed the joint probability distribution function of X and Y. It has these properties:

(i) $\quad 0 \le p(x_i, y_i) \le 1 \quad$ for all (x_i, y_j)

(ii) $\quad \sum_{j=1}^{\infty} \sum_{i=1}^{\infty} p(x_i, y_j) = 1$

The set of triples $\{x_i, y_j, p(x_i, y_j)\}$, $i, j, = 1, 2, 3, \cdots\cdots$ is sometimes called the joint probability distribution of (X, Y), and is often presented in a joint table as shown in Example 4.14 below:

EXAMPLE 4.14

Suppose X and Y have the following Bivariate distribution. The X values are given as $X = 0, 1, 2$ whilst the Y values are given as $Y = 1, 2$ and 3

x / y	0	1	2
1	0.1	0.1	0
2	0.1	0.4	0.1
3	0	0.1	0.1

Table 4.23

The joint probabilities $p(x, y) = P[X = x$ and $Y = y]$ as given in Table 4.23 above, are written as follows:

$p(0, 1) = P[X = 0 \text{ and } Y = 1] = 0.1$
$p(0, 2) = P[X = 0 \text{ and } Y = 2] = 0.1$
$p(0, 3) = P[X = 0 \text{ and } Y = 3] = 0$
$p(1, 1) = P[X = 1 \text{ and } Y = 1] = 0.1$
$p(1, 2) = P[X = 1 \text{ and } Y = 2] = 0.4$

$p(1, 3) = P[X = 1 \text{ and } Y = 3] = 0.1$
$p(2, 1) = P[X = 2 \text{ and } Y = 1] = 0$
$p(2, 2) = P[X = 2 \text{ and } Y = 2] = 0.1$
$p(2, 3) = P[X = 2 \text{ and } Y = 3] = 0.1$

This is a true probability distribution since, for each $p(x, y)$

(i) $0 \le p(x, y) \le 1$ and
(ii)

$$\sum_{j}\sum_{i} p(x_i, y_j)$$

$$= p(0,1) + p(0,2) + p(0,3) + p(1,1) + p(1,2) + p(1,3)$$
$$+ p(2,1) + p(2,2) + p(2,3) = 1$$

EXAMPLE 4.15(a)

An honest coin is tossed three times, and the following random variables are defined.

X = the number of tails on the last two tosses $\{X = 0, 1 \text{ or } 2\}$
Y = the number of tails on the first toss $\{Y = 0 \text{ or } 1\}$

The sample space S consists of eight equally sample points, each with probability $\dfrac{1}{8}$.

$$S = \left\{ \begin{matrix} HHH & HHT & HTH & THH & THT & TTH & HTT & TTT \\ a & b & c & d & e & f & g & h \end{matrix} \right\}$$

The corresponding values of X and Y are given in Table 4.24 below:

	S	$p(s)$	$X(s)$	$Y(s)$
a	HHH	$\frac{1}{8}$	0	0
b	HHT	$\frac{1}{8}$	1	0
c	HTH	$\frac{1}{8}$	1	0
d	THH	$\frac{1}{8}$	0	1

ℓ	THT	$\frac{1}{8}$	1	1
f	TTH	$\frac{1}{8}$	1	1
g	HTT	$\frac{1}{8}$	2	0
h	TTT	$\frac{1}{8}$	2	1

Table 4.24

The joint probabilities of the following events:

$[X = 0 \text{ and } Y = 0] = \{a\}$

$[X = 0 \text{ and } Y = 1] = \{d\}$

$[X = 1 \text{ and } Y = 0] = \{b, c\}$

$[X = 1 \text{ and } Y = 1] = \{e, f\}$

$[X = 2 \text{ and } Y = 0] = \{g\}$

$[X = 2 \text{ and } Y = 1] = \{h\}$

are determined and given in Table 4.25 below:

x y	0	1	2
0	$\frac{1}{8}$	$\frac{2}{8}$	$\frac{1}{8}$
1	$\frac{1}{8}$	$\frac{2}{8}$	$\frac{1}{8}$

Table 4.25: Joint probability density function of X and Y.

We note that

$$p(0,0) = \frac{1}{8}, p(0,1) = \frac{1}{8}, p(1,0) = \frac{2}{8}, p(1,1) = \frac{2}{8},$$
$$p(2,0) = \frac{1}{8}, p(2,1) = \frac{1}{8}$$

4.4.3. The Joint Cumulative Distribution Function

The probability of the form $F_{XY}(x, y) = P[X \leq x, \text{ and } Y \leq y]$ depends on both values x and y and hence is a function of two variables termed the joint distribution function of X and Y. For simplicity, we drop the subscripts from $F_{XY}(x, y)$, and write $F(x, y) = P[X \leq x, \text{ and } Y \leq y]$.

4.4.3.1 Definition (Joint Cumulative Distribution Function)

Let (X, Y) be a two dimensional random variable. The cumulative distribution function (cdf) F of the two dimensional random variable (X, Y) is defined by

$$F(x, y) = P[X \leq x, \ Y \leq y] = \sum_{x_i \leq x} \sum_{y_j \leq y} p(x_i, y_j)$$

Note that $F(x, y)$ has the following properties analogous to those discussed for the one-dimensional cumulative distribution function $F(x)$.

(a) $0 \leq F(x, y) \leq 1$, for $-\infty < x < +\infty$, $-\infty < y < +\infty$. This follows since $F(x, y)$ is a probability.

(b) If $x_1 \leq x_2$ and $y_1 \leq y_2$, then $F(x_1, y_1) \leq F(x_2, y_2)$. This follows since the region consisting of all points (x, y) satisfying $x \leq x_1$ and $y \leq y_1$, will lie inside the region $x \leq x_2$ and $y \leq y_2$. It follows that $[X \leq x_1 \text{ and } Y \leq y_1]$ is a subset of $[X \leq x_2 \text{ and } Y \leq y_2]$ and hence $F(x_1, y_1) \leq F(x_2, y_2)$.

(c)   (i) $\lim\limits_{\substack{x \to \infty \\ y \to \infty}} F(x, y) = F(\infty, \infty) = 1$

        (ii) $\lim\limits_{x \to -\infty} F(x, y) = F(-\infty, y) = 0$

        (iii) $\lim\limits_{y \to -\infty} F(x, y) = F(x, -\infty) = 0$

These properties seem reasonable since the event $[X \leq x \text{ and } Y \leq y]$ approaches a certain event if $x \to \infty$, $y \to \infty$ and approaches an impossible event if $x \to -\infty$, or if $y \to -\infty$.

EXAMPLE 4.16

We re-visit Example 4.15 above, and now consider the following events:

$[X \leq 0, \ Y \leq 0] = \{a\}$
$[X \leq 0, \ Y \leq 1] = \{a, d\}$
$[X \leq 1, \ Y \leq 0] = \{a, b, c\}$
$[X \leq 1, \ Y \leq 1] = \{a, b, c, d, e, f\}$
$[X \leq 2, \ Y \leq 0] = \{a, b, c, g\}$
$[X \leq 2, \ Y \leq 1] = \{a, b, c, d, e, f, g, h\} = S$

Because the above events are all combinations of mutually exclusive, equally likely elementary events, we obtain the table distribution function $F(x, y)$, as shown in Table 4.26 below:

x / y	0	1	2
0	$\frac{1}{8}$	$\frac{3}{8}$	$\frac{4}{8}$
1	$\frac{2}{8}$	$\frac{6}{8}$	1

Table 4.26: Joint cumulative distribution function $F(x, y)$ of X and Y

We note that $F(0,0) = \frac{1}{8}, F(0,1) = \frac{2}{8}, F(1,0) = \frac{3}{8}, F(1,1) = \frac{6}{8}, F(2,0) = \frac{4}{8}$ and $F(2,1)=P(S)=1$.

For example

$$F(1,1) = \sum_{i \leq 1} \sum_{j \leq 1} p(x,y) = p(0,0) + p(0,1) + p(1,0) + p(1,1) = \frac{1}{8}$$

$$+ \frac{1}{8} + \frac{2}{8} + \frac{2}{8} = \frac{6}{8}$$

4.4.4 Marginal Probability Distributions

Consider the two-dimensional random variable (X, Y) whose joint probability distribution $p(x, y)$ is known. Suppose that we are interested only in determining the individual probability distribution of X (i.e. $p(x)$ or of Y (i.e. $p(y)$, yet have to make use of the joint distribution of X and Y. How can we compute the distribution of X or of Y? This can best be illustrated with an example.

EXAMPLE 4.17

Consider two random variables X and Y with joint probability function, $P[X = x, \ Y = y]$ shown in the table below.

$\begin{array}{c}\ x \\ y\ \end{array}$	2	4	Sum Σ
1	0.42	0.28	0.70
3	0.10	0.10	0.20
5	0.08	0.02	0.10
	0.60	0.40	Σ: 1.0

Table 4.27

We have $p(2, 1) = 0.42$, $p(2, 3) = 0.10$, $p(2, 5) = 0.08$, $p(4, 1) = 0.28$, $p(4, 3) = 0.10$, $p(4, 5) = 0.02$. In addition to these entries of the Table, let us also compute the marginal totals, that is, the sum of the 2 columns and 3 rows of the table. For the X values: 2 and 4

$$P(X = 2) = P(X = 2, Y = 1) + P(X = 2, Y = 3) + P(X = 2, Y = 5)$$

$$= 0.48 + 0.10 + 0.08 = 0.6$$

$$P(X = 4) = P(X = 4, Y = 1) + P(X = 4, Y = 3) + P(X = 4, Y = 5)$$

$$= 0.28 + 0.10 + 0.02 = 0.4$$

Thus, in general for any given X

$$p(x) = \sum_{y} p(x, y) \quad \dots\dots\dots\dots\dots\dots\dots\dots\dots \quad 4.4.4.1$$

Similarly, for the Y values: 1, 3 and 5

$$p(Y = 1) = p(X = 2, Y = 1) + p(X = 4, Y = 1)$$
$$= 0.42 + 0.28 = 0.70$$

$$p(Y = 3) = p(X = 2, Y = 3) + p(X = 4, Y = 3)$$
$$= 0.10 + 0.10 = 0.20$$

$$p(Y = 5) = p(X = 2, Y = 5) + p(X = 4, Y = 5)$$
$$= 0.08 + 0.02 = 0.10 \text{ and generally for any given } y$$

$$p(y) = \sum_x p(x, y) \quad \ldots\ldots\ldots\ldots\ldots\ldots\ldots \quad 4.4.4.2$$

These probabilities of equations 4.4.4.1 and 4.4.4.2 appearing in the column and row margins represent the probability distribution of X and Y respectively. The word marginal merely describes how the distribution of X (or Y) may be calculated whenever Y (or X) is in play. A column or row sum is calculated and placed in the margin, whenever we have a two-dimensional discrete random variable (X, Y).

EXAMPLE 4.18

Suppose that T=(X-1)(Y-2)=XY-2X-Y+2, and (X,Y) have the following joint distribution given in Table 4.28 below:

Y \ X	0	1	2
0	0.1	.01	0.0
2	0.1	0.4	0.1
4	0.0	0.1	0.1

Table 4.28

(i)   Find the marginal probabilities of X and Y respectively, hence

(ii)   Compute the expectation of T, E(T).

SOLUTION:

$$E(T) = E(XY - 2X - Y + 2)$$
$$= E(XY) - 2E(X) - E(Y) + 2$$

The possible combinations of the product $(x \times y)$ is obtained in the table below.

$(x \times y))$	$p(x, y)$	$(xy)p(x, y)$
(0×0)	0.1	$0 \times 0.1 = 0$
(0×2)	0.1	$0 \times 0.1 = 0$
(0×4)	0	$0 \times 0 = 0$
(1×0)	0.1	$0 \times 0.1 = 0$
(1×2)	0.4	$2 \times 0.4 = 0.8$
(1×4)	0.1	$4 \times 0.1 = 0.4$
(2×0)	0	$0 \times 0 = 0$
(2×2)	0.1	$4 \times 0.1 = 0.4$
(2×4)	0.1	$8 \times 0.1 = 0.8$

Table 4.29

Case 1

$$E(xy) = \sum_x \sum_y (xy)p = (0 \times 0.1) + (0 \times 0.1) + (0 \times 0) + (0 \times 0.1) + (2 \times 0.4)$$
$$+(4 \times 0.1) + (0 \times 0) + (4 \times 0.1) + (8 \times 0.1)$$
$$= 0.0 + 0.0 + 0 + 0 + 0.8 + 0.4 + 0 + 0.4 + 0.8$$
$$= 2.4$$

To compute $E(X)$ and $E(Y)$, we first compute the marginal probabilities of X and Y respectively:

$$P(X = 0) = 0.2, \ p(X = 1) = 0.6, p(X = 2) = 0.2$$

$$\therefore \quad E(X) = \sum_x xp(x) = (0 \times 0.2) + (1 \times 0.6) + (2 \times 0.2)$$
$$= 0 + 0.6 + 0.4 = 1$$

Similarly, $(Y = 0) = 0.2$, $P(Y = 2) = 0.6$, $P(Y = 4) = 0.2$

$$\therefore \quad E(Y) = \sum_y yp(y) = (0 \times 0.2) + (2 \times 0.6) + (4 \times 0.2)$$
$$= 0 + 1.2 + 0.8 = 2$$

Thus
$$E(T) = E(XY) - 2E(X) - E(Y) + 2$$
$$= 2.4 - 2 - 2 + 2 = 0.4$$

Case 2
Alternatively, the values of $xy - 2x - y + 2$ is given in the table below

$(x \times y) - 2x - y + 2$		T	$p(x,y)$	
$(0 \times 0) - (2 \times 0) - 0 + 2$	=	2	$p(0,0)$	= 0.1
$(0 \times 2) - (2 \times 0) - 2 + 2$	=	0	$p(0,2)$	= 0.1
$(0 \times 4) - (2 \times 0) - 4 + 2$	=	-2	$p(0,4)$	= 0
$(1 \times 0) - (2 \times 1) - 0 + 2$	=	0	$p(1,0)$	= 0.1
$(1 \times 2) - (2 \times 1) - 2 + 2$	=	0	$p(1,2)$	= 0.4
$(1 \times 4) - (2 \times 1) - 4 + 2$	=	0	$p(1,4)$	= 0.1
$(2 \times 0) - (2 \times 2) - 0 + 2$	=	-2	$p(2,0)$	= 0
$(2 \times 2) - (2 \times 2) - 2 + 2$	=	0	$p(2,2)$	= 0.1
$(2 \times 4) - (2 \times 2) - 4 + 2$	=	2	$p(2,4)$	= 0.1

Table 4.30

The probability distribution of T is given in the table below:

T	-2	0	2
$P(T = t)$	0	0.8	0.2

Table 4.31

$$E(T) = \sum_t tp(t) = (-2 \times 0) + (0 \times 0.8) + (2 \times 0.2)$$

$$E(T) = 0 + 0 + 0.4$$

The above example illustrates that if $T = H(X,Y)$ is a function of two random variables, X and Y, then

(i) $E(T) = E\{H(X,Y)\} = \sum_x \sum_y H(x,y)p(x,y)$ as applied in case 1 above, or alternatively, applying the definition to the probability distribution of T, we have

(ii) $E(T) = \sum_t tp(t)$, as applied in case 2 above.

4.4.5 Conditional Distributions

Consider a random vector $\left(\underset{\sim}{X}, \; \underset{\sim}{Y} \right)$, comprising of two random variables X and Y. Suppose we now have information that a given event of positive probability, $P(Y = y_j)$, has occurred for which the random variable Y takes a single value $Y(s)$ in this set. We are now interested in the conditional distribution of the random *variable* X, given that Y takes on a value in some set. Thus, if (X, Y) has a discrete distribution in the x-y plane, with values (x_i, y_j) for $i = 1, 2, \cdots, n$ and $j = 1, 2, \cdots, m$: the conditional probabilities for X given $Y = y_j$ are computed as follows:

$$P(x_i/y_j) = P(X = i/Y = y_j) = \frac{P(X = x_i, Y = y_j)}{P(Y = y_j)} = \frac{p(x_i, y_j)}{p(y_j)} \quad \text{if } p(y_j) > 0$$

Similarly,

$$p(y_j/x_i) = P(Y = y_j/X = x_i) = \frac{P(X = x_i, Y = y_j)}{P(X = x_j)}$$

$$= \frac{p(x_i, y_j)}{p(x_i)} \quad \text{if } p(x_i) > 0$$

where $p(x_i, y_j)$ is the joint probability density function of X and Y, and $p(x_i), p(y_j)$ are the marginal probability functions of X and Y respectively. For a given i or j, $p(x_i/y_j)$ and $p(y_j/x_i)$ satisfy all the conditions for a probability distribution. We have $p(x_i/y_j) \geq 0$ and also

$$\sum_{i=1}^{\infty} p(x_i/y_j) = \sum_{i=1}^{\infty} \frac{p(x_i, y_j)}{p(y_j)} = \frac{p(y_j)}{p(y_j)} = 1$$

Similarly, $p(y_j/x_i) \geq 0$, and

$$\sum_{j=1}^{\infty} p(y_j/x_i) = \sum_{j=1}^{\infty} \frac{p(x_i, y_j)}{p(x_i)} = \frac{p(x_i)}{p(x_i)} = 1$$

Marginal probabilities can be expressed in terms of conditional probabilities. Thus,

$$P(X = x_i) = p(x_i) = \sum_j (x_i, y_j) = \sum_j p(x_i/y_j)p(Y = y_j) \qquad \ldots\ldots\ldots 4.4.5.1$$

Also

$$P(Y = y_j) = p(y_j) = \sum_i p(x_i, y_j)$$

$$= \sum_i p(y_j/x_i)p(X = x_i) \qquad \ldots\ldots\ldots 4.4.5.2$$

Equations 4.4.5.1 and 4.4.5.2 are referred to as the "Law of Total Probability".

EXAMPLE 4.19

The random vector (X, Y) has a discrete distribution of probabilities, defined by the accompanying Table 4.32 below:

X Y	1	2	3
2	$\frac{1}{12}$	$\frac{1}{6}$	$\frac{1}{12}$
3	$\frac{1}{6}$	0	$\frac{1}{6}$

4	0	$\frac{1}{3}$	0

Determine the following:

(a) $P(X = 1/X + Y \le 5)$
(b) $P(X = 2/Y = 2)$
(c) $P(Y = 2/X > 1)$
(d) $p(y/2)$ Table 4.32
(e) $p(x/2)$, $p(x/3)$ and $p(x/4)$; then verify that

$$P(X = x) = p(x) = \sum_y p(x/y)p(Y = y)$$

SOLUTION:

(a) The marginal probabilities:

$P(X = 1) = \frac{1}{12} + \frac{1}{6} + 0 = \frac{3}{12}$, $P(X = 2) = \frac{1}{6} + 0 + \frac{1}{3} = \frac{6}{12}$,

$P(X = 3) = \frac{3}{12}$

$P(Y = 2) = \frac{1}{12} + \frac{1}{6} + \frac{1}{12} = \frac{4}{12}$, $P(Y = 3) = \frac{4}{12}$, $P(Y = 4) = \frac{4}{12}$

$$P(X = 1/X + Y \le 5) = \frac{P(X = 1 \cap Y = 2) + P(X = 1 \cap Y = 3) + P(X = 1 \cap Y = 4)}{P(X + Y \le 5)}$$

$$P(X = 1 \cap Y = 2) + P(X = 1 \cap Y = 3) + P(X = 1 \cap y = 4)$$
$$= \frac{1}{12} + \frac{2}{12} + 0 = \frac{3}{12}$$

$$[P(X + Y) \le 5] = p(1,2) + p(1,3) + p(1,4) + p(2,2) + p(2,3)$$
$$+ p(3,2) = \frac{6}{12}$$

$$\therefore \quad P(X = 1/X + Y \le 5) = \frac{3}{12} \div \frac{6}{12} = \frac{1}{2}$$

(b) $P(X = 2/Y = 2) = \frac{P(X=2 \cap Y=2)}{P(Y=2)} = \frac{^2/_{12}}{P(Y=2)}$

$$P(Y = 2) = \sum_x p(X/2) = p(1,2) + p(2,2) + p(3,2) = \frac{4}{12}$$

$$\therefore \quad P(X = 2/Y = 2) = \frac{2}{12} \div \frac{4}{12} = \frac{1}{2}$$

(c)

$$P(Y = 2/X > 1) = \frac{P(Y = 2 \cap X > 1)}{P(X > 1)} = \frac{P(Y = 2 \cap X = 2) + P(Y = 2 \cap X = 3)}{P(X = 2) + p(X = 3)}$$

But

$$P(X = 2) = \sum_y p(2, Y) = p(2, 2) + p(2, 3) + p(2, 4) = \frac{6}{12}$$

$$P(X = 3) = \sum_y p(3, Y) = p(3, 2) + p(3, 3) + p(3, 4) = \frac{3}{12}$$

$$\therefore \quad P(Y = 2/X > 1) = \frac{\frac{2}{12} + \frac{1}{12}}{\frac{6}{12} + \frac{3}{12}} = \frac{3}{9} = \frac{1}{3}$$

(d)

$$P(Y/2) = \frac{P(Y \cap X = 2)}{P(X = 2)}$$

Computed separately for $y = 2, 3, 4$

$$P(Y = 2/X = 2) = \frac{P(Y = 2 \cap X = 2)}{P(X = 2)} = \frac{\frac{2}{12}}{\frac{6}{12}} = \frac{1}{3}$$

$$P(Y = 3/X = 2) = \frac{p(Y = 3 \cap X = 2)}{p(X = 2)} = \frac{0}{6/12} = 0$$

$$P(Y = 4/X = 2) = \frac{P(Y = 4 \cap X = 2)}{P(X = 2)} = \frac{\frac{4}{12}}{\frac{6}{12}} = \frac{2}{3}$$

This distribution is given in the table below:

Y	2	3	4
$p(Y/2)$	$\frac{1}{3}$	0	$\frac{2}{3}$

Table 4.33

(e)

$$P(X/2) = P(X = x/Y = 2) = \frac{P(X = x \cap Y = 2)}{P(Y = 2)}$$

Computed separately for $X = 1, 2, 3$

$$P(X = 1/Y = 2) = \frac{P(X = 1, Y = 2)}{P(Y = 2)} = \frac{\frac{1}{12}}{\frac{4}{12}} = \frac{1}{4}$$

$$P(X = 2/Y = 2) = \frac{P(X = 2, Y = 2)}{P(Y = 2)} = \frac{\frac{2}{12}}{\frac{4}{12}} = \frac{1}{2}$$

$$P(X = 3/Y = 2) = \frac{P(X = 3, Y = 2)}{P(y = 2)} = \frac{\frac{1}{12}}{\frac{4}{12}} = \frac{1}{4}$$

Also

$$P(X/3) = P(X = x/Y = 3) = \frac{P(X = x \cap Y = 3)}{P(Y = 3)}$$

Computed separately for $x = 1, 2, 3$ and

$$P(X/4) = P(X = x/Y = 4) = \frac{P(X = x \cap Y = 4)}{P(Y = 4)}$$

Computed separately for $x = 1, 2, 3$ and

The distributions are given in the table below:

x	1	2	3
$p(X = x/Y = 2)$	$\frac{1}{4}$	$\frac{1}{2}$	$\frac{1}{4}$
$p(X/3)$	$\frac{1}{2}$	0	$\frac{1}{2}$
$p(X/4)$	0	1	0

Table 4.34

$$P(X = x) = \sum_y p(X/Y)p(Y = y)$$

$$X = 1: \quad P(X = 1) = \sum_y p(X = 1/Y)p(Y = y)$$

$$= P(X = 1/2)P(Y = 2) + P(X = 1/3)P(Y = 3)$$
$$+ P(X = 1/4)P(Y = 4)$$

$$= \left(\frac{1}{4} \times \frac{4}{12}\right) + \left(\frac{1}{2} \times \frac{4}{12}\right) + \left(0 \times \frac{4}{12}\right) = \frac{3}{12}$$

$$X = 2: \quad P(X = 2) = \sum_y p(X = 2/Y)p(Y = y)$$

$$= p(X = 2/2)p(Y = 2) + p(X = 2/3)p(Y = 3) + p(X = 2/4)p(Y = 4)$$

$$= \left(\frac{1}{2} \times \frac{4}{12}\right) + \left(0 \times \frac{4}{12}\right) + \left(1 \times \frac{4}{12}\right) = \frac{6}{12}$$

$$X = 3: \quad P(X = 3) = \sum_y p(X = 3/Y)p(Y = y)$$

$$= p(X = 3/2)p(Y = 2)$$
$$+ p(X = 3/3)p(Y = 3) + p(X = 3/4)p(Y = 4)$$

$$= \left(\frac{1}{4} \times \frac{4}{12}\right) + \left(\frac{1}{2} \times \frac{4}{12}\right) + \left(0 \times \frac{4}{12}\right) = \frac{3}{12}$$

4.4.6 Conditional Expectation

If (X, Y) is a two dimensional discrete random variable, we define the conditional expectation of X for a given $Y = y_j$ as $E(X/y_j) = \sum_{i=1}^{\infty} x_i p(x_i/y_j)$. The conditional expectation of Y for a given X is defined as $E(Y/x_i) = \sum_{j=1}^{\infty} y_j p(y_j/x_i)$. Since $p(x/y)$ represents the conditional probability mass function (pmf) of X for given $Y = y$,

$E(X/y)$ is the expectation of X conditioned on the event $(Y = y)$. In general $E(X/y)$ is a function of y and hence is a random variable. Similarly, $E(Y/x)$ is a function of x and is also a random variable. Since $E(Y/X)$ and $E(X/Y)$ are random variables, it will be meaningful to speak of their expectations.
That is,

$$E_Y[E(X/Y)] = E(X) \qquad \text{............} \qquad 4.4.6.1$$

and $\qquad E_X[E(Y/X)] = E(Y) \qquad \text{.........} \qquad 4.4.6.2$

In equation 4.4.6.1, the inner expectation is taken with respect to the conditional distribution of X given Y equals y, whilst the outer expectation is taken with respect to the probability distribution of Y. Similarly, for equation 4.4.6.2, the inner expectation is taken with respect to the conditional distribution of Y given X equals x, whilst the outer expectation is taken with respect to the probability distribution of X.

EXAMPLE 4.20
Suppose that packets consisting of varying quantity of items are supplied to a store each day. If W is the number of items in the packet, the probability distribution of the random variable W is given as follows:

w	5	10	15	20	25
$P(W = w)$	0.02	0.05	0.2	0.35	0.30

Table 4.35

The probability that any particular item is faulty is the same for all items and equals 0.15. If Q is the number of faulty items each day, what is the expected value of Q?

For given $W = w$, Q has a Binomial distribution. Since W is itself a random variable, we have
$E(Q) = E_W[E(Q/W)]$. However, $E(Q/W) = 0.15W$ since for given W, Q has a Binomial distribution. Hence, $E(Q) = E(0.15W) = 0.15E(W)$. But

$$E(W) = \sum wP(w)$$
$$= (5 \times 0.02) + (10 \times 0.05) + (15 \times 0.2) + (20 \times 0.35) + (25 \times 0.30)$$
$$= 0.1 + 0.5 + 3.0 + 7.0 + 7.5 = 18.1$$

$\therefore$ $E(Q) = 0.15E(W) = 0.15 \times 18.1 = 2.715$

4.4.7 Conditional Variance

Given a joint distribution for (X, Y), the variance of the conditional variable $X/Y = y$, (or X/y) is the conditional variance,

$$Var(X/y) = E\{[X - E(X/y)]^2/y\} = E(X^2/y) - [E(X/y)]^2$$

This is a function of the value y given as the observed value of Y. Applying the parallel axis theorem to the distribution of X/y:

$$E\{[X - E(X)]^2/y\} = Var(X/y) + [E(X/y) - E(X)]^2$$

Averaging in y with respect to the distribution of Y yields

$$E\{[X - E(X)]^2/y\} = E[Var(X/y)] + E\{[E(X/y) - E(X)]^2\}$$

The last term can be interpreted as a variance, the variance of $h(Y)$, where $h(y) = E(X/y)$ is a function of y, and $E[h(y)] = E_Y[E(X/Y)] = E(X)$. Thus

$$Var\ X = E[Var(X/Y)] + Var[E(X/Y)]$$

or in words, the variance of X is the mean of its conditional variance plus the variance of its conditional mean.

EXAMPLE 4.21

Let (X, Y) have the discrete distribution given in the following table of probabilities.

$_Y\backslash^X$	0	2	4
1	$\frac{1}{4}$	0	$\frac{1}{4}$
3	$\frac{1}{12}$	$\frac{1}{3}$	$\frac{1}{12}$

Verify that $E(Var\ X/Y) =$ Table 4.36

SOLUTION:

$$E_Y[Var\ (X/Y)] = \sum_y Var\ (X/y)p(Y = y)$$
$$= Var(X/Y = 1)p(Y = 1) + Var(X/Y = 3)p(Y = 3)$$

$$Var(X/y = 1) = E(X^2/y = 1) - [E(X/y = 1)]^2$$

$$E(X^2/y = 1) = \sum_x x^2 p(x/y = 1)$$
$$= 0^2 p(0/y = 1) + 2^2 p(2/y = 1) + 4^2 p(4/y = 1)$$

$$= 0^2 \frac{p(X = 0, Y = 1)}{p(Y = 1)} + 2^2 \frac{p(X = 2, Y = 1)}{p(Y = 1)} + 4^2 \frac{p(X = 4, Y = 1)}{p(Y = 1)}$$

$$= \left(0 \times \frac{1}{2}\right) + (4 \times 0) + \left(16 \times \frac{1}{2}\right) = 8$$

$$E(X/y = 1) = \sum_x xp(x/y = 1)$$
$$= 0\ p(0/y = 1) + 2\ p(2/y = 1) + 4\ p(4/y = 1) = 2$$

$$\therefore \quad \{E(X/y = 1)\}^2 = 4$$

Hence $Var(X/y = 1)p(Y = 1) = (8 - 4) \times \frac{2}{4} = 2$

Also
$$Var(X/y = 3) = E(X^2/y = 3) - [E(X/y = 3)]^2$$
$$E(X^2/y = 3) = \sum_x x^2 p(x/y = 3)$$
$$= 0^2 p(0/y = 3) + 2^2 p(2/y = 3) + 4^2 p(4/y = 3) = \frac{16}{3}$$

$$E(X/y = 3) = \sum_x xp(x/y = 3)$$
$$= 0\ p(0/y = 3) + 2\ p(2/y = 3) + 4\ p(4/y = 3)$$
$$= \frac{6}{3}$$

$$\therefore \quad [E(X/y = 3)]^2 = \frac{36}{9} = \frac{12}{3}$$

$$\therefore \quad Var(X/y = 3) = \left(\frac{16}{3} - \frac{12}{3}\right) = \frac{4}{3}$$

Hence

$$[Var(X/y = 3)]p(Y = 3) = \frac{4}{3} \times \frac{6}{12} = \frac{2}{3}$$

$$\therefore \quad E(Var\ X/Y) = \sum_y Var(X/y)p(Y = y) = 2 + \frac{2}{3} = \frac{8}{3}$$

$$Var(X) = E(X^2) - [E(X)]^2$$

$$E(X^2) = \sum_x x^2 p(x) = \left(0^2 \times \frac{4}{12}\right) + \left(2^2 \times \frac{4}{12}\right) + \left(4^2 \times \frac{4}{12}\right) = \frac{80}{12}$$

$$E(X) = \sum_x x\,p(x) = \left(0 \times \frac{4}{12}\right) + \left(2 \times \frac{4}{12}\right) + \left(4 \times \frac{4}{12}\right) = \frac{24}{12} = 2$$

$$\therefore \quad [E(X)]^2 = 4$$

Hence

$$Var(X) = \frac{80}{12} - 4 = \frac{80 - 48}{12} = \frac{32}{12} = \frac{8}{3}$$

$$E(Var\ X/y) = Var(X) = \frac{8}{3}$$

4.5. Covariance and Correlation

The term covariance itself comes from the notion that this quantity attempts to measure the degree to which certain variables are related. Let

X, Y denote two random variables that have joint probability density function (pdf), $p(x, y)$. We try to develop a measure of how well these two variables are linearly related. If X tends to be large when Y is large, and small when Y is small, then the covariance will be positive; if large values of X tend to correspond to small values of Y, (or vice versa), the covariance will be negative; if knowing that X is large does not give much information as to the tendency of Y, the covariance will be close to zero. Our measure of how the variables move together should not depend on where the variables happen to be centred (when plotted in the x-y plane). We could take deviations from their means and obtain $[X - E(X)]$ and $[Y - E(Y)]$ respectively.

A good measure of how X and Y vary together can be obtained by summing the products $[X - E(X)][Y - E(Y)]$ for each pair of (X, Y) values, attaching the associated probability value to each set, this is called covariance, that is

$$COV(X, Y) = \sum_x \sum_y [X - E(X)][Y - E(Y)]p(x, y)$$

$$= E\{[X - E(X)][Y - E(Y)]\}$$

$$= E(XY) - E(X)E(Y) \qquad \ldots\ldots\ldots \quad 4.5.1$$

Notice that $COV(X, X) = Var(X)$ and $COV(Y, Y) = Var(Y)$. Also, if either X or Y is a constant (not a random variable), then $COV(X, k_1) = COV(k_2, Y) = 0$ ie. covariance between a random variable and a constant is zero.

The covariance as given in equation 4.5.1 depends upon the units in which X and Y are measured, in this sense it is a poor measure of coherence. For instance, if X and Y were the weights of individuals measured in grams, the covariance will be $1000 \times 1000 = 10^6$ (one million) times bigger than its value measured in kilograms. To obtain a measure of coherence that does not have this anomaly, a new measure referred to as "correlation coefficient" is used, denoted and defined as follows:

$$\rho_{XY} = \frac{COV(X,Y)}{\sqrt{Var(X)}\sqrt{Var(Y)}} = \frac{E\{[X - E(X)][Y - E(Y)]\}}{\sqrt{Var(X)}\sqrt{Var(Y)}}$$

$$= \frac{E(XY) - E(X)E(Y)}{\sqrt{Var(X)}\sqrt{Var(Y)}}$$

Here, a change in the units in X would introduce a factor $\sqrt{Var(X)}$ as well as in $E\{[X - E(cX)][Y - E(Y)]\}$, and these could cancel, similarly, a scale change in Y would introduce a factor $\sqrt{Var(Y)}$ as well as in $E\{[X - E(cX)][Y - E(Y)]\}$, and these would cancel.

The value ρ_{XY} is always bounded: $-1 \le \rho \le 1$. Whenever X and Y have a perfect positive linear relation [i.e. if all the (X, Y) points lie on a straight line with positive slope], then ρ takes on the limiting value of $+1$; similarly, if there is a perfect negative linear relation, [i.e. if all the (X, Y) points lie on a straight line with negative slope], then ρ takes on the limiting value of -1.

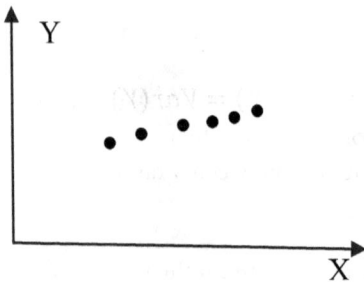

Figure 4.19 : $\rho = +1$
Perfect positive correlation

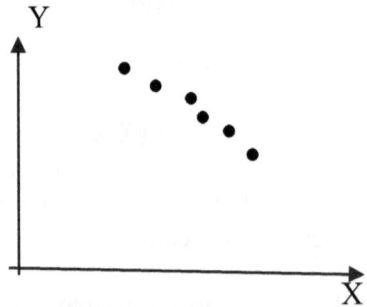

Figure 4.20 : $\rho = -1$ Perfect negative correlation

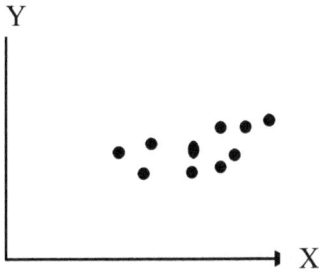

Figure 4.21:
$0 < \rho < 1$ positive correlation

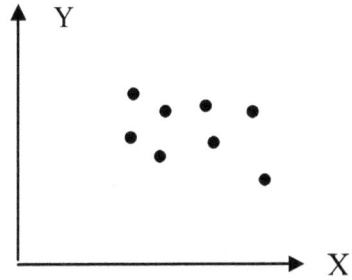

Figure 4.22:
$0 < \rho < 1$ positive correlation

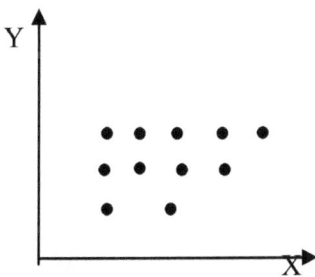

Figure 4.23:
$\rho = 0$ no correlation
coefficient

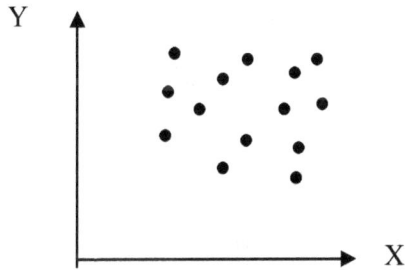

Figure 4.24:
$\rho = 0$ no correlation coefficient

REMARKS

(i)   If X and Y are two random variables and, a and b are constants then

(a) $V(aX + bY) = a^2 V(X) + b^2 V(Y) + 2ab\, COV(X, Y)$ where, $COV(X, Y)$ is the covariance between X and Y

(b) $V(aX - bY) = a^2 V(X) + b^2 V(Y) - 2ab\, COV(X, Y)$

(ii)   Let $X_1, X_2, \cdots X_n$, be n random variables, then
$V(X_1 + X_2 + \cdots X_n) = V(X_1) + V(X_2) + \cdots + V(X_n) + 2\sum_{i<j} COV(X_i, X_j)$

(iii) $V(aX + k) = a^2 V(X) + V(k) + 2a\, COV(X, k)$
$= a^2 V(X) + 0 + 0$

Variance of a constant is zero, and also, the covariance between a random variable and a constant is zero.

EXAMPLE 4.22

The random variable X and Y have standard deviations $D(X) = 1$ and $D(Y) = 2$ respectively. The correlation coefficient $\rho_{XY} = 0.05$. Compute the variance:
$V(5X - 4Y)$.

SOLUTION.
$V(aX - bY) = a^2V(X) + b^2V(Y) - 2ab\, COV(X,Y)$ where a, b are constants.

$$\therefore \quad V(5X - 4Y) = 5^2V(X) + 4^2V(Y) - 2(5 \times 4)COV(X,Y)$$

Since $\rho_{XY} = \dfrac{COV(X, Y)}{\sqrt{V(X)}\sqrt{V(Y)}}$, we have $COV(X,Y) = \rho_{XY}\sqrt{V(X)}\sqrt{V(Y)} = 0.05 \times 1 \times 2 = 0.1$

$D(X) = \sqrt{V(X)} = 1 \qquad \Rightarrow V(X) = 1$

$D(Y) = \sqrt{V(Y)} = 2 \qquad \Rightarrow V(Y) = 4$

$\therefore \quad V(5X - 4Y) = (25 \times 1) + (16 \times 4) - 2(5 \times 4) \times 0.1$
$$= 25 + 64 - 4 = 85$$

EXAMPLE 4.23(a)

Consider the random variable X and Y. If $E(X) = 4, E(X^2) = 20$, $E(Y) = 3, E(Y^2) = 18$ and $E(XY) = 10$. Compute the correlation coefficient ρ_{XY}.

SOLUTION:

$$\rho_{XY} = \frac{E(XY) - E(X)E(Y)}{\sqrt{V(X)}\sqrt{V(Y)}} = \frac{10 - (4 \times 3)}{\sqrt{V(X)}\sqrt{V(Y)}}$$

$V(X) = E(X^2) - [E(X)]^2 = 20 - 4^2 = 4 \quad \Rightarrow \quad \sqrt{V(X)} = 2$

$V(Y) = E(Y^2) - [E(Y)]^2 = 18 - 3^2 = 9 \quad \Rightarrow \quad \sqrt{V(Y)} = 3$

Hence $\quad \rho_{XY} = \dfrac{10-12}{2 \times 3} = -\dfrac{2}{6} = -\dfrac{1}{3} \cong -0.3333$

EXAMPLE 4.23(b)

The random variables X and Y have means $E(X) = 3$, and
$E(Y) = 2$; and standard deviations $\sigma_X = 1$ and $\sigma_Y = 2$ respectively.
The covariance of X and Y is equal to 0.05. Compute the mean and
variance of the random variables:

(i) $\qquad\qquad\qquad\qquad\qquad P = 5X - 4Y$
(ii) $\qquad\qquad\qquad\qquad\qquad Q = 5X + 4Y$

SOLUTION:
Mean of P: $\quad E(P) = E\{5X - 4Y\}$
$\qquad\qquad\qquad\quad = 5E(X) - 4E(Y)$
$\qquad\qquad\qquad\quad = (5 \times 3) - (4 \times 2)$
$\qquad\qquad\qquad\quad = 15 - 8 = 7$

Mean of Q: $\quad E(Q) = E\{5X + 4Y\}$
$\qquad\qquad\qquad\quad = 5E(X) + 4E(Y)$
$\qquad\qquad\qquad\quad = (5 \times 3) + (4 \times 2)$
$\qquad\qquad\qquad\quad = 15 + 8 = 23$

NOTE.

If $R = aX \pm bY$, $\quad V(R) = a^2 V(X) + b^2 V(Y) \pm 2ab\, Cov(X, Y)$

Variance of P: $\quad V(P) = V(5X - 4Y)$
$\qquad\qquad\qquad\qquad = 5^2 V(X) + 4^2 V(Y) - 2(5 \times 4) \times 0.05$
$\qquad\qquad\qquad\qquad = (25 \times 1) + (16 \times 4) - (2 \times 20 \times 0.05)$
$\qquad\qquad\qquad\qquad = 25 + 64 - 2 = 87$

Variance of Q: $\quad V(Q) = V(5X + 4Y)$
$\qquad\qquad\qquad\qquad = 5^2 V(X) + 4^2 V(Y) + 2(5 \times 4) \times 0.05$
$\qquad\qquad\qquad\qquad = (25 \times 1) + (16 \times 4) + (2 \times 20 \times 0.05)$
$\qquad\qquad\qquad\qquad = 25 + 64 + 2 = 91$

4.6. Independent Random Variables

We define the independence of two random variables in terms of the independence of two events. The random variables X and Y are called independent if for every x and y, the event $(X = x)$ and $(Y = y)$ are independent. This property of independence has the following consequences:

(i) X and Y are independent if and only if
$$p(X = x/Y = y) = p(X = x) \qquad \ldots\ldots\ldots\ldots \qquad 4.6.1$$
i.e. $p(x/y) = p(x)$: the conditional probability that the event $= x$ occurs given that event $Y = y$ has already occurred, is simply the probability of the event $X = x$ occurring. In other words, the prior occurrence of the event $Y = y$ does not in any way affect the probability distribution of the event $X = x$ occurring.

(ii)   We know by definition that, $p(x/y) = \frac{p(x,y)}{p(y)}$ if $p(y) > 0$

and $p(y/x) = \frac{p(x,y)}{p(x)}$, if $p(x) > 0$

Independence of X and Y would mean that

$$\text{and} \quad \left. \begin{array}{l} p(x/y)p(y) = p(x)p(y) = p(x,y) \\ p(y/x)p(x) = p(y)p(x) = p(x,y) \end{array} \right\}$$
4.6.2

for all x and y. In words, two events X and Y are independent of each other, if and only if their joint probability density function $p(x, y)$, equals the product $p(x)p(y)$, of their individual marginal probability functions.

(iii)   If two events X and Y are independent, their correlation coefficient $\rho_{XY} = 0$. The converse is not necessarily true. By definition of ρ_{XY}:

$$\rho_{XY} = \frac{\text{COV}(X, Y)}{\sqrt{V(X)}\sqrt{V(Y)}} = \frac{E(XY) - E(X)E(Y)}{\sqrt{V(X)}\sqrt{V(Y)}} = 0$$

This would mean that, covariance of X and Y, equals zero.
That is,

$COV(X, Y) = E(XY) - E(X)E(Y) = 0$. Hence,

$$E(XY) = E(X)E(Y) \qquad \ldots\ldots\ldots\ldots\ldots \quad 4.6.3$$

Equations 4.6.1, 4.6.2, 4.6.3 are the main consequences of the random variables X and Y being independent of each other. In most applications, it suffices to use any of the three equations to verify that, any two random variables, (X, Y) having a joint distribution are independent of each other.

EXAMPLE 4.24

Two independent random variables X and Y have means and variances: $(X) = 5, V(X) = 3, \ E(Y) = 1$ and $V(Y) = 2$ respectively. Compute the mean and variance of the random variable $Z = 2X - 3Y - 7$.

SOLUTION:
$$
\begin{aligned}
E(Z) &= E(2X - 3Y - 7) \\
&= 2E(X) - 3E(Y) - E(7) \\
&= (2 \times 5) - (3 \times 1) - 7 = 0
\end{aligned}
$$

$$
\begin{aligned}
V(Z) &= 2^2 V(X) + 3^2 V(Y) + (-1)^2 V(7) - 2(2 \times 3)COV(X, Y) \\
&\quad - (2 \times 2)COV(X, 7) \\
&\quad -(2 \times 3)COV(Y, 7) \\
&= (4 \times 3) + (9 \times 2) + 0 - (12 \times 0) - (4 \times 0) - (6 \times 0) \\
V(Z) &= 12 + 18 = 30
\end{aligned}
$$
NOTE:
[X and Y are independent; $\therefore$ $COV(X, Y) = 0$. Also $COV(X, 7) = 0$, $COV(Y, 7) = 0$: covariance between a random variable and a constant equals zero. $V(7) = 0$: variance of a constant value equals zero].

EXAMPLE 4.25

Two random variables X and Y have joint probability distribution $P(X = x, Y = y)$, given in Table 4.37 below: We wish to determine whether X and Y are independent.

x \ y	2	4
1	0.42	0.28
3	0.10	0.10
5	0.08	0.02

Table 4.37

X \ Y	2	4	$P(Y = y)$		
1	0.42	0.28	0.7	←	$P(Y = 1)$
3	0.10	0.10	0.2	←	$P(Y = 3)$
5	0.08	0.02	0.1	←	$P(Y = 5)$
$P(X = x)$	0.60	0.40	$\sum : 1$		
	↑	↑			
	$P(X = 2)$	$P(X = 4)$			

Table 4.38

SOLUTION:
The marginal probabilities $P(X = x)$ and $P(Y = y)$ for values of x and y are given in Table 4.38 above. Table 4.39 below gives the joint probabilities and the corresponding products of the marginal probabilities.

	Joint Probabilities $p(x, y)$	Product of marginal probabilities $p(x)p(y)$	Is $p(x, y) = p(x)p(y)$?
(a)	$p(2, 1) = 0.42$	$p(X = 2)p(Y = 1)$ $= 0.6$ $\times 0.7$ $= 0.42$	Yes $p(2, 1) = p_x(2)p_y(1)$
(b)	$p(2, 3) = 0.10$	$p(X = 2)p(Y = 3)$ $= 0.6$ $\times 0.2$ $= 0.12$	No $p(2, 3) \neq p_x(2)p_y(3)$

(c)	$p(2,5) = 0.08$	$p(X = 2)p(Y = 5)$ $= 0.6$ $\times 0.1$ $= 0.06$	No $p(2,5) \neq$ $p_x(2)p_y(5)$
(d)	$p(4,1) = 0.28$	$p(X = 4)p(Y = 1)$ $= 0.4$ $\times 0.7$ $= 0.28$	Yes $p(4,1) =$ $p_x(4)p_y(1)$
(e)	$p(4,3) = 0.10$	$p(X = 4)p(Y = 3)$ $= 0.4$ $\times 0.2$ $= 0.08$	No $p(4,3) \neq$ $p_x(4)p_y(3)$
(f)	$p(4,5) = 0.02$	$p(X = 4)p(Y = 5)$ $= 0.4$ $\times 0.1$ $= 0.04$	No $p(4,5) \neq$ $p_x(4)p_y(5)$

Table 4.39

CONCLUSION:

X and Y are not independent. For X and Y to be independent, $p(x,y)$ must be equal to $p(x)p(y)$ for all combinations of the values of X and Y.

EXAMPLE 4.26

The random variables X and Y can assume only the values 0 and 1. If $E(X) = \frac{1}{4}$, $E(Y) = \frac{1}{3}$ and $E(XY) = \frac{1}{12}$. Compute the joint and marginal probabilities of X and Y.

SOLUTION:

$$E(X) = \sum_x x\, p(X = x)$$

i.e. $\frac{1}{4} = 0 \times p(X = 0) + 1 \times p(X = 1)$ giving $p(X = 1) = \frac{1}{4}$

$$\therefore \ P(X = 0) = 1 - p(X = 1) = \frac{3}{4}$$

Similarly,

$$E(Y) = \sum_{y} y \, p(Y = y)$$

i.e. $\frac{1}{3} = 0 \times p(Y = 0) + 1 \times p(Y = 1)$ giving $p(Y = 1) = \frac{1}{3}$

$$\therefore \ P(Y = 0) = 1 - p(Y = 1) = \frac{2}{3}$$

Note that $E(XY) = \frac{1}{12} = E(X)E(Y) = \left(\frac{1}{4} \times \frac{1}{3}\right)$ implying that X and Y are independent. It is also true that $p(x, y) = p(x)p(y)$ holds for all combinations of x and y.

$$p(0, 0) = P(X = 0)P(Y = 0) = \frac{3}{4} \times \frac{2}{3} = \frac{6}{12} = \frac{1}{2}$$

$$p(0, 1) = P(X = 0)P(Y = 1) = \frac{3}{4} \times \frac{1}{3} = \frac{3}{12} = \frac{1}{4}$$

$$p(1, 0) = P(X = 1)P(Y = 0) = \frac{1}{4} \times \frac{2}{3} = \frac{2}{12} = \frac{1}{6}$$

$$p(1, 1) = P(X = 1)P(Y = 1) = \frac{1}{4} \times \frac{1}{3} = \frac{1}{12}$$

This information is presented in the table below.

X \ Y	0	1	$p(Y = y)$
0	$\frac{6}{12}$	$\frac{2}{12}$	$\frac{2}{3}$
1	$\frac{3}{12}$	$\frac{1}{12}$	$\frac{1}{3}$
$p(X = x)$	$\frac{3}{4}$	$\frac{1}{4}$	$\Sigma: 1$

Table 4.40

4.7. Some Important Discrete Random Variables

4.7.1 Bernoulli Trial.

A fair coin is tossed once, and the outcome of this random experiment is observed: it is either 'Head' or 'Tail'. Suppose our main interest is to observe whether a Head shows up. It would be convenient to classify the outcome in two mutually exclusive ways (success or failure), and define a random variable on this outcome. Let the random variable say X, assume the value 1 if "Head shows up" and consider the experiment as being successful, or X may assume the value 0 if "Tail shows up" and the experiment is considered a failure. We can write it formally as

$$X = \begin{cases} 1 \text{ if success(Head shows up)} \\ 0 \text{ if failure (Tail shows up)} \end{cases}$$

Associated with each value of X, is a certain probability:
Let $p = p(\text{Head}) = p(X = 1)$ and $(1 - p) = p(\text{Tail}) = p(X = 0)$. The value $(1 - p)$ is sometimes denoted as q, and $p + q = 1$. The probability function of this single toss of a coin can be represented simply as shown in table below:

X	0	1
$p(X = x)$	$1 - p$	p

Table 4.41

Or it can be given by the formula $p(X = k) = p^k q^{1-k}$, for $k = 0, 1$. Individual trials of experiments of the above nature are referred to as Bernoulli trials. There are many random variables of this type. A few examples are experiments for which the outcomes can be classified as
(i)   Dead or alive
(ii)   Defective or Not defective
(iii)   Black or white
(iv)   Yes or no
(v)   Good or bad
and so on.

4.7.1.1 Mean and Variance of a Bernoulli Trial

For an individual Bernoulli trial, the mean $E(X)$ of the random variable X, is given as

$$E(X) = \sum_x x\, p(x) = 0\, p(X = 0) + 1\, p(X = 1)$$

$$E(X) = 0 + p = p$$

The variance, $V(X) = E(X^2) - [E(X)]^2$

$$E(X^2) = \sum_x x^2 p(x) = 0^2\, p(X = 0) + 1^2\, p(X = 1)$$

$$= 0 + p = p$$
$$\therefore\ V(X) = E(X^2) - [E(X)]^2 = p - p^2 = p(1-p) = pq$$

Where $q = 1 - p$ and $p + q = 1$.
Thus, for a Bernoulli trial, the mean of the random variable, $E(X) = p$, and the variance $V(X) = pq$.

EXAMPLE 4.27.

In an experiment of tossing a fair die, let A be the fixed event 'two' occurs. Then
$$P(A) = \frac{1}{6} = p \text{ Hence}$$

$$X = \begin{cases} 1, if\ a\ two\ ocurs \\ 0, if\ a\ two\ does\ not\ occurs \end{cases}$$

Then $E(X) = p = \frac{1}{6}$ and $Var(X) = p(1-p) = \frac{1}{6}\left(\frac{5}{6}\right) = \frac{5}{36}$

Standard deviation of X, $\sigma_x = \sqrt{\left(\frac{5}{36}\right)} = \frac{\sqrt{5}}{6}$

4.7.2 The Binomial Distribution

If the Bernoulli experiment is repeated over and over again, we obtain a sequence of independent Bernoulli trials. Assume that the probability of success say p is the same for each trial over the sequence of trials, thus the probability of failure, $(1-p) = q$ is also constant over the sequence. Assume further that, the classification of any particular trial (for example, success or failure) is independent of the classification of any other trial. Using these assumptions, we may speak of the trials as Bernoulli trials. (This does not mean that the same results are obtained for each trial, but simply that the experiments are the same). For the type of distribution we are discussing, it is important that we specify, beforehand, the number of times the experiment is to be repeated (that is, the total number of trials, say n, is to be finite and its value specified prior to performing the experiment). We shall then speak of n independent "trials" each resulting in either "success" or "failure". The random variables $X_1, X_2, \cdots\cdots, X_n$ which assign to the individual trials values 1 or 0 (depending on whether we have a success or a failure), keep count of the event, say A of interest, and their sum over n trials is precisely the number of "successes" among those trials. It is necessary to give a simple illustration.

EXAMPLE 4.28

Toss a coin three times. Each toss is considered as a Bernoulli trial. The trials are independent of each other, and the probability of head or tail remains constant over the sequence of three trials. The sample space

$S = \{HHH,\ HHT,\ HTH,\ THH,\ TTH,\ THT,\ HTT,\ TTT\}$

There are 8 individual outcomes in the sample space. Our interest is focused not on the individual outcomes of S. Rather we simply wish to know how many heads (successes) are found (irrespective of the order in which they occur).

Case1.

Suppose we are interested in counting 3 heads (3 successes) occurring in 3 tosses of the coin. The event $A_1 = \{H, H, H\}$. We define the random variables,

$X_1 = 1$: [Head, (success on first toss]
$X_2 = 1$ [Head, (success on second toss]
$X_3 = 1$ [Head, (success on third toss]

$\therefore$ total number of successes, $k = X_1 + X_2 + X_3 = 1 + 1 + 1 = 3$ Heads.

The sample space S, shows that the event (3 heads) can be accomplished in only one way. Hence the probability of three heads $= p \times p \times p = p^3 = 1p^3q^0 = \frac{1}{8}$

Case 2

Suppose we are interested in counting 2 heads (2 successes) in the three trials. The event

$A_2 = \{HHT, HTH, TH\}$. The result 2 heads can be accomplished in any of 3 mutually exclusive ways.

(a) $X_1 = 1$ [Head (success) on first toss]
$X_2 = 1$ [Head (success) on second toss]
$X_3 = 0$ [Tail (Failure) on the third toss], giving

$X_1 + X_2 + X_3 = k = 2$ heads, with probability
$p \times p \times (1 - p) = p^2(1 - p) = p^2q = \frac{1}{8}$

(b)   Similarly, (HTH): success on first toss, failure on second toss and success on third toss has probability equal to $p(1 - p)p = p^2(1 - p) = p^2q = \frac{1}{8}$ and

(c) (THH): Failure on first toss, success on second and third tosses, has probability equal to $(1 - p)p^2 = p^2q = \frac{1}{8}$

The required probability of obtaining 2 successes in 3 trials is, therefore, given as

$$3p^2(1 - p)^{3-2} = 3p^2(1 - p) = 3p^2q = \frac{3}{8}$$

Case 3

The probability of obtaining 1 head (one success) in the 3 trials can be derived as

$$3pq^2 = 3p(1 - p)^2 = 3p^1(1 - p)^{3-1} = \frac{3}{8}$$

Case 4

Finally, the probability of obtaining no head (i.e. no success) in the three trials, is equal to

$$(1-p)^3 = q^3 = p^0 q^3 = 1p^0 q^3 = \frac{1}{8}$$

Note that the sum of the probabilities equals 1, i.e.

$$1p^3 q^0 + 3p^2 q^1 + 3p^1 q^2 + 1p^0 q^3 \qquad \dots \quad 4.7.2.1$$
$$\frac{1}{8} \quad + \quad \frac{3}{8} \quad + \quad \frac{3}{8} \quad + \quad \frac{1}{8}$$

for the sum can be written as $(p+q)^3 = \left(\frac{1}{2}+\frac{1}{2}\right)^3$. The coefficients 1, 3, 3, 1 which appear in the terms of equation 4.7.2.1 above, are called Binomial coefficients in the expression of the Binomial Theorem, often met in elementary calculus.

The total number of successes, say $X = k$, is called a Binomial random variable. In general, the event $X = k$ successes includes $\binom{n}{k}$ outcomes, each with probability $p^k (1-p)^{n-k}$. The Binomial random variable X, has parameters n and p. Thus, if X has a Binomial distribution, we can write for short $X \sim \text{Bin}(n, p)$, or $X \sim b(n, p)$ where n denotes the fixed number of trials and p the probability of success.

In essence consider an experiment, E and let A be some fixed event associated with E. Suppose that $P(A) = p$ and hence $P(A^c) = 1 - p$. Consider n independent repetitions of E and assume $P(A) = p$ remains the same for all repetitions. Let X, the random variable, be defined as follows:
X = number of times the event A occurs
We call X a Binomial random variable with parameters n and p. The r.v, X takes possible values 0, 1, 2,…, n.

Note that, the Binomial distribution is the approximated model for a r.v X which measures the number of occurrences of a given attribute in a sample of n 'individuals' selected at random from a given population.

4.7.2.1 Mean and Variance of a Binomial random variable.

Let X be a Binomial variable on n repetitions. Then

$P(X = k \text{ successes}) = P(X = k) = \binom{n}{k} p^k (1-p)^{n-k}$, $k = 0, 1, 2, \cdots, n$.

i.   Legitimate

ii.   The mean $E(X)$ of the Binomial random variable is given as $E(X) = np$

iii.   and the variance

$V(X) = npq$.

This result will be discussed in later chapters.

PROOF:

i.   For legitimacy, Clearly, $P(X = k) \geq 0 \; \forall x$ and

$$\sum_{k=0}^{n} P(X = k) = \sum_{k=0}^{n} \binom{n}{k} p^k (1-p)^{n-k}$$
$$= (p + 1 - p)^n$$
$$= 1$$
$$= 1$$

By the binomial theorem which states that the expansion of

$$(a + b)^n = \sum_{k=0}^{n} \binom{n}{k} a^k (b)^{n-k} \text{ where } n \in Z^+$$

ii.

$$E(X) = \sum_{k=0}^{n} k P(X = k)$$

$$= \sum_{k=0}^{n} k \binom{n}{k} p^k (1-p)^{n-k}$$

$$= \sum_{k=0}^{n} k \frac{n!}{k!\,(n-k)!} p^k (1-p)^{n-k}$$

$$= \sum_{k=1}^{n} \frac{n!}{(k-1)!\,((n-k)!} p^k (1-p)^{n-k}$$

$$= \sum_{k=1}^{n} \frac{n(n-1)!}{(k-1)!\,(n-k)!} p^k (1-p)^{n-k}$$

$$= n \sum_{k=1}^{n} \frac{(n-1)!}{(k-1)!\,((n-k)!} p^k (1-p)^{n-k}$$

Let $s = k-1 \Rightarrow k = s+1$, s = 0 when k = 1 and s = n - 1 when k=n

$$E(X) = n \sum_{s=0}^{n-1} \frac{(n-1)!}{s!\,(n-(s+1))!} p^{s+1} (1-p)^{n-(s+1)}$$

$$= np \sum_{s=0}^{n-1} \binom{n-1}{s} p^s (1-p)^{n-(s+1)}$$

$$= np(p+1-p)^{n-1}$$
$$= np(1)^{n-1}$$
$$= np$$

iii.　　Variance of X, σ^2

$$Var(X) = E(X^2) - \left(E(X)\right)^2 = E(X^2) - (np)^2 \; from \; (ii)$$

But

$$E(X^2) = \sum_{k=0}^{n} k^2 P(X=k) = \sum_{k=0}^{n} k^2 \binom{n}{k} p^k (1-p)^{n-k}$$

$$= \sum_{k=0}^{n} k^2 \frac{n!}{k!\,(n-k)!} p^k (1-p)^{n-k}$$

$$= \sum_{k=0}^{n} k^2 \frac{n!}{k!\,(n-k)!} p^k (1-p)^{n-k}$$

$$= \sum_{k=0}^{n} [k(k-1)+k] \frac{n!}{k!\,(n-k)!} p^k (1-p)^{n-k}$$

$$= \sum_{k=0}^{n} k(k-1) \frac{n!}{k!\,(n-k)!} p^k (1-p)^{n-k}$$

$$+ \sum_{k=0}^{n} k \frac{n!}{k!\,(n-k)!} p^k (1-p)^{n-k}$$

$$= n(n-1) \sum_{k=0}^{n} \frac{(n-2)!}{(k-2)!\,(n-k)!} p^k (1-p)^{n-k} + E(X)$$

$$= n(n-1) \sum_{s=0}^{n-2} \binom{n-2}{s} p^{s+2} (1-p)^{n-(s+2)} + np,$$

On letting $S = k - 2$

$$= n(n-1)p^2 \sum_{s=0}^{n-2} \binom{n-2}{s} p^s (1-p)^{n-(s+2)} + np$$

$$= n(n-1)p^2 (p+1-p)^{n-2} + np$$

$$= n(n-1)p^2 + np$$

therefore

$$Var(X) = n(n-1)p^2 + np - (np)^2$$
$$= np - np^2$$
$$= np(1-p)$$
$$= npq$$

EXAMPLE 4.29

A box consists of 10 white and 20 black marbles. One selects at random 10 marbles with replacement). Compute the probability of obtaining at least 2 white marbles.

SOLUTION:

The probability of a white marble, $p = \dfrac{10}{30} = \dfrac{1}{3}$

Probability of a black marble, $(1-p) = \dfrac{20}{30} = \dfrac{2}{3}$

Ten marbles were selected, i.e. $n = 10$.

Probability of obtaining at least 2 white is the same as the probability of obtaining 2 or more white marbles, which is the same as $1 -$ Prob(no white) $-$ Prob(1 white).

$$P(X = 0: \text{no white}) = \binom{10}{0}\left(\frac{1}{3}\right)^0 \left(\frac{2}{3}\right)^{10} = \left(\frac{2}{3}\right)^{10} \cong 0.01734$$

$$P(X = 1: 1 \text{ white}) = \binom{10}{1}\left(\frac{1}{3}\right)^1\left(\frac{2}{3}\right)^9 = 10\left(\frac{2^9}{3^{10}}\right) \cong 0.08671$$

∴ probability of at least 2 white $= 1 - 0.01734 - 0.08671 =$
$0.89595 \approx 0.90$.

EXAMPLE 4.30

The proportion of defective items in a certain manufacturing process is
0.20. If 5 items are selected at random from the process, find the
(a) expected number of good items
(b) probability that none is defective
(c) Probability that at least three are defective.

SOLUTION

(a)   Let X be the number of good items. Then $X \sim b(5; 0:80)$.
Therefore
$$E(X) = np = 5(0:8) = 4$$
(b)   Let Y be the number of defective items. Then $Y \sim b(5; 0:20)$
$$P(Y = 0) = \binom{5}{0}(0.2)^0(0.8)^5 = 0.3277$$
(c)   In the notation of (b)
$$P(Y \geq 3) = P(Y = 3) + P(Y = 4) + P(Y = 5)$$

$$= \binom{5}{3}(0.2)^3(0.8)^2 + \binom{5}{4}(0.2)^4(0.8)^1$$
$$+ \binom{5}{5}(0.2)^5(0.8)^1$$
$$= 00579$$

Alternatively,
$$P(Y \geq 3) = 1 - P(Y \leq 2)$$

$$P(Y \leq 2) = \binom{5}{0}(0.2)^0(0.8)^5 + \binom{5}{1}(0.2)^1(0.8)^4$$
$$+ \binom{5}{2}(0.2)^2(0.8)^3 = 0.9421. Therefore$$
$$P(Y \geq 3) = 1 - 0.9421 = 0.0579$$

Calculation in R

(b) $P(Y = 0) = choose(5,0) * 0.2^0 * 0.8^5$ or

dbinom(0,5,0.2)

(c)

$P(Y \geq 3) => dbinom(3,5,0.2) + dbinom(4,5,0.2)$
$+ dbinom(5,5,0.2)$

OR

$P(Y \geq 3) = 1 - P(Y \leq 2) = 1 - pbinom(2,5,0.2)$

4.7.2.2. The Binomial Table

In actual practice, binomial probabilities are rarely calculated directly. They have been tabulated extensively for various values of n and p which were generated by computers.

The cumulative binomial distribution of $X \sim b(n; p)$ is given as

$$P(X \leq x) = \sum_{r=0}^{x} \binom{n}{r} p^r q^{n-r} = \sum_{r=0}^{x} b(r; n; p)$$

It is denoted as B$(x; n; p)$. Thus B$(x; n; p) = \sum_{r=0}^{x} b(r; n; p)$

In R, $P(X = x)$ can be found using the R command $dbinom(x, n, p)$ where as $P(X \leq x)$ is found using the R command $pbinom(x, n, p)$

EXAMPLE 4.31
If $X \sim b(10; 0:3)$, find using the binomial tables
(i) $P(X \leq 7)$
(ii) $P(X > 3)$
(iii) $P(X = 5)$

SOLUTION
(i) $P(X \leq 7) = B(7; 10; 0:3) = 0.9984$
In R, use $pbinom(7,10,0.3)$

(ii) $P(X > 3) = 1 - P X(\leq 3)$
$$= 1 - B(3; 10; 0:3)$$
$$= 0.3504$$
In R use, $1 - pbinom(3,10,0.3)$

(iii) $P(X = 5) = P(X \leq 5) - P(X \leq 4)$
$$= B(5; 10; 0:3) - B(4; 10; 0:3)$$
$$= 0.1029$$

In R use, $pbinom(5,10,0.3) - pbinom(4,10,0.3$ Or $dbinom(5,10,0.3)$

4.7.2.3 Normal Approximation To Binomial Distribution

If n is small, the individual terms of the Binomial distribution are relatively easy to compute. However, if n is relatively large, these computations become rather cumbersome. Fortunately, the Binomial probabilities have been tabulated. A method of approximating such probabilities with the aid of the Central Limit Theorem is explained in section 6.1.8 of chapter 6, where we treat the Normal distribution for large populations.

4.7.3 The Hypergeometric Distribution

Suppose that we have a lot of N items, r of which are defective and $(N - r)$ of which are non defective. Suppose that we choose at random, n items from the lot $(n \leq N)$ without replacement.

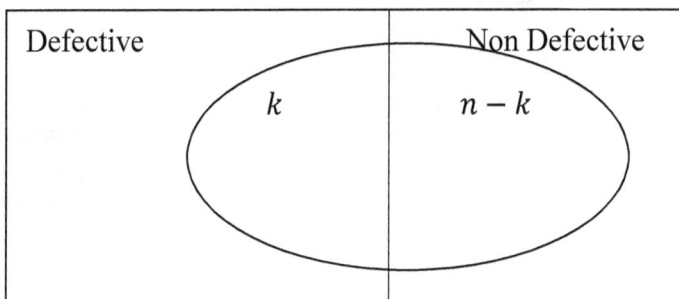

| Defective | Non Defective |
| k | $n - k$ |

Figure 4.25

Let X be the number of defectives found. The random variables $X = k$ if and only if we obtain precisely k defective items (from the r defectives in the lot) and precisely $(n - k)$ non defectives [from the $(N - r)$ non defectives in the lot]. We have

$$p(X = k) = \frac{\binom{r}{k}\binom{N-r}{n-k}}{\binom{N}{n}}, \quad k = 0, 1, 2, \cdots, n$$

4.7.3.1

as the ratio of the number of sample points containing exactly k defectives to the total number of sample points.

A discrete random variable having the probability distribution of equation 4.7.3.1 is said to have the Hypergeometric distribution and it is denoted as $X \sim H(N; r; n)$.

NOTE: With the Binomial distribution (see section 4.6.2) the probability of exactly k defectives among the n parts is given by

$$p(X = k) = \binom{n}{k} p^k (1 - p)^{n-k}, \quad k = 0, 1, 2, \cdots, n$$

This is used when sampling is done with replacement or when sampling from a "large" population.

The Hypergeometric distribution gives the probability of exactly k defectives in a sample of size n when sampling is done without replacement from a finite population. The important distinction between these distributions is that the events "the ith part chosen as defective" are independent if the sampling procedure does not deplete the population as the sample is drawn (sampling from an infinite population or sampling with replacement) as compared to the case in which the sampling procedure does deplete the population (sampling without replacement from a finite population).

4.7.3.1 Theorem (The Mean and the Variance of the Hypergeometric Distribution)

Let X have a hypergeometric distribution as given by equation 4.6.4 (a) above. Let $= \frac{r}{N}$, $q = 1 - p = 1 - \frac{r}{N}$. Then we have

(a) $E(X) = np$

(b) $V(X) = npq \frac{N-n}{N-1}$

(c) $p(X = k) \approx \binom{n}{k} p^k q^{n-k}$ for large N.

Some comments on the above results

$$Var(X) \sim npq \text{ as } N \rightarrow \infty \text{ 1 since } lim_{N \rightarrow \infty} \frac{N-n}{N-1} = 1$$

Thus, the above results show that for sufficiently large N, the distribution of X may be approximated by the Binomial distribution. This is because for large N relative to n, sampling without replacement is approximately the same as sampling with replacement which produces the Binomial distribution since $\frac{k}{N}$ is a constant at each choice (in this case, experiment). [Rule of thumb is that n should not exceed 5% of N].

EXAMPLE 4.32

A box consists of 2 black and 8 white marbles. Two marbles are selected at random from the box (without replacement). Compute the probability that both marbles are
(a)   Black
(b)   Have different colours

SOLUTION:

$$P(X = k) = \frac{\binom{r}{k}\binom{N-r}{n-k}}{\binom{N}{n}}, \quad k = 0, 1, 2$$

$N = 2$ Black $+ 8$ White $= 10$ Marbles

r = black Marbles = 2
$N - r$ = white marbles = 8
Sample n = 2 marbles

(a) $\quad P(X = 2 \text{ Black}) = \frac{\binom{2}{2}\binom{8}{0}}{\binom{10}{2}} = \frac{1}{45}$

(b) $\quad P(X = 1 \text{ White} + 1 \text{ Black}) = \frac{\binom{2}{1}\binom{8}{1}}{\binom{10}{2}} = \frac{(2)(8)}{45} = \frac{16}{45}$

EXAMPLE 4.31
A population consists of 4 families. The numbers of individuals in these families are 3, 4, 5 and 6 respectively. Four individuals are chosen at random (without replacement) from the population. Compute the probability that the sample consists of a member from each family.

SOLUTION:

$$P(X = \text{one from each family}) = \frac{\binom{3}{1}\binom{4}{1}\binom{5}{1}\binom{6}{1}}{\binom{18}{4}} = \frac{3 \times 4 \times 5 \times 6}{12 \times 15 \times 17}$$

$$= \frac{2}{17}$$

Using R:
Suppose that a population of size N consist of r units with the attribute and $N - r$ without it. If a sample size n is taken, without replacement, and X is the number of items with the attribute in the sample, by using R,

$P(X = k)$, can be found with the command $dhyper(k, r, N - r, n)$
and
$P(X \leq x)$, s given by the command $phper(k, r, N - r, n)$

EXAMPLE 4.32

From a group of 20 PhD statisticians, 10 are randomly selected for employment. What is the probability that the 10 selected include all the 5 best statisticians in the group of 20?

SOLUTION
For this example, N = 20; n = 10; and r = 5. We seek P(X = 5), where X denotes the number of best statisticians among the ten selected. Then

$$P(X = 5) = \frac{\binom{5}{5}\binom{20-5}{10-5}}{\binom{20}{10}}$$

$$= \frac{21}{1292}$$

$$= 0.0162$$

4.7.4   The Geometric Distribution

The Geometric distribution assumes, as in the case of the Binomial distribution, that the individual Bernoulli repetitions of the experiment E, are independent of each other. For each trial, the probability p of success is constant; hence the probability $1 - p = q$ also remains unchanged.

There is, however, a departure from the assumptions leading to the Binomial distribution. That is, the number of repetitions, n, is not predetermined in a Geometric distribution; it is a random variable. The experiment is repeated until the first success (for instance, the occurrence of first Head in repeated tosses of a fair coin) occurs. Suppose we define the random variable X as the number of trials required up to and including the first success. Then X assumes the possible values $1, 2, ...$ Since $X = k$ if and only if the first $(k - 1)$ repetitions of E result in failure while the kth trial results in success, we have

Event X: $\underbrace{\text{(failure)} \cap \text{(failure)} \cap \text{(failure)} \cdots \cap \text{(failure)}}_{(k-1)\text{times}} \cap \text{(success)}$

and

$$p(X = k) = (1 - p)(1 - p)(1 - p) \cdots (1 - p)(p) = (1 - p)^{k-1}p = q^{k-1}p \quad ....$$

For $k = 1, 2, \cdots$

A random variable having probability density function such as equation 4.7.4.1 is said to have a Geometric distribution.

$P(X = k) = q^{k-1}p, \quad k = 1, 2, \cdots$ defines a legitimate probability distribution. Obviously, $P(X = k) \geq 0, \ and.$

$$\sum_{k=1}^{\infty} P(X = k) = \sum_{k=1}^{\infty} q^{k-1}p = p \sum_{k=1}^{\infty} (q^0 + q^1 + q^2 + q^3 + \cdots)$$

$$= p(1 + q + q^2 + \cdots) = p\left[\frac{1}{1-q}\right] = 1$$

4.7.4.1 The Mean (or Expected Value), $E(X)$

The expected value,

$$E(X) = \sum_{k=1}^{\infty} kp(X = k) = \sum_{k=1}^{\infty} kpq^{k-1} = p \sum_{k=1}^{\infty} kq^{k-1} = p \sum_{k=1}^{\infty} \frac{d}{dq} q^k$$

And thus,

$$p\frac{d}{dq} \sum_{k=1}^{\infty} q^k = p\frac{d}{dq}\left[\frac{q}{1-q}\right] = \frac{1}{p}$$

The interchange of differentiation and summation is justified here since the series converges for $|q| < 1$.

4.7.4.2 The Variance, $V(X)$

A similar computation shows that,

$$V(X) = E(X^2) - [E(X)]^2$$

$$= \sum_{=1}^{\infty} k^2 p(X = k) - \left[\frac{1}{p}\right]^2$$

$$= \sum_{k=1}^{\infty} k^2 pq^{k-1} - \left[\frac{1}{p}\right]^2$$

$$= \sum_{k=1}^{\infty} \{k(k-1) + k\}pq^{k-1} - \left[\frac{1}{p}\right]^2 = \sum_{k=1}^{\infty} k(k-1)\, pq^{k-1}$$

$$+ \sum_{k=1}^{\infty} k\, pq^{k-1} - \left[\frac{1}{p}\right]^2$$

$$= pq \sum_{k=1}^{\infty} k(k-1)q^{k-2} + \frac{1}{p} - \left[\frac{1}{p}\right]^2 = pq \sum_{k=1}^{\infty} \frac{d^2}{dq^2}(q^k) + \frac{1}{p} - \left[\frac{1}{p}\right]^2$$ On interchanging the order of summation and differentiation gives

$$V(X) = pq \frac{d^2}{dq^2} \sum_{k=1}^{\infty} (q^k) + \frac{1}{p} - \left[\frac{1}{p}\right]^2 = pq \frac{d^2}{dq^2}\left(\frac{q}{1-q}\right) + \frac{1}{p} - \left[\frac{1}{p}\right]^2$$

$$pq = \frac{d}{dq}(1-q)^{-2} + \frac{1}{p} - \left[\frac{1}{p}\right]^2$$

$$= pq \left(\frac{2}{(1-q)^3}\right) + \frac{1}{p} - \left[\frac{1}{p}\right]^2 = \frac{2q}{p^2} + \frac{1}{p} - \left[\frac{1}{p}\right]^2$$

$$= \frac{2q}{p^2} + \frac{-(1-p)}{p^2} = \frac{q}{p^2}$$

We shall discuss in later chapters other techniques for deriving these results.

4.7.4.3 Forgetfulness Property

Let X represent the number of trials until the first success in a sequence of Bernoulli trials.

Suppose that we have observed a fixed number, k, of these trials and found them all to be failures. We are now concerned with the extra number of trials, say T, that must occur until the first success. Then $X = k + T$, or $T = X - k$. The conditional probability of T, given that $X > k$ (i.e. the first k trials until the first success is greater than k) can be derived as follows:

$$P[T = t/X > k] = P[X - k = t/X > k]$$

$$= P[X = k + t/X > k]$$

$$= \frac{P[X = k + t \text{ and } X > k]}{P[X > k]}$$

for $t = 1, 2, 3, \cdots$

The event $\{[X = k + t] \cap [X > k]\} = \{X = k + t\}$

$$P[T = t/X > k] = \frac{p[X = k + t]}{P[X > k]} = \frac{q^{k+t-1}p}{\sum_{j=k+1}^{\infty} q^{j-1}p} = \frac{q^{k+t-1}p}{q^k p \left(\frac{1}{1-q}\right)}$$

$$= q^{t-1}p$$

.... 4.7.4.3.1

Thus, we observe that conditional on $X > k$, the number of trials resulting until the first success, $T = X - k$, has the same probability density function $q^{t-1}p$, $t = 1, 2, 3, \cdots$ as X had originally(see for example equation 4.7.4.1 of section 4.7.4 above). This is termed the forgetfulness (or lack of memory) property of the Geometric probability density function. If a sequence of "failures" is observed in a sequence of Bernoulli trials, one need not remember how long the sequence is to determine the probabilities of additional trials needed until the first "success".

EXAMPLE 4.31

A lot of manufactured goods are sampled by testing items sequentially until the first defective is found. If k or more items are tested before the first defective is found, the lot is accepted as meeting specifications.

(a) If $k = 4$ and the lot contains 10% defective items, what is the probability that it will be accepted?

(b) What value should be given to k if one wishes that a lot having 40% defectives be rejected with probability of at least 0.80?

SOLUTION:

(a) $p(\text{defective}) = 0.1$ $p(\text{not defective}) = 0.9$

$$\therefore \ P(X = k) = (1 - q)^3 q = p^3 q = (0.9)^3(0.1) = 0.0729$$

(b) Rejecting lot with probability of at least 0.80 is the same as accepting the lot with probability of at most 0.2. Required probability
$$= P(X = k) = p^{k-1}q = (0.6)^{k-1}(0.4) \leq 0.2$$

i.e. $k - 1 = \dfrac{\ln(0.5)}{\ln(0.6)} \cong 1.357$ giving $k \leq 2.$

Using R
Computation of probabilities and cumulative probabilities associated with a geometric random variable can be evaluated using various software packages.

$P(X = x)$, probability associated with a geometric random variable X with success probability p can be found using the R command
$$dgeom(x - 1, p)$$
$P(X \leq x)$, the cumulative probability of a geometric distribution with success probability p is evaluated in R using the command
$$pgeom(x - 1, p)$$
Note that $(x - 1)$ is the number of failures before the first success and x is the number of trial on which the first success occurs.

4.7.5   The Poisson Density Function

If X is a discrete random variable with range $0, 1, 2, \ldots$ and

$$P_k = P(X = k) = \begin{cases} \dfrac{e^{-\lambda}\lambda^k}{k!} & ,for \ k = 1,2 \ldots \\ 0, & otherwise \end{cases}$$

We say that X has a Poisson distribution $P_0(\lambda)$, with parameter $\lambda > 0$. The probability function $P(X = k)$ is in fact a valid probability density function. That is,

(a) $\qquad 0 \le P_k \le 1$

(b) $\qquad \displaystyle\sum_{k=0}^{\infty} P_k = 1$

Each term in the formula for P_k is positive, and

$$\sum_{k=0}^{\infty} P_k = \sum_{k=0}^{\infty} \lambda^k \frac{e^{-\lambda}}{k!} = e^{-\lambda}\left[1 + \frac{\lambda}{1} + \frac{\lambda^2}{2!} + \frac{\lambda^3}{3!} + \cdots\right]$$

The sum in the bracket is the infinite series expansion of e^{λ}, hence

$$\sum_{k=0}^{\infty} P_k = e^{-\lambda} \sum_{k=0}^{\infty} \frac{\lambda^k}{k!} = e^{-\lambda} e^{\lambda} = 1$$

4.7.5.1 The Mean (or Expected Value) ,E(X), of the Poisson Distribution

The mean $E(X)$, of the Poisson probability density function is derived as:

$$E(X) = \sum_{k=0}^{\infty} kP(X = k) = \sum_{k=0}^{\infty} k \frac{e^{-\lambda}\lambda^k}{k!} = \sum_{k=1}^{\infty} \frac{e^{-\lambda}\lambda^k}{(k-1)!}$$

Letting $r = k - 1$, $E(X)$ becomes

$$E(X) = \sum_{r=0}^{\infty} \frac{e^{-\lambda}\lambda^{r+1}}{r!} = \lambda \sum_{r=0}^{\infty} e^{-\lambda}\frac{\lambda^r}{r!} = \lambda$$

4.7.5.2 The Variance, V(X), of the Poisson Distribution

The variance of X is given as:

$$V(X) = E(X^2) - [E(X)]^2$$

$$= \sum_{k=0}^{\infty} k^2 \frac{e^{-\lambda}\lambda^k}{k!} - \left[\sum_{k=0}^{\infty} k \frac{e^{-\lambda}\lambda^k}{k!}\right]^2$$

$$= \sum_{k=1}^{\infty} k \frac{e^{-\lambda}\lambda^k}{(k-1)!} - \lambda^2$$

Again letting $r = k - 1$, we obtain

$$E(X^2) = \sum_{r=0}^{\infty} (r+1) \frac{e^{-\lambda}\lambda^{r+1}}{r!} = \lambda \sum_{r=0}^{\infty} r \frac{e^{-\lambda}\lambda^r}{r!} + \lambda \sum_{r=0}^{\infty} \frac{e^{-\lambda}\lambda^r}{r!}$$

$$= \lambda^2 \sum_{r=1}^{\infty} \frac{e^{-\lambda}\lambda^{r-1}}{(r-1)!} + \lambda \sum_{r=0}^{\infty} \frac{e^{-\lambda}\lambda^r}{r!}$$

$$= \lambda^2 + \lambda$$

$$\therefore \quad V(X) = E(X^2) - [E(X)]^2 = \lambda^2 + \lambda - \lambda^2 = \lambda$$

4.7.5.3 The Poisson Distribution as an approximation to the Binomial Distribution

The discussion of this section leads us to ask the following question: what happens to the Binomial probabilities $\binom{n}{k} p^k q^{n-k}$ if $n \to \infty$ and $p \to 0$ in such a manner that np remains constant equal to, say, λ? The following calculations help provide an answer to this question. The Binomial probability $P(X = k)$ is given as

$$P(X = k) = \binom{n}{k} p^k q^{n-k} = \frac{n!}{k!\,(n-k)!} p^k q^{n-k}$$

$$= n(n-1)(n-2) \cdots (n-k+1) p^k (1-p)^{n-k}$$

Let $np = \lambda$. Hence, $p = \frac{\lambda}{n}$ and $(1 - p) = 1 - \frac{\lambda}{n}$. Replacing all terms involving p by their equivalent expression in terms of λ, we obtain

$$P(X = k) = \frac{n(n - 1) \cdots (n - k + 1)}{k!} \left(\frac{\lambda}{n}\right)^k \left(\frac{n - \lambda}{n}\right)^{n-k}$$

$$= \frac{\lambda^k}{k!} \left[(1) \left(1 - \frac{1}{n}\right)\left(1 - \frac{2}{n}\right) \cdots \left(1 - \frac{k-1}{n}\right)\right] \left[1 - \frac{\lambda}{n}\right]^{n-k}$$

$$= \frac{\lambda^k}{k!} \left[(1) \left(1 - \frac{1}{n}\right)\left(1 - \frac{2}{n}\right) \cdots \left(1 - \frac{k-1}{n}\right)\right] \left[\left(1 - \frac{\lambda}{n}\right)^n \left(1 - \frac{\lambda}{n}\right)^{-k}\right]$$

Letting $n \to \infty$ in such a way that $np \to \lambda$ remains constant, each term in the product

$1 \left(1 - \frac{1}{n}\right) \cdots \left(1 - \frac{k-1}{n}\right)$ approaches 1, also $\left(1 - \frac{\lambda}{n}\right)^n \left(1 - \frac{\lambda}{n}\right)^{-k} \to$
$e^{-\lambda}(1) = e^{-\lambda}$. Hence, the

limit as $n \to \infty$ with $np = \lambda$ (and as $p = \frac{\lambda}{n} \to 0$),

$$\binom{n}{k} p^k q^{n-k} \to \frac{\lambda^k e^{-\lambda}}{k!}$$

Thus, the Poisson probability function can be used as a convenient approximation to the Binomial probability density function in the case of large n and small p.

$$\binom{n}{k} p^k q^{n-k} \cong \frac{e^{-\lambda} \lambda^k}{k!}$$

Where $\lambda = np = $ constant

NOTE.

The Poisson probability density function may be used in a number of applications. For example, the random variable, X, may denote the number of alpha particles emitted by a radioactive substance that enter a prescribed region during a prescribed interval of time. With a suitable value of λ, it is found that X may be assumed to have a Poisson distribution.

Again, let X denote the number of defects on a manufactured article, such as a vehicle door. Upon examining many of these doors, it is found, with an appropriate value of λ, that X may be said to have a Poisson distribution.

The number of motor car accidents in some unit of time (or the number of insurance claims in some unit of time) is often assumed to be a random variable which has a Poisson distribution. Each of these instances can be thought of as a process that generates a number of changes (accidents, claims, etc) in a fixed interval (of time or space and so on). If a process leads to a Poisson distribution, that process leads to a Poisson process. The Poisson process is an example of continuous parameter Stochastic Processes which will be treated in Chapter 8.

EXAMPLE 4.32

If X has a Poisson distribution with parameter λ, and if $P(X = 0) = 0.2$. Evaluate $P(X > 2)$.

SOLUTION:

$$P(X > 2) = \sum_{k=3}^{\infty} P(X = k) = \sum_{k=3}^{\infty} \frac{e^{-\lambda}\lambda^k}{k!}$$

$$= 1 - \sum_{k=0}^{2} \frac{e^{-\lambda}\lambda^k}{k!} = 1 - P(X = 0) - P(X = 1) - P(X = 2)$$

$$P(X = 0) = \frac{e^{-\lambda}\lambda^0}{0!} = e^{-\lambda} = 0.2 \qquad \Rightarrow \lambda = \ln 5 \cong 1.6094$$

$$P(X = 1) = \frac{e^{-\lambda}\lambda^1}{1!} = \lambda e^{-\lambda} = 0.2\ln 5 \cong 0.3219$$

$$P(X = 2) = \frac{e^{-\lambda}\lambda^2}{2!} = \frac{0.2(\ln 5)^2}{2} \cong 0.2590$$

$$\therefore \quad P(X > 2) = 1 - 0.2 - 0.3219 - 0.2590 = 0.2191$$

EXAMPLE 4.33

Suppose that the probability that an item produced by a particular machine is defective equals 0.2. If 10 items produced from this machine are selected at random, what is the probability that not more than one defective is found? Use the binomial and Poisson distributions and compare the answers.

SOLUTION:
Probability of at most one defective is $P(X = 0) + P(X = 1)$

Binomial: $P(X = 0) = \binom{10}{0}(0.2)^0(0.8)^{10} = (0.8)^{10} \cong$ 0.1073742

$$P(X = 1) = \binom{10}{1}(0.2)^1(0.8)^9 = 2(0.8)^9 \cong 0.2684355$$

$$\therefore \quad P(X = 0) + P(X = 1) = 0.37581$$

Poisson: $\quad \lambda = np = 2$

$$P(X = 0) = \frac{\lambda^0 e^{-\lambda}}{0!} = e^{-2}$$

$$P(X = 1) = \frac{\lambda^1 e^{-\lambda}}{1!} = 2e^{-2}$$

$$\therefore \quad P(X = 0) + P(X = 1) = 3e^{-2} \cong 3(0.1353353) \cong 0.406$$

Using R

Computation of probabilities and cumulative probabilities associated with a Poisson random variable can also be evaluate using various software packages.

$P(X = x)$, probability associated with a Poisson random variable with mean λ can be found using the R command $dpois(x, \lambda)$, $P(X \leq x)$, the cumulative probability of a Poisson distribution with mean λ is evaluated in R using the command $ppois(x, \lambda)$.

EXAMPLE 4.34
At a busy traffic junction, the probability, p of an individual car having an accident is 0.0001. During a certain part of the day, 1000 cars pass through the junction. What is the probability of two or more of these cars being involved in an accident within this period?

SOLUTION.
Let X be the number of accidents. Then
$X \sim b(1000; 0.0001). Therefore$
$$P(X \geq 2) = 1 - P(X = 0) - P(X = 1)$$
$$= 1 - (0.9999)^{1000} - 1000(0.0001)(0.9999)^{999}$$
which is much tedious even with an ordinary calculator.
By the Poisson approximation to the Binomial, $np = \lambda =$
$1000(0.0001) = 0.1$
$$P(X \geq 2) = 1 - (1 + 0.1)e^{-0.1} = 0.00467$$

Using R we have
$$P(X \geq 2) = 1 - P(X \leq 1)$$
$$= 1 - ppois(0.1, 1)$$
$$= 0.00467$$

4.7.6 The Negative Binomial (or Pascal) Probability Distribution

Suppose that independent trials, each having probability p; $0 < p < 1$ of being a success are performed until we obtain a total of k success. If we let X be the number of trials required, then

$$P(X = x) = \binom{x-1}{k-1} p^k (1-p)^{x-k}, \qquad x = k, k+1, k+2, ...$$

Explanation:
In order for the kth success to occur at the xth trial, there must be $k-1$ successes in the first $(x-1)$ trials and then x^{th} trial must be a success. The probability of the first $(x-1)$ events is

$$P(X = x) = \binom{x-1}{k-1} p^{k-1}(1-p)^{x-k}$$

and the probability of the last event is p. Thus, by independence, we

$$P(X = x) = \binom{x-1}{k-1} p^{k-1}(1-p)^{x-k} * p$$

$$= \binom{x-1}{k-1} p^k(1-p)^{x-k}, \qquad x = k, k+1, k+2, ...$$

NOTATION:

When X has the negative binomial distribution with parameters k and p, we denote it by $X \sim b^-(k; p)$ and it's p.m.f $P(X = x)$ is denoted by $X \sim b^-(x; k; p)$. That is,

$$X \sim b^-(x; k; p) = \binom{x-1}{k-1} p^k(1-p)^{x-k}, \qquad x$$
$$= k, k+1, k+2, ...$$

Given that $X \sim b^-(x; k; p)$, then $E(X) = \frac{k}{p}$ and $Var(X) = \frac{kq}{p^2}$

Note also that $\binom{x-1}{k-1} = \frac{k}{x} \binom{x}{k}$

4.7.6.1 Relationship between the Binomial and Negative Binomial random variables

$$b^-(x; k; p) = \frac{k}{x} b(k; x, p)$$

PROOF:

$$b^-(x; k; p) = \binom{x-1}{k-1} p^k(1-p)^{x-k}$$
$$= \frac{k}{x} \binom{x}{k} p^k(1-p)^{x-k} = \frac{k}{x} b(k; x, p)$$

Using R:

If k = 2, 3, 4, … and X has a negative binomial distribution with success probability p,
$P(X = x)$ can be found using the R command $dnbinom(x - k, k, p)$.
Also $P(X \leq x)$, is found by using the R command $pnbinom(x - k, k, p)$

EXAMPLE 4.33
A fair coin is tossed until we obtain three 'heads'. Find the probability that the third head is obtained on the tenth toss by using the binomial tables.

SOLUTION

Let X be the number of tosses up to and including the third head then
$X \sim b^-(10; 3; 0:5)$. The required probability is
$$P(X = 10) = b^-(10; 3; 0:5)$$
$$= \frac{3}{10} b(3; 10; 0:5)$$
$$= \frac{3}{10} [B(3; 10; 0:5) - B(2; 10; 0:5)]$$
$$= 0.0352$$

Calculation in R:
$$P(X = 10), \qquad k = 3, \qquad p = 0.5$$
$$dnbinom(5 - 3,3,0.2)$$
$$= 0.0352$$

EXAMPLE 4.34

A geological study indicates that an exploratory oil well drilled in a particular region should strike oil with probability 0.2. Find the probability that the third oil strike comes on the fifth well drilled.

SOLUTION

Assuming independent drilling and probability 0.2 of striking oil with any one well, let X denote the number of trials on which the third oil strike occurs. Then it is reasonable to assume that X has a negative binomial distribution with p = 0:2. Because we are interested in $k = 3$ and $x = 5$,

$$P(X = 5) = b^-(5; 3; 0.2)$$
$$= \binom{4}{2} 0.2^3 (0.8)^2$$
$$= 6(0.008)(0.64)$$
$$= 0.0307$$

Calculation in R:

$$P(X = 5), \quad k = 3, \quad p = 0.2$$
$$dnbinom(5 - 3, 3, 0.2) = 0.03072$$

EXERCISE 4

4.1 The random variable X assumes values of 0, 1 and 10 only. If $E(X) = 2$ and
$$V(X) = 13.$$
Derive the probability distribution function for X.

4.2 X and Y are independent random variables such that, X assumes the value 0 with probability q_1 and the value 1 with probability p_1, where $p_1 + q_1 = 1$; while Y assumes the values 0 and 1 with probabilities q_2 and p_2 respectively, where
$p_2 + q_2 = 1$. Compute the mean of the random variable $Z = X + Y$.

4.3 The two faces of a coin M_1 are marked with numbers 1 and 2 respectively, while those of another coin M_2 are marked 2 and 3. Let X and Y denote the results obtained on both coins respectively when they are tossed once.
(a)   Derive the probability distribution function for $U = X + Y$ and $V = X - Y$ respectively.
(b)   Determine whether U and V are independent.

4.4 Two symmetric dice, T_1 and T_2, are thrown once, and the sum of the numbers showing on their faces noted. Compute the probability that this sum equals at least

8 given that it is equal to at most 11.

4.5 X and Y are independent random variables such that X assumes the values 0, 3, 6 with probabilities $\frac{1}{6}, \frac{2}{3}$ and $\frac{1}{6}$ respectively; while Y assumes values 0, 1, 10 with probabilities $\frac{1}{2}, \frac{1}{3}$ and $\frac{1}{6}$ respectively. Compute the mean and standard deviation

of $Z = X + Y$.

4.6 X and Y are two random variables such that $E(X) = 2$, $E(X^2) = 6$, $E(Y) = 3$,

$E(Y^2) = 10$ and $E(XY) = 7$. Compute $E(X - Y)$ and $V(X - Y)$.

4.7 Two random variables are such that $E(X) = 4$, $E(X^2) = 20$, $E(Y) = 3$, $E(Y^2) = 18$

and $E(XY) = 10$. Compute $E(X + Y)$ and $V(X + Y)$.

4.8 X and Y are two independent random variables given in the tables below:

x	0	2	4
$P(X = x)$	0.3	0.5	0.2

y	0.5	1	2
$P(Y = y)$	0.3	0.2	0.5

Compute   (a) $E(XY)$ (b) $E(X/Y)$

4.9 The random variable X has cumulative distribution function, $F_X(x)$, given below:

$$F_X(x) = \begin{cases} 0 & \text{if} \quad x < 2 \\ 0.4 & \text{if} \quad 2 \le x < 5 \\ 1 & \text{if} \quad x \ge 5 \end{cases}$$

Derive the probability mass function for X.

4.10. Let X be a discrete random variable whose cumulative distribution is

$$F(x) = \begin{cases} 0, for\ x < -3 \\ \dfrac{1}{6}, for -3 \le x < 6 \\ \dfrac{1}{2}, for\ 6 \le x < 10 \\ 1, for\ 10 \le x \end{cases}$$

a) Find the probability mass function of X.
b) Find;
i. $P(X \le 4)$
ii. $P(-5 < X \le 4)$
iii. $P(X = -3)$
iv. $P(X = 4)$

4.11 The cumulative distribution function for the random variable X is given as,

$$F_X(x) = \begin{cases} 0 \ \ \text{if} \ \ \ \ x < 3 \\ \dfrac{1}{4} \ \ \text{if} \ \ 3 \le x < 5 \\ 1 \ \ \text{if} \ \ \ \ x \ge 5 \end{cases}$$

Compute the mean and standard deviation of X.

4.12 Let X and Y have the joint distribution defined by the following table of probabilities

Y \ X	1	2	3
2	$\dfrac{1}{12}$	$\dfrac{1}{6}$	$\dfrac{1}{12}$
3	$\dfrac{1}{6}$	0	$\dfrac{1}{6}$
4	0	$\dfrac{1}{3}$	0

Determine the following:
(a)   The marginal probabilities
(b) $P(X = Y)$
(c) $P(X = 2 \text{ or } Y = 4)$
(d) $P(X + Y \leq 4)$

4.13 Let (X, Y) have the discrete distribution given in the following table of probabilities:

Y \ X	0	2	4
1	$\frac{1}{4}$	0	$\frac{1}{4}$
3	$\frac{1}{12}$	$\frac{1}{3}$	$\frac{1}{12}$

(a)   Verify that $E_X(VarY/X) = VarY$
(b)   Compute the correlation coefficient ρ_{XY}
(c)   Compute the covariance of X and Y given that $X \neq 4$

4.14 Consider the discrete random variables (X, Y) with joint probabilities given in the table below:

Y \ X	-1	0	1
0	0.1	0.1	0.1
2	0.1	0.2	0.1
4	0.1	0.1	0.1

(a)   Compute the correlation coefficient ρ_{XY}.
(b)   Determine whether X and Y are independent.

4.15 The random vector (X, Y) has a discrete distribution defined by the table given below:

Y \ X	1	2	4
2	$\frac{1}{12}$	$\frac{1}{6}$	$\frac{1}{12}$
3	$\frac{1}{6}$	0	$\frac{1}{6}$
4	0	$\frac{1}{3}$	0

Determine the following:
(a) $P(Y = 2/X + Y \leq 5)$
(b) $P(Y = 2/X = 2)$
(c) $P(X = 1/Y > 2)$
(d) $p(x/2)$
(e) $p(y/1)$, $p(y/2)$ and $p(y/3)$: then verify that
$P(Y = y) = p(y) = \sum_x p(y/x)p(X = x)$

4.16. The random variable X counts the number of heads in two tosses of a fair coin.

Compute the mean and variance of X.

4.17. The random variable X has a Binomial distribution with parameters given as

$$X \sim \text{Bin}\left(2, \frac{1}{3}\right)$$

Compute $E\left\{\frac{1}{1+X}\right\}$

4.18. A box consists of white and black marbles in the ratio $1:4$. Eight marbles are selected at random (with replacement) from the box. Compute the probability
that at least three black marbles are selected.

4.19. The random variable $X \sim \text{Bin}(150, \ p)$. Compute $p(10 < X < 65)$ if
(a) $p = 0.4$
(b) $p = 0.6$

4.20. A random sample of 4 pupils is selected (without replacement) from a class consisting of 12 boys and 9 girls. Compute the probability that the sample consists of at least 2 boys.

4.21. Suppose that X has a Poisson distribution.
(a) If $P(X = 2) = \frac{2}{3}P(X = 1)$,

Evaluate $P(X = 0)$ and $P(X = 3)$.

     (b) If $P(X = 1) = P(X = 2)$, find $P(X = 4)$.

4.22. An insurance company has discovered that only about 0.1 percent of the population is involved in a certain type of accident each year. If its policy holders were randomly selected from the population, what is the probability that not more than 5 of its clients are involved in such an accident next year?

4.23. The random variables X and Y have the joint distribution specified by the following table:

Y \ X	0	1	2	3
1	0	0	p	0
2	0	p^2	p^2	0
3	0	p^2	p^2	0
4	p	0	0	p

(a)   Find the value of p
(b)   Are X and Y independent?
(c)   Find $E(X)$ and $V(X)$
(d)   Find $E(Y)$ and $V(Y)$
(e)   Evaluate $P(Y = 2/X + Y = 3)$

4.24. (a) The discrete random variable X has cumulative distribution function $F(x)$ given
    as

$$F(x) = \begin{cases} 0 & if & x < 6 \\ 0.25 & if & 6 \le x < 10 \\ 1 & if & x \ge 10 \end{cases}$$

    Obtain the probability distribution of X, hence compute
    (i) $E(X)$
    (ii) $V(X)$
    (b) The random variables X and Y are such that $V(X) = 1$ and $V(Y) = 2$. If the correlation coefficient between X and Y, that is $\rho(X, Y) = 0.05$. Obtain $V(X + Y)$

4.25. (a) let X be a geometrically distributed random variable i.e.
$P(X = x) = q^{x-1}p,$

$x = 1, 2, \ldots$. Where $p + q = 1$. Show that this distribution lacks memory: (i.e. Show that for positive integers a and b, $P(X > a + b/X > b) = P(X > a)$).

(b) Let X_1 and X_2 be two independent observations of the geometric distribution X

in (a) above. Compute the probability that,

(i) $P(X_1 > X_2)$
(ii) $P(X_1 = X_2)$
(iii) $P(X_1 > X_2/X_1 + X_2 = 4)$

4.26. (a) Two sides X and Y of a rectangle are independent random variables with $E(X) = 3$ and $E(Y) = 3, V(X) = 4, \ V(Y) = 4$. Let u denote the area of the rectangle and v denote the circumference. Compute the correlation coefficient between the area $u = XY$ and the circumference $v = 2(X + Y)$ of the rectangle.

Hint: Correlation coefficient between two random variables u and v is given by

$$\rho = \frac{E(uv) - E(u)E(v)}{\sqrt{\text{Var}(u)}\sqrt{\text{Var}(v)}}$$

(b) A population consists of five elements A, B, C, D, E with the following values of variable X.

	A	B	C	D	E
x	1	5	7	9	11

We take with equal probabilities and without replacement, three elements from the population. Let $\tilde{X}$ be the median in this sample. Give the probability distribution of $\tilde{X}$.

4.27. Two articles among 10 (ten) articles are defective. Determine the possible values and the corresponding probabilities for the random variables:

X: The number of defectives in a random selection of 1 (article) from the lot, and

Y: The number of defective articles in a random selection of 4.

4.28. (a) Let X and Y have the discrete distribution given in the following table of probabilities:

Y \ X	0	2	4
1	$\frac{1}{4}$	0	$\frac{1}{4}$
3	$\frac{1}{12}$	$\frac{1}{3}$	$\frac{1}{12}$

Obtain (i) $E(X/Y)$ (ii) $E(X/Y = 2)$ (iii) verify that $E_Y[E(X/Y)] = E(X)$

(b) A bivariate random variable (X, Y) has a discrete distribution defined by the table of probabilities given below:

Y \ X	1	2	3
2	$\frac{1}{12}$	$\frac{1}{6}$	$\frac{1}{12}$
3	$\frac{1}{6}$	0	$\frac{1}{6}$
4	0	$\frac{1}{3}$	0

Determine the following:

(i) $P(Y/X = 1)$
(ii) $P(Y/X = 2)$
(iii) $P(Y/X = 3)$ and verify that
(iv) $P(Y = y) = \sum_x P(Y = y/X = x)\, P(X = x)$

4.29 (a) The six sides of a die are numbered 1, 3, 3, 4, 4, 6 respectively. The die is thrown twice;

Let X denote the outcome of the first throw and Y the outcome of the second throw.

(i) Obtain the probability distribution of the random variable $Z = X - Y$

(ii) Compute $E(Z)$ and $V(Z)$

(b) The discrete random variable X has cumulative distribution function $F(x)$ given as

$$F(x) = \begin{cases} 0 & \text{if} & x < 2 \\ 0.4 & \text{if} & 2 \le x < 5 \\ 1 & \text{if} & x \ge 5 \end{cases}$$

Obtain the probability distribution of X, hence compute $E(x)$ and $V(x)$.

4.30. (a) The random variable X and Y have joint distribution specified by the following table:

	X	
Y	0	1
1	0.1	0.2
2	0.2	0.1
3	0.3	0.1

(i)   Obtain the marginal probabilities for X and Y
(ii) $E(X)$ and $E(Y)$
(iii)   Compute $E(Y/X = 0)$
(iv)   Compute $E(Y/X = 1)$
(v)   Compute $E(X/Y = 1)$
(vi)   Compute $E(X/Y = 2)$
(vii)   Compute $E(X/Y = 0)$

(b) Let the random variable Y have a Binomial distribution as given below:
     Calculate
(i) $P(Y = 4)$, where $Y \sim \text{Bin}(7, 0.3)$
(ii)   Suppose that X ha s a Poisson distribution.
If $P(X = 2) = \frac{2}{3} P(X = 1)$, Evaluate $P(X = 0)$.

4.31. (a) Let X be a discrete random variable with a Poisson distribution, assuming the possible values $0, 1, 2, \dots, n$. If $P(X = k) = \dfrac{e^{-\lambda} \lambda^k}{k!}$

for $k = 1, 2, \dots n$ and $\lambda > 0$.

Prove that

$$\sum_{k=0}^{\infty} P(X = k) = 1$$

(b) Suppose that X has a Poisson distribution. If $P(X = 3) = \frac{5}{6}P(X = 2)$, evaluate
$P(X = 0)$ and $P(X = 1)$

(c) If $X \sim \text{Bin}\left(8, \frac{1}{3}\right)$, find

    (i) $P(X = 0)$ (ii) $P(X = 1)$ (iii) $P(X \geq 2)$

CHAPTER 5
CONTINUOUS RANDOM VARIABLES

5.1. One Dimensional Case

Suppose that the range space of the random variable X is made up of very large but finite number of values: say all values x in the interval $0 \leq x \leq 1$ of the form $0, 0.01, 0.02, \ldots 0.98, 0.99, 1.00.$ with non-negative number

$p(x_i) = P[X = x_i], i = 1, 2, \ldots,$ whose sum equals 1. It might be mathematically easier to idealize the above probabilistic description of X by supposing that X can assume all possible values $0 \leq x \leq 1$. If we do this, what happens to the point $p(x_i) = P[X = x_i]$? Since the possible values of X are non-countable, we cannot really speak of the ith value of X, and hence $p(x_i) = P[X = x_i]$ becomes meaningless. What we are trying to say is that, a continuous random variable has a probability of zero of assuming exactly any of its values. Consequently, its probability distribution cannot be given in tabular form, as was the case, in previous chapters, of a discrete random variable. This may seem startling, but it becomes more plausible when we consider a particular example. Let us discuss a random variable whose values are the heights of all people over 20 years of age. Between any two values, say 62.4 and 63.6 inches, or even 62.99 and 63.04 inches, there are an infinite number of heights, one of which is 63 inches. The probability of selecting a person at random exactly 63 inches tall and not one of the infinitely large set of heights so close to 63 inches that you cannot humanly measure the difference is extremely remote, and thus we assign a probability of zero to the event. It follows that the following probabilities are all the same if X is a continuous random variable:

$P(a \leq X \leq b) = P(a \leq X < b) = P(a < X \leq b) = P(a < X < b).$
That is, it does not matter whether we include an endpoint of the interval or not.

While the probability distribution of a continuous random variable cannot be presented in tabular form, it does have a formula. We replace the function $p(x_i)$, defined, only for the discrete case, for $x_1, x_2, x_3 \ldots$ by the functional notation $f(x_i)$. In dealing with continuous random variables, $f(x)$ is usually called the probability density function of the random variable X. Since X is defined over a continuous sample space, the graph of $f(x)$ will be continuous and may, for example, take one of the forms shown in Figure 5.1 below:

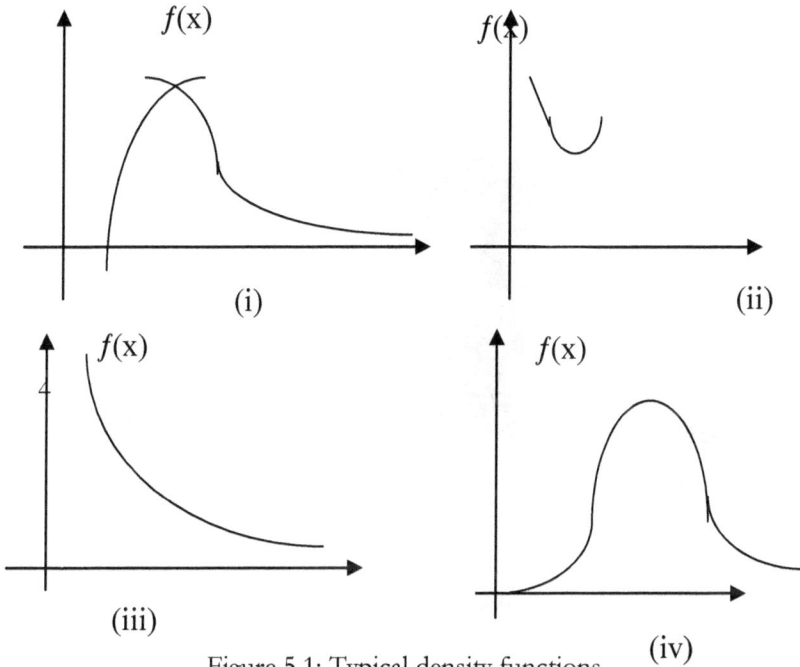

Figure 5.1: Typical density functions

A probability density function is constructed so that the area under the curve bounded by the X-axis is equal to one when computed over the range of X for which f(x) is defined. If X assumes values only in some finite interval

[a, b], we may simply set $f(x) = 0$ for all $X \notin [a, b]$. If $f(x)$ is represented as shown in figure 5.2 below, then the probability that X

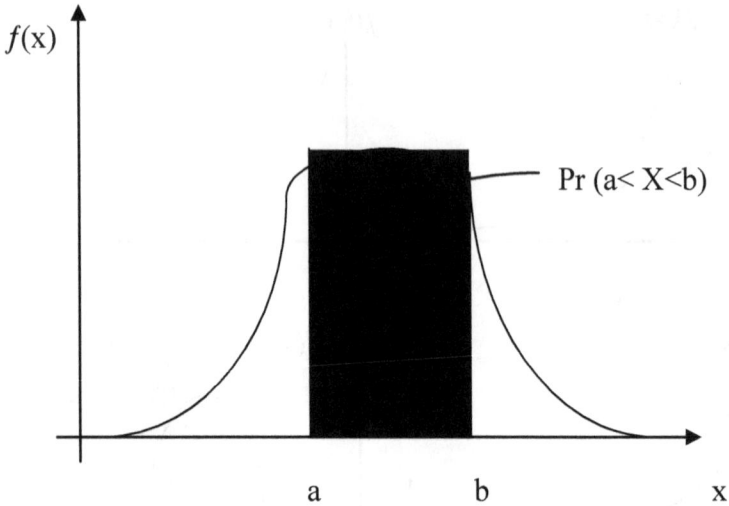

Figure 5.2

assumes a value between a and b is equal to the shaded area under the density function between the ordinates x = a and x = b. The exact area can be obtained only by methods of integral calculus.

The probability density function, $f(x)$, does not represent the probability of anything. Only when the function is integrated between two limits does it yield a probability. We can, however, give an interpretation of $f(x) \, \Delta x$ as follows: From the Mean Value Theorem of calculus it follows that

$$P(x \leq x + \Delta x) = \int_{x}^{x+\Delta x} f(s)d(s) = \Delta x f(\gamma), \qquad x \leq \gamma \leq x + \Delta x.$$

If Δx is small, $f(x)\Delta x$ equals approximately $P(x \leq X \leq x + \Delta x)$. (If f is continuous from the right, this approximation becomes more accurate as $\Delta x \to 0$).

5.1.1   Definition of Probability Density Function

X is said to be a continuous random variable if there exists a function f called the probability density function (pdf) of X, satisfying the following conditions:

(a) $f(x) \geq 0$, for all x

(b) $\int_{-\infty}^{\infty} f(x) dx = 1$

(c)   For any a, b with $-\infty < a < b < +\infty$, we have

$$P(a \leq X \leq b) = \int_a^b f(x) dx$$

As a consequence of the above probabilistic description of X, any specified value of X, say x_0, we have $P(X = x_0) = 0$, since

$P(X = x_o) = \int_{x_o}^{x_o} f(x) dx = 0$. This result may seem quite contrary to

our intuition. We must realize, however, that if we allow X assume all values in some interval, then probability zero is not equivalent with impossibility. Hence in the continuous case, $P(A) = 0$ does not imply $A = \Phi$ (the empty set).

REMARK: $P(a \leq X \leq b)$ *represents the area under the graph of the pdf between x=a and x=b*

EXAMPLE 5.1

Let X b a continuous random variable with probability density function
$$f(x) = \begin{cases} 2x \ for \ 0 < x < 1 \\ 0 \ elsewhere \end{cases}$$
a)   Show that f is a legitimate pdf.
b)   Compute:
i) $P(X \leq 0.4)$,
ii) $P(0.2 < X < 0.4)$
iii) $P(X \leq \frac{1}{2} : \frac{1}{3} \leq X \leq \frac{2}{3})$

SOLUTION

a)   *Clearly f(x) $\geq$0 for every x. this is because for any value of x, 0<x<1, f(x)* $=2x \geq 0$. *next, we need to show that* $\int_{-\infty}^{\infty} f(x) dx = 1$

$\int_{-\infty}^{\infty} f(x) dx = \int_{-\infty}^{0} 0 dx + \int_0^1 2x dx + \int_1^{\infty} 0 dx$

Since f(x) is not defined over the interval $-\infty < x < 0$ *and* $1 < x < \infty$

$$\int_{-\infty}^{\infty} f(x)dx = \int_0^1 2x dx$$
$$= [x^2]^1_0$$
$$= 1\text{-}0$$
$$1$$

Hence f(x) is legitimate.

b) *Computing*

i) $\quad P\ (X \leq 0.4) = \int_{-\infty}^{0.4} f(x)dx$

$$= \int_{-\infty}^0 0 dx + \int_0^{0.4} 2x dx$$
$$= [x^2]^{0.4}_0 = 0.16$$

ii) $\quad P\ (0.2 < X < 0.4) = \int_{0.2}^{0.4} 2x dx$

$$= \int_{0.2}^{0.4} 2x dx$$
$$= [x^2]^{0.4}_{0.2} = 0.12$$

iii) $\quad$ *Required probability P (X<1)*

P (X<1) = P (X≤1) - P (X=1)

But

$P\ (X=1) = \int_1^1 \frac{3}{4}(2x - x^2)dx$
$$= 0$$

Hence P (X<1) = P (X≤1) = $\frac{1}{2}$

5.2. Cumulative Distribution Function (cdf)

5.1.2   Definition

Let X be a continuous random variable with probability density function (pd*f*) *f*. We define F to be the cumulative distribution function of the random variable X (abbreviated cd*f*) where

$$F(x) = P(X \leq x) = \int_{-\infty}^{x} f(s)ds$$

NOTE: (Because x is used as a dummy in defining the cdf, another dummy, s, is needed in the integral to distinguish between upper limit and the variable of integration). This relation gives F(x) in terms of f(x) is shown in Figure 5.3 below.

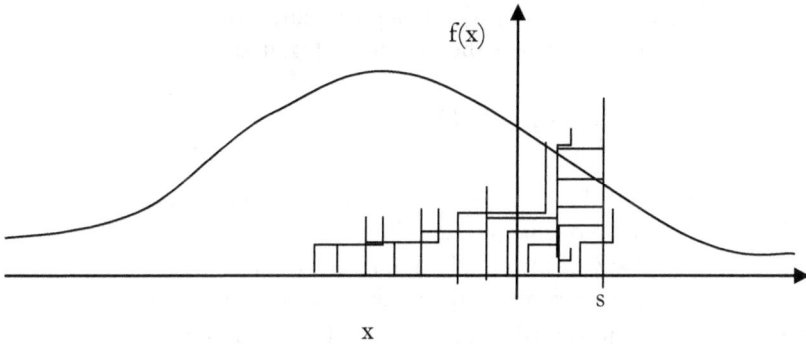

Figure 5.3

Shaded area = $F(x) = P[X \leq s] = \int_{-\infty}^{s} f(x)dx$

5.2.2 Properties of the cd$f$: F

(a)   The function F is non decreasing, giving the cumulated probability from the left up to the point x; is a non-decreasing function of x; for probability (which is non-negative) can only be added on, going from left to right. That is, if $x_1 \leq x_2$, we have $F(x_1) \leq F(x_2)$. The function starts at $-\infty$ with value zero and rises to the value 1 at $+\infty$:

(b) $0 \leq F(x) \leq 1$

(c) $F(x) \geq 0$

(d) $\lim_{x \to -\infty} F(x) = \int_{-\infty}^{-\infty} f(s)\, ds = F(-\infty) = 0$

(e) $\lim_{x \to +\infty} F(x) = \int_{-\infty}^{+\infty} f(s)ds = F(+\infty) = 1$

(f) $f(x) = \frac{d}{dx} F(x)$ for all x at which F is differentiable.

It is important that the reader clearly distinguishes between the random variable (X) and the value assumed by the random variable.

Because the distribution function F(x) is a probability, for any given x, it is expressible as an area, or definite integral, of the density function.

$F(x) = P[X \leq x]$ = area under $f(x)$ to the left of x.

$$= \int_{-\infty}^{x} f(s)ds$$

The integral giving the cd$f$. (F(x)) in terms of the density $f(x)$ is a particular indefinite integral of $f(x)$, sharing with any indefinite integral the property that its derivative is the function integrated:

$$f(x) = F'(x) = \frac{d}{dx}F(x) = \frac{d}{dx}\int_{-\infty}^{x} f(s)ds$$

(this fact is the essential content of what is called in calculus the Fundamental Theorem of Integral Calculus). Thus the slope of F(x) at any point is the density of the distribution at that point. This distribution concentrates probability heavily at points where F(x) is steep [large density $f(x)$] and spreads it thinly where F(x) rises slowly (small density).

EXAMPLE 5.2

Suppose that a continuous random variable has cd$f$ F given by

$$F(x) = \begin{cases} 0, & x \leq 0 \\ 1 - e^{-x}, & x > 0 \end{cases}$$

Then the derivative of F(x) with respect to x, i.e.

$$F'(x) = \frac{d}{dx}F(x) = f(x) = e^{-x},$$

and thus the pd$f$. f, is given by

$$f(x) = \begin{cases} e^{-x}, & x \geq 0 \\ 0, & elsewhere \end{cases}$$

EXAMPLE 5.3

Suppose a continuous r.v X has pdf

$$f(x) = \begin{cases} 2x, 0 \leq x \leq 1 \\ 0, elsewhere \end{cases}$$

Find the cdf $F(x)$ if X and sketch it.

SOLUTION

The cdf $F(x)$ of X is given by:

$$Fx(X) = P(X \le x) = \int_{-\infty}^{x} f(x)dx$$

Step 1: if x<0, then

$$F(x) = \int_{-\infty}^{x} f(d)dt = \int_{-\infty}^{0} 0\,dt = 0$$

Step 2: if $0 \le x \le 1$), then

$$F(x) = \int_{-\infty}^{x} f(x)dx$$

$$= \int_{-\infty}^{0} 0dt + \int_{0}^{x} 2tdt$$

$$= 0 + x^2 = x^2$$

Step 3: if 1<x, then

$$F(x) = \int_{-\infty}^{x} f(t)dt$$

$$= \int_{-\infty}^{0} 0dt + \int_{0}^{1} 2tdt + \int_{1}^{x} 0dt$$

$$= 0 + 1 + 0 = 1$$

Conclusion: the cumulative distribution function

of X is:

$$F(x) = \begin{cases} 0, for\ x < 0 \\ x^2, for\ 0 \le x \le 1 \\ 1, 1 < x \end{cases}$$

5.3. Expected Value of one Dimensional Continuous Random Variable

The expected value or mean or average or expectation of a discrete random variable was defined as the weighted sum of its possible values, the weights being the corresponding probabilities. A continuous random variable has too many possible values for this definition to work, but the same idea will give the proper definition if the "summation" is properly carried out. The probability element $f(x)dx$ can be used as a weight for the value x, the products of value and probability element being integrated to give the expected value or mean value. As in the case of discrete distribution, the formula is precisely the formula for centre of gravity of a continuous mass distribution with mass density function $f(x)$, normalized or scaled so that the total is 1.

It is clear from the analogy of mass distributions that the expected value or centre of gravity of a symmetric distribution is the centre of symmetry. That is if c is such a point:

$$f(x - c) = f(x + c) \text{ for all x, then the mean value}$$
$$E(X) = c.$$

5.3.1.1. Definition

Let X be a continuous random variable with pdf f. The expected value of X is defined as

$$E(X) = \int_{-\infty}^{+\infty} x f(x) dx$$

Again it may happen that this (improper) integral does not converge. Hence we say that $E(X)$ exists if and only if

$$\int_{-\infty}^{\infty} |x| f(x) dx \text{ is finite.}$$

EXAMPLE 5.4

Suppose that X has the density

$$f(x) = \begin{cases} 2x \text{ , } if \ 0 < x < 1 \\ 0 \text{ , } elsewhere \end{cases}$$

The expected value is computed as follows:

$$E(X) = \int_{-\infty}^{\infty} x f(x) dx$$
$$P(x \leq X \leq x + \Delta x) = \int_{x}^{x+\Delta x} f(s) ds = \Delta x f(\gamma),$$
$$= 0 + \int_{0}^{1} x.2 x \, dx + 0$$

$$= \left[\frac{2x^3}{3} \right]_0^1 = \frac{2}{3}$$

Note that the expected value E(x) is not, incidentally, the same as the median, which divides the area into two equal parts. The median $\widetilde{m}$ is computed as follows:

$$P\left(X \le \widetilde{m}\right) = P\left(x \ge \widetilde{m}\right) = \frac{1}{2}$$

i.e $\int_0^{\widetilde{m}} f(x)dx = \int_{\widetilde{m}}^1 f(x)dx = \frac{1}{2}$

for this condition to be satisfied,

$\int_0^{\widetilde{m}} 2x \ dx = \int_{\widetilde{m}}^1 2x \ dx = \frac{1}{2}$ would mean that the median $\widetilde{m} = \frac{1}{\sqrt{2}}$

5.3.2. Properties of the Expected Value

We shall list some important properties of the expected value of a random variable which will be useful for subsequent work. In each case we shall assume that all the expected values to which we refer exists.

5.3.2.1 Property 1
If $X = k$ where k is a constant then $E(X) = E(k) = k$
(i.e. the expected value of a constant is equal to the constant).

Proof:

$$E(X) = \int_{-\infty}^{+\infty} kf(x)dx = k \int_{-\infty}^{+\infty} f(x)dx = k$$

5.3.2.2 Property 2

Suppose k is a constant and X is a random variable.
Then

$$E(kX) = kE(X)$$

Proof: $E(kX) = \int_{-\infty}^{+\infty}(kX)f(x)dx = k \int_{-\infty}^{+\infty} xf(x)dx = kE(X)$

5.3.2.3 Property 3: Linearity Property.

Suppose k_1 and k_2 are constants, and X and Y are two random variables.
Then
$E(k_1 + k_2Y) = k_1E(X) + k_2E(Y)$ and

$$E\ (k_1X - k_2Y)\ =\ k_1E\ (X) - k_2E\ (Y)$$

The proof of 5.1.3.2.3 is left to the reader.

5.4. The Variance of one Dimensional Continuous Random Variable

5.4.1 Definition (Variance of a Continuous Random Variable)

Let X be a continuous random variable with pdf., f. We define the variance of X, denoted by $V(X)$ or $\sigma_x{}^2$, as follows:
$$V(X)\ =\ E\ [X - E(X)]^2.$$
The evaluation of $V(X)$ may be simplified as
$$V(X)\ =\ E\ [X - E(X)]^2\ =\ E\{X^2 - 2\,X\,E(X) + [E(X)]^2\}$$
$$=\ E(X^2) - 2E(X)E(X) + [E(X)]^2$$
$$=\ E(X^2) - [E(X)]^2$$
$$=\int_{-\infty}^{\infty} x^2\ f(x)dx - \left[\int_{-\infty}^{\infty} x f(x)dx\right]^2$$

EXAMPLE 5.5

Suppose that X is a continuous random variable with pdf:
$$f\ (x) = \begin{cases} 1 + x\ ,\ if -1 \le x \le 0 \\ 1 - x\ ,\ if\ \ \ 0 \le x \le 1 \end{cases}$$
(see Figure 5.4 below)

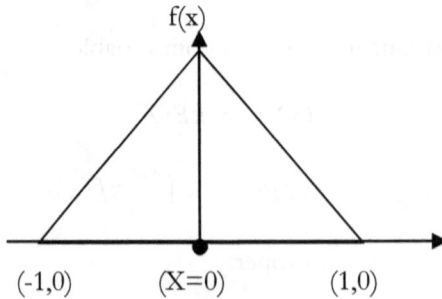

$f(x)$

(-1,0)　　　(X=0)　　　(1,0)

Figure 5.4

Because of the symmetry of the pdf about $X = 0$, $E(X) = 0$

NOTE: Suppose that a continuous random variable has a pdf which is symmetric about $X = 0$. That is $f(-x) = f(x)$ for all x; then provided that $E(X)$ exists, $E(X) = 0$, which is an immediate consequence of the definition of $E(X)$. This may be extended to an arbitrary point of symmetry $x = a$, in which case

$$E(X) = a.$$

To compute the variance $V(X)$,

$$E\left(x^2\right) = \int_{-1}^{0} x^2 \left(1 + x\right) dx + \int_{0}^{1} x^2 \left(1 - x\right) dx = \frac{1}{6}$$

Hence $V(X) = E\left(X^2\right) - \left[E(X)\right]^2 = \frac{1}{6}$

5.5. Function of a One Dimensional Continuous Random Variable

Suppose X is a continuous random variable with pdf. f, and H is a continuous function. Then $Y = H(X)$ is a continuous random variable. In order to obtain the pdf of Y, we proceed as follows:

(a)   Obtain $F(y)$ the cdf of Y, where $F(y) = P(Y \leq y)$, by finding the event A (in the range space of X) which is equivalent to the event $(Y \leq y)$.

(b)   Determine $F(y)$ with respect to y in order to obtain $f(y)$.

(c)   Determine those values of y in the range space of Y for which $f(y) > 0$.

EXAMPLE 5.6

Suppose that X has pdf:

$$f(x) = \begin{cases} 2x, & if\ 0 < x < 1 \\ 0, & elsewhere \end{cases}$$

Let $Y = H(X) = 3X + 1$.   Hence the cumulative distribution function of Y,

$$F(y) = P(Y \le y) = P(3X + 1 \le y) = P\left(X \le \frac{y-1}{3}\right) = \int_0^{\frac{y-1}{3}} 2x\,dx$$

$$= \left[\frac{y-1}{3}\right]^2$$

Thus $f(y) = F'(y) = \frac{d}{dy}F(y) = \frac{2}{3}(\frac{y-1}{3})$. Since $f(x) > 0$ for $0 < x < 1$, we find that $f(y) > 0$ for $1 < y < 4$

There is another slightly different way of obtaining the same result. Consider

again $F(y) = P(Y \le y) = P[X \le \frac{y-1}{3}] = F\left(\frac{y-1}{3}\right)$, where F is the cdf. of X; that is

$F(x) = P(X \le x)$. In order to evaluate the derivative of F, $F'(y)$, we use the chain rule for differentiation as follows:

$$\frac{dF(y)}{dx} = \frac{dF(y)}{dr} \times \frac{dr}{dy}, \quad \text{where } r = \frac{y-1}{3} \cdot \text{Hence,}$$

$$F'(y) = F'(r) * \frac{1}{3} = 2\left(\frac{y-1}{3}\right) * \frac{1}{3} = \frac{2}{3}\left(\frac{y-1}{3}\right), \text{ as before}$$

Theorem 5.5.1

Let X be a continuous random variable with pdf., f, where $f(x) > 0$ and for any constants a and b; $a < x < b$. Suppose that $Y = H(x)$ is a strictly monotone (increasing or decreasing) function of X. Assume that this function is differentiable (and hence continuous) for all x. Then the random variable $Y = H(x)$ has a pdf. given by

$$f(y) = f(x)\left|\frac{dx}{dy}\right|, \text{ where } x \text{ is expressed in terms of } y. \text{ If } H \text{ is}$$
increasing, then f is

non-zero for those values of y satisfying $H(a) < y < H(b)$. If H is decreasing, then f is non-zero for those values of y satisfying $H(b) < y < H(a)$.

Proof.

(a) Assume that H is a strictly increasing function Hence,

$$F(y) = P(Y \le y) = P[H(x) \le y] = P[X \le H^{-1}(y)] = F[H^{-1}(y)]$$

Differentiating F(y) with respect to y we obtain , using the chain rule for

derivatives: $\dfrac{dF(y)}{dy} = \dfrac{dF(y)}{dx}\cdot\dfrac{dx}{dy}, where\ x = H^{-1}(y)$. Thus,

$$F(y) = \frac{dF(x)}{dx}\frac{dx}{dy} = f(x)\frac{dx}{dy}.$$

(b)   Assume that H is a decreasing function. Therefore,

$$F(y) = P[Y \le P[H(x) \le y] = P[X \ge H^{-1}(y)] = 1 - P[X \le H^{-1}(y)] = 1 - F[H^{-1}(y)$$

Proceeding as above, we may write,

$$\frac{dF(y)}{dy} = \frac{dF(y)}{dx}\frac{dx}{dy} = \frac{d}{dx}[1 - F(x)]\frac{dx}{dy} = -f(x)\frac{dx}{dy}$$

NOTE: The algebraic sign in (b) above is correct since, if y is a decreasing function of x, x is a decreasing function of y and hence $\dfrac{dx}{dy} < 0$. Thus, by using the absolute value sign around $\dfrac{dx}{dy}$,we my combine the result of (a) and (b) of the Theorem.

EXAMPLE 5.7

Suppose that: $f(x) = \begin{cases} \frac{1}{2}, if\ -1 < x < 1 \\ 0,\ elsewhere \end{cases}$

Let $H(x) = x^2$. This is obviously not a monotone function over the interval [-1,1]. We obtain the pdf. of $Y = X^2$ as follows:

$$F(y) = P[Y \le y] = P[X^2 \le y] = p[-\sqrt{y} \le X \le \sqrt{y}] = F(\sqrt{y}) - F(-\sqrt{y}), \text{ where F is the}$$

cdf. of the random variable X. Therefore the pdf. is derived as:

$$f(y) = F'(y) = \frac{f(\sqrt{y})}{2\sqrt{y}} - \frac{f(-\sqrt{y})}{-2\sqrt{y}} = \frac{1}{2\sqrt{y}} \left[f(\sqrt{y}) - f(-\sqrt{y}) \right].$$

Thus,

$$f(y) = \frac{1}{2\sqrt{y}} \left(\frac{1}{2} + \frac{1}{2} \right) = \frac{1}{2\sqrt{y}}, \quad for \quad 0 < y < 1.$$

5.5.1 Expected Value of a Function of Random Variable

5.5.1.1 Definition (Expected Value of a Function of a Continuous Random Variable)

Let X be a continuous random variable and Y a function of X, ie. $Y = H(X)$. If X has pdf. given as f, we define $E(y) = \int_{-\infty}^{\infty} y f(y) dy$.

NOTE: The above definition is completely consistent with the previous definition given for the expected value of a random variable. In fact, the definition simply represents restatement in terms of the variable Y. One disadvantage of applying it in order to obtain the expected $E(Y)$, is that the probability distribution of Y (that is, the probability distribution over the range space R_Y) is required. We discussed earlier methods by which we may obtain either the point probabilities $f(y_i)$, or f the pdf. of Y. However, the question arises as to whether we can obtain $E(Y)$ without first finding the probability distribution of Y, simply from the knowledge of the probability distribution of X. This is of course possible as the following theorem indicates.

5.5.1.2 Theorem

Let X be a continuous random variable and let $Y = H(X)$. The expected value $E(Y) = E[H(X)] = \int_{-\infty}^{\infty} H(X) f(x) dx$.

NOTE: The theorem makes the evaluation of E(Y) much simpler, for it means that we need not find the probability of Y in order to evaluate E(Y). The knowledge of the probability distribution of X suffices.

EXAMPLE 5.8

Suppose that X is a continuous random variable with the following pdf:

$$f(x) = \begin{cases} e^{\frac{x}{2}}, & if \ x \le 0 \\ e^{-\frac{x}{2}}, & if \ x > 0 \end{cases}$$

Let Y=$|X|$ to obtain E(Y) we may proceed in one of two ways.

(a)   E(Y)=$\int_{-\infty}^{\infty} |x| f(x) dx = \frac{1}{2}[\int_{-\infty}^{0}(-x)e^{\frac{x}{2}} dx + \int_{0}^{\infty}(x)e^{-\frac{x}{2}} dx =$
$\frac{1}{2}[1+1] = 1$.

(b)   To evaluate $E(Y)$ using the definition, we need to obtain the pdf. of Y=$|X|$, ie. f(y). Let F be the cdf. of Y. Then,

$$F(y) = P[Y \le y] = P[|X| \le y] = P[-y \le X \le y]$$
$$= 2P[0 \le X \le y],$$

since the pdf is symmetric about zero. Therefore,

$$F(y) = 2\int_{0}^{y} f(x) dx = 2\int_{0}^{y} \frac{e^{-x}}{2} dx = -e^{-y} + 1$$

Thus we have for f, the pdf of Y, $f(y) = F'(y) = ye^{-y}$ for $y \ge$ 0.   Hence,
$E(Y) = \int_{0}^{\infty} yf(y) dy = \int_{0}^{\infty} ye^{-y} dy = 1$,   as obtained before.

5.6. Uniformly Distributed Random Variable

5.6.1 Definition (Uniformly Distributed Random Variable).
Suppose that X is a continuous random variable assuming all values in the interval$[a, b]$, where both a and b are finite. If the pdf of X is given by

$$f(x) = \begin{cases} \dfrac{1}{b-a}, a \le x \le b \\ 0, \ elsewhere \end{cases}$$

We say that X is uniformly distributed over the interval $[a, b]$ (see Figure 5.5 below).

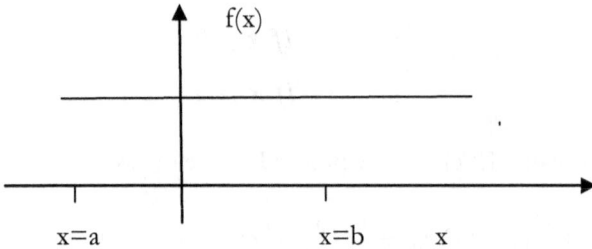

Figure 5.5

NOTE **5.6.1.2**:

(i)   Uniformly distributed random variable has pdf. which is constant over the interval of definition.   In order to satisfy the condition $\int_{-\infty}^{\infty} f(x)dx = 1$, this constant must be equal to the reciprocal of the length of the interval.

(ii)   A uniformly distributed random variable represents the continuous analogue to equally likely outcomes in the following sense: For any subinterval $[c, d]$, where $a \leq c \leq d \leq b$, $P(c \leq X \leq d)$ is the same for all intervals having the same length. That is,

$$P(c \leq X \leq d) = \int_{c}^{d} f(x)dx = \frac{d - c}{b - a}$$

and thus depends only on the length of the interval and not on the location of that interval.

(iii)   We can now make precise the intuitive notion of choosing a point

in an interval, say, $[a, b]$. By this we shall simply mean that the x-coordinate of the chosen point say x, is uniformly distributed over [a,b].

EXAMPLE 5.8

A point X, assumed to be uniformly distributed, is chosen on the line segment$[0,2]$. What is the probability that the chosen point lies between 1 and $\frac{3}{2}$?

SOLUTION:

Letting X represent the co-ordinate of the chosen point. According to Note above 5.5.1.2, the pdf. of X is given by:

$$f(x) = \begin{cases} \frac{1}{2}, & 0 < x < 1 \\ 0, elsewhere \end{cases}$$

Hence by 5.1.5.2(ii) above,

$$P[1 \leq X \leq \tfrac{3}{2]} = \int_1^{\frac{3}{2}} f(x)\, dx = \int_1^{\frac{3}{2}} \tfrac{1}{2}\, dx = \tfrac{1}{4}$$

5.6.2 Cumulative Distribution Function

We now obtain an expression for the cumulative distribution function (cdf.), $F(x)$, of a uniformly distributed random variable.

$$F(x) = P(X \leq x) = \int_{-\infty}^{x} f(s)\, ds$$

$$F(x) = \begin{cases} 0, & if\, x < a \\ \frac{x-a}{b-a}, & if\ a \leq x < b \\ 1, & if\ x \geq b \end{cases}$$

The graph of the CDF is shown in Figure 5.6 below:

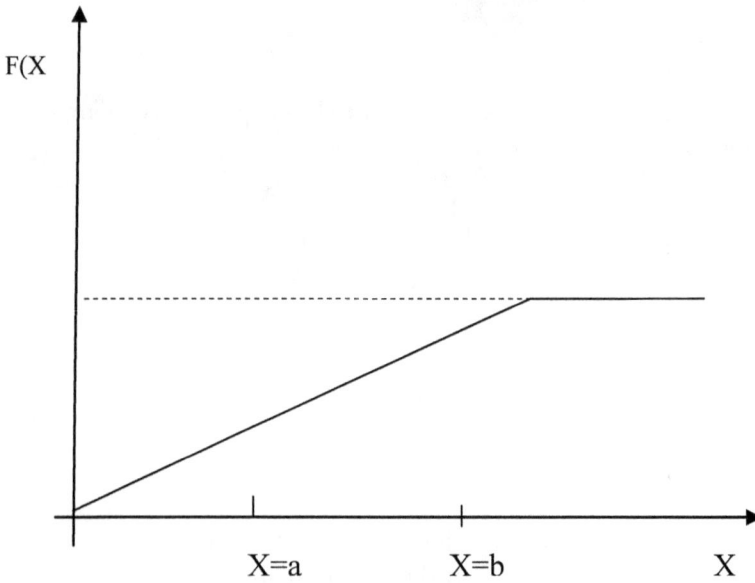

Figure 5.6

5.6.3 Expected Value and Variance

5.6.3.1 Theorem

Let X be uniformly distributed over the interval $[a, b]$. Then

$$E(X) = \frac{a + b}{2}$$

Proof:

The pdf of X is given by $f(x) = \begin{cases} \frac{1}{b-a}, & if \ a \leq x \leq b \\ 0, & elsewhere \end{cases}$

Hence $E(X) = \int_a^b \frac{x}{b-a} dx = \frac{1}{b-a}\left[\frac{x^2}{2}\right]_a^b = \frac{a+b}{2}.$

(Observe that this represents the midpoint of the interval $[a, b]$, as we would intuitively expect)

5.6.3.2 Theorem

Let X be uniformly distributed over the interval $[a, b]$. The variance

$$V(X) = \frac{(b-a)^2}{12}$$

Proof:

To compute $V(X)$, we evaluate $E(X^2) = \int_a^b x^2 \frac{1}{b-a} dx = \frac{b^2 - a^2}{3(b-a)}$

Hence $V(X) = E(X^2) - [E(X)]2 = \frac{(b-a)^2}{12}$

NOTE:

The result is intuitively meaningful. It states that the variance of X does not depend on values of a and b individually but only on $(b - a)^2$, that is, on the square of their difference. Hence, two random variables each of which is uniformly distributed over some interval (not necessarily the same), will have equal variances so long as the lengths of the intervals are the same.

5.7. Two Dimensional (Bivariate) Continuous Random Variable

The concepts discussed above for the one-dimensional case, also hold for higher dimensional random variables. In particular, for the two dimensional case, the joint variation of continuous random variables is described by a joint density function, $f(x, y)$. As in the case of the one-dimensional variable, the density is non-negative and its integral over the plane is equal to 1(one).

We say that (X, Y) is a bivariate continuous r.v. if (X, Y) can assume all (possible) values in some unaccountable set of the Euclidean plane. For example, (X, Y) is a continuous r,v, if

$\{(X, Y) = (x, y)/ \quad a \le x \le b, c \le y \le d\}$This is represented in Figure 5.7 below

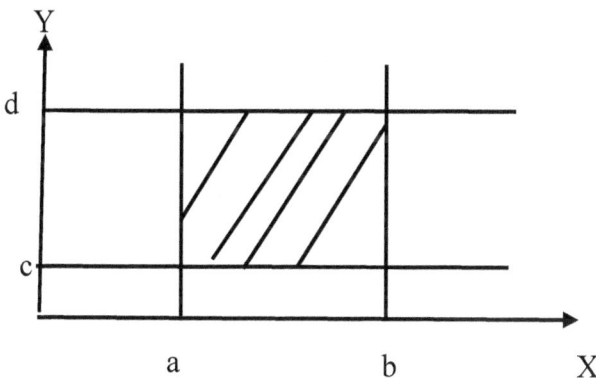

Figure 5.7

5.7.1.   Definition (Joint Probability Density Function)

Let (X, Y) be a two-dimensional continuous random variable (bivariate) taking all possible values in some region R of the Euclidean plane. The joint pdf of (X, Y) is a function $f(X, Y)$ satisfying the following conditions:

(a) $f(x, y) \geq 0 \ \forall \ (x, y) \in R$

(b) $\iint_R f(x, y) dx dy = 1$

This integral is a double integral over the plane of values of the pair (X, Y). Probabilities for the regions of these values are calculated in terms of the probability element $f(x, y) dx dy$. by integrating over the region of interest.

EXAMPLE 5.9

A two-dimensional r.v. (X, Y) has the joint pdf given by

$$f(x, y) = \begin{cases} 1, & 0 \leq x \leq 1, 0 \leq y \leq 1 \\ 0, & \textit{otherwise} \end{cases}$$

Show that this joint pdf is legitimate.

SOLUTION

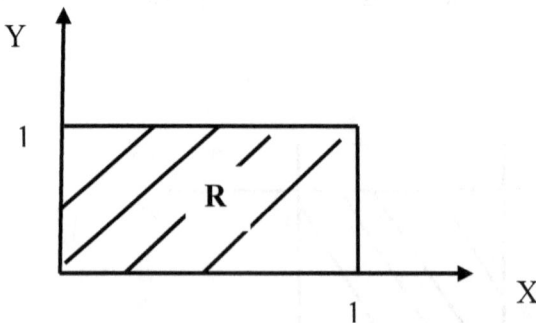

Figure 5.9

(i)   Clearly $f(x, y) \geq 0 \; \forall (x, y) \in R$

(ii)   Next, from Figure 5.9 we have

$$\iint_R f(x, y) dx dy = \int_{y=0}^{1} \int_{x=0}^{1} 1 \, dx dy$$

$$= \int_{y=0}^{1} \left[\int_{x=0}^{1} 1 \, dx \right] dy$$

$$= \int_{0}^{1} 1 \, dy = 1$$

Calculation of Probabilities

$$P(a_1 \leq X \leq b_1, a_2 \leq Y \leq b_2)$$

$$= \begin{cases} \displaystyle\int_{a_2}^{b_2} \int_{a_1}^{b_1} f(x, y) dx dy \,, & \text{if } (X, Y) \text{ is continuous} \\[2em] \displaystyle\sum_{y=a_2}^{b_2} \sum_{x=a_1}^{b_1} P(X = x, Y = y), & \text{if } (X, Y) \text{ is discrete} \end{cases}$$

NOTE

Part of the probability may involve an impossible event so always draw the region for the calculation on the sample space, then do your calculations over the resultant relevant region.

EXAMPLE 5.10

A bivariate continuous random variable (X, Y) has joint pdf

$$f(X, Y) = \begin{cases} x^2 + \dfrac{xy}{3}, & 0 \leq x \leq 1, 0 \leq y \leq 1 \\[1.5em] 0, & \text{otherwise} \end{cases}$$

(a)   Find $P(X + Y \leq 1)$ (b) $P(X + Y \geq 1)$

SOLUTION

(a)

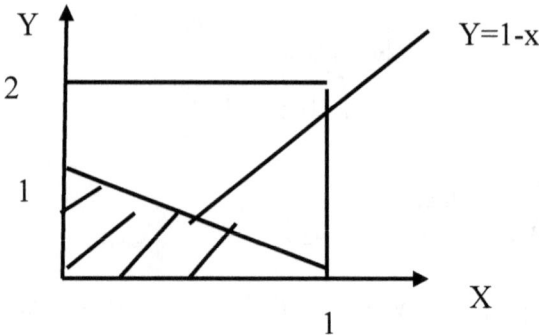

Figure 5.10

From figure 5.10 above,

$$P(X + Y \leq 1) = \int_{x=0}^{1} \int_{y=0}^{1-x} \left(x^2 + \frac{xy}{3}\right) dy dx = \frac{7}{72},$$

Alternatively

$$P(X + Y \leq 1) = \int_{y=0}^{1} \int_{x=0}^{1-y} \left(x^2 + \frac{xy}{3}\right) dx dy = \frac{7}{72}$$

Or,

$$P(X + Y \leq 1) = 1 - P(X + Y > 1)$$

$$= \int_{x=0}^{1} \int_{y=1-x}^{2} \left(x^2 + \frac{xy}{3}\right) dy dx$$

$$= \frac{7}{72}$$

Or,

$$P(X + Y \leq 1) = 1 - P(X + Y > 1)$$

$$= 1 - \left[\int_{y=0}^{2} \int_{x=0}^{1} \left(x^2 + \frac{xy}{3}\right) dx dy + \int_{y=0}^{1} \int_{x=1-y}^{1} \left(x^2 + \frac{xy}{3}\right) dx dy \right]$$

$$= \frac{7}{72}$$

(b)
$$P(X + Y \geq 1) = 1 - P(X + Y < 1)$$
$$= 1 - \frac{7}{72}$$
$$= \frac{65}{72}$$

5.7.2 Definition (Expected Value and Variance)

Let (X, Y) be a two-dimensional continuous random variable. If Z is a function of X and Y, the value of Z is computed, much as was the function of a single variable, as the weighted sum of the values over the whole plane where the weight is the probability element. We define the expected value, E(Z), as: $E(Z) = \int_{-\infty}^{\infty} zf(z)dz$. and the Variance, V(z), as $V(Z) = [E(Z)^2] - [E(Z)]^2 = \int_{-\infty}^{\infty} z^2 f(z)dz - \{\int_{-\infty}^{\infty} zf(z)dz\}^2$.

5.7.2.1 Theorem

Let (X, Y) be a two-dimensional continuous random variable and let $Z = H(X, Y)$ with joint pdf., f. We have $E(Z) = E[H(X, Y)] = \iint_{-\infty}^{\infty} H(X, Y)f(x, y)dxdy$.

NOTE: Again, as in the one-dimensional case, this is an extremely useful result since it states that we need not find the probability distribution of the random variable Z in order to evaluate its expectation $E(Z)$. We can find $E(Z)$ directly from the knowledge of the joint distribution of (X, Y). With these definitions for the distribution of the two random variables (X, Y), one can obtain the following extensions of ideas and formulae given earlier for the discrete case:

5.7.2.2 Additivity of the Expected value
$$xf(x, y)dxdy + yf(x, y)dxdy$$

$$E(X + Y) = E(X) + E(Y) \quad \text{ie.}$$

$$\iint_{-\infty}^{\infty}(x+y)f(x,y)dxdy =$$
$$-\infty\infty xf x, ydxdy +-\infty\infty yf x, ydxdy$$

5.7.3 Condition for Independence Illustrated by the Probability Density Function)

If X and Y are independent, then $f(x,y)dxdy = f_x(x)dx.f_y(y)dy$

5.7.4 Covariance of X and Y

$$Covariance(X,Y) = Cov(X,Y) = \sigma_{xy} = E\big[(X-\mu_x)(Y-\mu_y)\big]$$
$$= E(XY) - E(X)E(Y)$$

5.7.5 Variance of a Sum

$$Var(X+Y) = Var(X) + Var(Y) + 2cov(X,Y)$$

$$Var(X-Y) = Var(X) + Var(Y) - 2cov(X,Y)$$

If X and Y are independent, we have $Cov(X,Y) = 0$ and $V(X \pm Y) = V(X) + V(Y)$

NOTE: **5.7.5.1** The Condition for Independence of X and Y implies that:

$$Cov(X,Y) = E(XY) - E(X)E(Y)$$

i.e $E(XY) = E(X)E(Y)$.

Proof:

$$E(XY) = \int_{-\infty}^{\infty}\int_{-\infty}^{\infty} xy\, f(x,y)dxdy = \int_{-\infty}^{\infty}\int_{-\infty}^{\infty} xy f_x(x)\, f_Y(y)dxd$$
$$= \int_{-\infty}^{\infty} x f_X(x)dx \int_{-\infty}^{\infty} y f_Y(y)dy$$
$$= E(X)E(Y)$$

NOTE:

In the above proof, $f_X(x)$ and $f_Y(y)$ are the marginal density functions of X and Y respectively.

5.7.6 Marginal Density Functions

Let (X, Y) be a two-dimensional continuous random variable with joint density function $f(x, y)$.

We obtain the marginal density functions $f_X(x)$ and $f_Y(y)$ of X and Y respectively as follows:-

$f_x(x) = \int_{-\infty}^{\infty} f(x, y)dy$ = marginal density function of X.

and

$f_y(y) = \int_{-\infty}^{\infty} f(x, y)dx$ = marginal density function of Y.

5.7.7. Definition A bivariate continuous r.v. (X, Y) taking values in the Euclidean plane R is said to be Uniformly distributed if its joint pdf $f(x, y)$ is given by

$$f(x, y) = \begin{cases} k, \forall \ (x, y) \in R \\ 0. otherwise \end{cases}$$

Where k is a constant. It is easy to show that $k = \dfrac{1}{area\ of\ R}$. The import of the underlined phrase "Uniformly distributed" is that probability over sub-regions of R depend only on their area and not on their locations.

EXAMPLE 5.11

Suppose that a bivariate continuous r.v. (X, Y) is uniformly distributed over the region R indicated below.

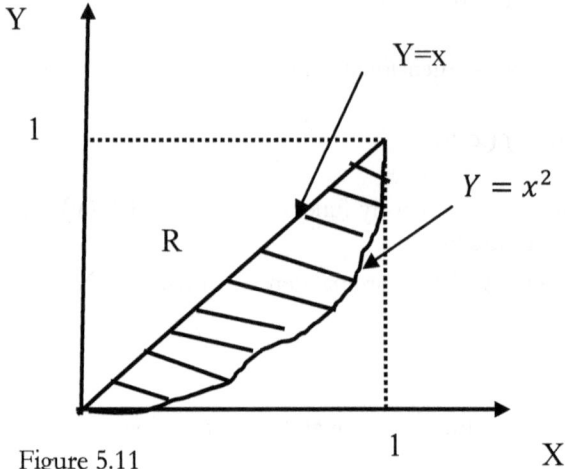

Figure 5.11

Find the marginal pdfs of X and Y

SOLUTION
From Figure 5.11 we have:

$$Area\ of\ R = \int_0^1 (x - x^2)dx$$

$$= \left[\frac{x^2}{2} - \frac{x^3}{3}\right]_0^1$$

$$= \frac{1}{6}$$

Therefore the pdf $f(x, y) = \begin{cases} 6\ if\ (x, y) \in R \\ 0. \quad ortherwise \end{cases}$

The marginal pdf of X is

$$g(x) = \int_{y=x^2}^{y=x} 6dy$$

$$= [6y]_{x^2}^{x}$$
$$= 6x - 6x^2$$

i.e.

$$g(x) = \begin{cases} 6x(1-x), 0 \le x \le 1 \\ \\ 0. \quad ortherwise \end{cases}$$

The marginal pdf of Y is

$$h(y) = \int_{x=y}^{x=\sqrt{y}} 6dx$$
$$= 6(\sqrt{y} - y)$$

i.e.

$$h(y) = \begin{cases} 6(\sqrt{y} - y), 0 \le y \le 1 \\ \\ 0. \quad ortherwise \end{cases}$$

EXAMPLE 5.12

Suppose that the two-dimensional random variable (X, Y) is uniformly distributed over the triangular region. $R = \{(x, y) | 0 < x < y < 1\}$ as shown in Figure 5.12 below.

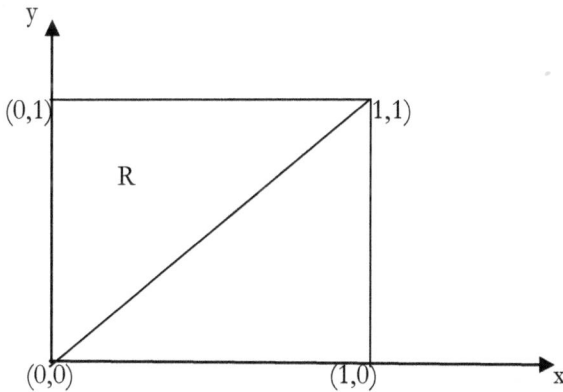

Figure 5.12

From Figure 5.12 the area the triangle R is given by $A = \frac{1}{2}(1)(1) = \frac{1}{2}$. Therefore the pdf,

$$f(x,y) = \begin{cases} \dfrac{1}{R}, & (x,y) \in R \\ 0, & otherwise \end{cases}$$

Hence, the joint pdf is given as

$$f(x,y) = \begin{cases} 2, & (x,y) \in R \\ 0, & elsewhere \end{cases}$$

Thus, the marginal pdf's of X and Y are:

$$f(x) = \int_{x}^{1} 2dy = 2(1-x), \qquad 0 \le x \le 1$$

and $f_Y(y) = \int_{0}^{y} 2dx = 2y$, $\qquad 0 \le y \le 1$

Therefore,

$$E(X) = \int_{0}^{1} x \cdot 2(1-x)dx = \frac{1}{3}$$

$$E(Y) = \int_{0}^{1} y.2y\, dy = \frac{2}{3}$$

$$E(X^2) = \int_{0}^{1} x^2 \cdot 2(1-x)dx = \frac{1}{6}$$

$$E(Y^2) = \int_{0}^{1} y^2\, 2y\, dy = \frac{1}{2}$$

$$V(X) = E(X^2) - [E(X)]^2 = \frac{1}{18}$$

$$V(X) = E(Y^2) - [E(Y)]^2 = \frac{1}{18}$$

Hence $E(XY) = \int_{0}^{1}\int_{0}^{y} xy(2)\, dx\, dy = \frac{1}{4}$

$$Cov(X,Y) = E(XY) - E(X)E(Y) = \frac{1}{4} - \frac{2}{9} = \frac{1}{36}$$

5.7.8 Conditional Distributions

Let (X, Y) have a continuous bivariate distribution and a value Y=y is observed. If Y is continuous, the probability that Y takes on a particular value is 0; an attempt at defining conditional probability as it was defined before in the discrete case yields.

$$P(X \le x/Y = y) = \frac{P(X \le x \, and \, Y = y)}{P(Y = y)} = \frac{0}{0}.$$

which leads nowhere. And yet, since one can observe a single value Y = y, it is certainly desirable to have a model with conditional probabilities of events given Y = y, and to be able to derive these from the model for the random (X,Y).

Since (X, Y) has a continuous distribution, and a value Y = y is observed, the conditional distribution of X certainly should be continuous along the line of

Y – values defined by the given information. Moreover, the density of that distribution should have the property (by analogy with the discrete case) that its weighted integral gives the marginal or unconditional density of X:

$$\int_{-\infty}^{\infty} f(x/y)f_Y(y)dy = f_X(x)$$

Where, again, a simplified notation has been used for the conditional density.

$$f(x/y) \equiv f_{X/Y=y}(x)$$

This desirable property for $f(x/y)$ can be reconciled with the definition of a marginal density in terms of the joint density:

$$f_X(x) = \int_{-\infty}^{\infty} f(x,y)dy$$

by defining

$$f(x/y) = \frac{f(x,y)}{f_Y(y)}$$

The function $f(x/y)$ so defined is (as a function of x) a bona fide density function, being non-negative with integral

$$\int_{-\infty}^{\infty} f(x/y)dx = \int_{-\infty}^{\infty} \frac{f(x,y)}{f_Y(y)}dx = 1.$$

A similar definition for $f(y/x)$ is given as

$$f(y/x) = \frac{f(x,y)}{f_X(x)}, where$$

$$f(y/x) = f_{Y/X=x}(y)$$

is a function of y, and $f_X(x) = \int_{-\infty}^{\infty} f(x,y)dy$, so that $\int_{-\infty}^{\infty} f(y/x)dy = -\infty \infty f(x,y) dy f X(x) = 1$

The integral over a set of X (or Y) values defines the conditional probability of that set,

$$P(X \text{ in } A/Y = y) = \int_A f(x/y)dx \text{ or}$$

$$P(Y \text{ in } A/X = x) = \int_A f(y/x)dy.$$

If A is in particular, the interval $(-\infty, x)$ or $(-\infty, y)$ one obtains the conditional cumulative density function (cdf),

$$F(x/y) = F_{X/Y=y}(x) = \int_{-\infty}^{x} f(u/y)du$$

or

$$F(y/x) = F_{Y/X=x}(y) = \int_{-\infty}^{y} f(u/x)du$$

EXAMPLE 5.13

A random vector (X, Y) has a density that is zero outside the triangle with vertices (0, 0), (0, 1) and (1, 0) and is constant within the triangle:

$$f(x,y) = \begin{cases} 2, & if \ x+y \leq 1, \ x \geq 0 \ and \ y \geq 0 \\ 0, & otherwise \end{cases}$$

The marginal density for x is the area under the cross section at X = x as a function of x.

$$f_x(x) = \int_{-\infty}^{\infty} f(x,y)dy = \int_0^{1-x} 2dy = 2(1-x), \quad for \ 0 \leq x \leq 1.$$

The marginal density for y is the area under the cross section at $Y = y$ as a function of y.

$$f_y(y) = \int_{-\infty}^{\infty} f(x, y)dx = \int_0^{1-y} 2dx = 2(1 - y), \quad for\ 0 \le y \le 1.$$

The conditional density for Y given, $X = x$, is then

$$f(y/x) = \frac{f(x,y)}{f_X(x)} = \frac{2}{2(1-x)} = \frac{1}{1-x}, \quad 0 \le y \le 1 - x$$

The value is zero outside the indicated range. That is, the density of Y given $X = x$ is constant on the interval $[0, 1 - x]$.

Similarly, the conditional density of X given $Y = y$ is given as

$$f\left(x/y\right) = \frac{f\left(x, y\right)}{f_Y\left(y\right)} = \frac{2}{2\left(1 - y\right)} = \frac{1}{1 - y}, \quad 0 < x \le 1 - y$$

The value being zero outside the indicated range, and is constant on the interval
$[0, 1 - y]$.

EXAMPLE 5.14
Let (X, Y) have the distribution defined by the density function

$$f_{X,Y}(x, y) = \begin{cases} e^{-x-y}, & for\ x > 0, y > 0 \\ 0, & otherwise \end{cases}$$

The marginal distribution for Y is the integral of this on x:

$$f_y(y) = \int_{-\infty}^{\infty} e^{-x-y}dx = \int_0^{\infty} e^{-x-y}dx = e^{-y}, \quad for\ y > 0.$$
The marginal distribution for X is the integral of this on y.

$$f_X(x) = \int_{-\infty}^{\infty} e^{-x-y}\,dy = \int_0^{\infty} e^{-x-y}\,dy = e^{-x}, \quad for\ x > 0.$$
The conditional density of X given $Y = y$ is then

$$f(x/y) = \frac{e^{-x-y}}{e^{-y}}, \quad for\ x > 0.$$
The conditional density of Y given $X = x$ is given as

$$f(y/x) = \frac{e^{-x-y}}{e^{-x}} = e^{-y}, \quad for\ y > 0.$$

It will be seen in the next section that the fact that this conditional density of x (or y) is independent of y (or x) and is indeed equal to the marginal density of X is of particular significance.

5.7.8.1 Condition for Independent Random Variables

It can happen, as in Example 5.14 above, that the conditional distribution of one variable in the pair (X,Y) is independent of any condition imposed on the other.
The random variables X and Y are then said to be independent random variables.
In particular, suppose that for every event A in the value space of X and for each value y of Y,

P(X in A/Y = y) = P(X in A).

Then for every event B in the value space of Y,

$$P(X \text{ in } A \text{ and } Y \text{ in } B) = \int_B P(X \text{ in } A/Y = y) f_Y(y) dy$$

$$= P(X \text{ in } A) \int_B f_Y(y) dy$$

$$= P(X \text{ in } A) P(Y \text{ in } B)$$

Then the experiment that consists of observing a value of X and the experiment that consists of observing a value of Y are independent experiments.
If the above factorisation is valid for all events A and B, it must hold, in particular, for
$\{X \leq x\}$ and $\{Y \leq y\}$: so that the cumulative functions
$$F_{XY}(x, y) = F_X(x) . F_Y(y)$$
Differentiation of these relations for cumulative distribution functions yields the following relation for densities:
$$f_{XY}(x, y) = f_X(x) . f_Y(y)$$
But then whether f denotes density or probability,

$$f(x/y) = \frac{f(x, y)}{f_Y(y)} = \frac{f_X(x) f_Y(y)}{f_Y(y)} = f_X(x),$$

which means that the conditional distribution of X given $Y = y$ is independent of y. Therefore, this defining criterion for independence and the various factorisation conditions for probabilities, cumulative distribution functions, and density or probability functions are equivalent. It does not matter which is taken as defining independence of X and Y.

EXAMPLE 5.15

Let (X, Y) have the joint density e^{-x-y}, for positive x and y, as in Example 5.14 above. It was seen in that example that
$f(X/y) = e^{-x}$, for $x > 0$, which is exactly the marginal density of X. Moreover, the marginal density of Y is e^{-y}, for $y > 0$, so that
$$f(x, y) = e^{-x-y} = e^{-x}e^{y} = f_X(x)f_Y(y)$$

The component variables X and Y are independent.

EXAMPLE 5.16

Consider again the random vector (X, Y) of Example 5.13 above where the distribution has a constant density in the triangle bounded by the coordinate axes and by $x + y \leq 1$ in the plane of values of the random vector. This does not have the factorization property defining independence of X and Y. The marginal densities are (as seen in Example 5.13.).

$$f_X(x) = 2(1-x), \qquad f_Y(y) = 2(1-y)$$
for $0 < x < 1$ and $0 < y < 1$ respectively, whereas the joint density $f_{XY}(x, y)$ is 2 on the given triangle and zero (0) outside.

5.8. Conditional Expectation

Just as we defined the expected value of a random variable X (in terms of its probability distribution) as $\int_{-\infty}^{\infty} x f(x)dx$, so we can define the conditional expectation of a random variable (in terms of its conditional probability distribution) as follows:

5.8.1 Definition (Conditional Expectation)

Let (X, Y) be a two-dimensional continuous random variable. We define the conditional expectation of X for given $Y = y$ as

$$E(X/Y = y) = \int_{-\infty}^{\infty} x f(x/y)dx$$

where $f(x/y)$ is the conditional expectation pdf of X given $Y = y$

and the conditional expectation of Y for given $X = x$ as

$$E(X/X = x) = \int_{-\infty}^{\infty} y f(y/x)dy$$

where $f(y/x)$ is the conditional expectation pdf of Y given $X = x$

NOTE:

(i)   It is important to realize that in general $E(X/y)$ is a function of y

and hence is a random variable. Similarly $E(Y/x)$ is a function of x and is also a random variable. [Strictly speaking, $E(X/y)$ is the value of the random variable $E(X/Y)$].

(ii)   Since $E(Y/X)$ and $E(X/Y)$ are random variables, it will be meaningful to speak of their expectations. Thus we may consider $E_Y[E_x(X/Y)]$, for instance. It is important to realize that the inner expectation is taken with respect to the conditional distribution of X given Y equals y, while the outer expectation is taken with respect to the probability distribution of Y.

(iii) $E(X/Y)$ is called the regression function (curve) of X on Y; $E(Y/X)$ is called the regression function (curve) of Y on X.

EXAMPLE 5.16

The joint density function of (X, Y) is $f(x, y) = \dfrac{e^{-x/y}e^{-y}}{y}, 0 \le x \le \infty, 0 \le y \le \infty$

(i)   Find $E(X/Y = y)$ and   (ii) $E(X^2/Y = y)$

SOLUTION

(i)   We have

$$g(x/y) = \frac{f(x,y)}{h(y)}$$

$$= \frac{\dfrac{e^{-x/y}e^{-y}}{y}}{\displaystyle\int_0^\infty \dfrac{e^{-x/y}e^{-y}}{y}dx}$$

$$= \frac{\dfrac{e^{-x/y}e^{-y}}{y}}{\left[-e^{-x/y}e^{-y}\right]_{x=0}^{\infty}}$$

$$= \frac{\dfrac{e^{-x/y}e^{-y}}{y}}{e^{-y}}$$

$$= \frac{e^{-x/y}}{y}$$

i.e. $g(x/y) = \dfrac{e^{-x/y}}{y}$

Therefore

$E(X/Y) = \int_0^\infty \dfrac{xe^{-x/y}}{y}dx = \left[-e^{-x/y}\right]_{x=0}^{\infty} + \int_0^\infty e^{-x/y}dx$, on integrating

by parts

$$= 0 + \left[-ye^{-x/y}\right]_{x=0}^{\infty} = y$$

(ii)   Next,

$E(X^2/y) = \int_0^\infty \dfrac{x^2 e^{-x/y}}{y}dx = \left[-x^2 e^{-x/y}\right]_{x=0}^{\infty} + \int_0^\infty 2xe^{-x/y}dx$, on

integrating by parts

$$= 0 + 2y\int_0^\infty \dfrac{xe^{-x/y}}{y}dx$$
$$= 2y^2$$

5.8.2 Theorem

$$E_Y[E(X/Y)] = E(X)$$

and

$$E_X[E(Y/X)] = E(Y)$$

Proof:

By definition,

$$E\left(X/y\right) = \int_{-\infty}^{\infty} x f\left(x/y\right)dx = \int_{-\infty}^{\infty} x \cdot \frac{f\left(x,y\right)}{f\left(y\right)}dx$$

where f(x, y) is the joint pdf. of (X, Y) and f(y) is the marginal pdf of Y.

Hence $E_Y\left[E\left(X/Y\right)\right] = \int_{-\infty}^{\infty} \cdot E\left(X/y\right)f\left(y\right)dy$

$$= \int_{-\infty}^{\infty}\left[\int_{-\infty}^{\infty} x\frac{f\left(x,y\right)}{f\left(y\right)}dx\right]f\left(y\right)dy \cdot$$

If all the expectations exist, it is permissible to write the above iterated integral with the order of integration reversed. Thus

$$E_Y\left[E\left(X/Y\right)\right] = \int_{-\infty}^{\infty} x\left[\int_{-\infty}^{\infty} f\left(x,y\right)dy\right]dx$$

$$= \int_{-\infty}^{\infty} x f\left(x\right)dx = E\left(X\right).$$

A similar argument may be used to establish that

$$E_X[E(Y/X)] = E(Y).$$

This is left as an exercise for the student.

EXAMPLE 5.17

A $r.v.\,X$ has a poison distribution with parameter λ. Given $X = k$, a r.v. Y has a binomial distribution with parameters k and p. Find the expected value of the r.v. Y.

SOLUTION:
$$E(Y) = E_y\{E_x(Y/X)\}$$
$$= E(Xp) = pE(X) \text{ since p is a constant}$$
$$= p\lambda$$

EXAMPLE 5.18

For any non-negative integer valued $r.v.\,X$ given $X = k$, Y is a r.v. that has a binomial distribution with parameter k and p.

Find $E(Y)$ if X has

(i)   a binomial distribution with parameters M and q
(ii)   pmf $P(X = u) = p(1 - p)^u, u = 0,1,2,3, ...$
(iii)   pmf $P(X = u) = p(1 - p)^{u-1}, u = 1,2,3, ...$

SOLUTION

$$E(Y) = E_y\{E_x(Y/X)\} = E(Xp) = pE(X)$$

(i) $E(X) = Mq$, therefore $E(Y) = pMq$

(ii) $E(X) = \frac{1-p}{p}$, therefore $E(Y) = p\frac{1-p}{p} = 1 - p$

(iii) $E(X) = \frac{1}{p}$, therefore $E(Y) = p\frac{1}{p} = 1$

5.8.3   Condition for Independent Random Variables

Suppose that X and Y are independent random variables. Then

$$E(X/Y) = E(X) \text{ and } E(Y/X) = E(Y).$$

EXAMPLE 5.13

Given $f(x,y) = \begin{cases} 6(1-x-y) \text{ for } 0<y<1-x, \ x>0 \\ 0, \ elsewhere \end{cases}$

The conditional density

$$f(x/y) = \frac{f(x,y)}{f_Y(y)} = \frac{f(x,y)}{\int_{-\infty}^{\infty} f(x,y)dx} = \frac{2(1-x-y)}{(1-y)^2},$$

for $0 < x < 1 - y$.

The conditional mean, given Y = y is then the integral of x with respect to this conditional density:

$$E(X/Y=y) = \int_{-\infty}^{\infty} x \, f(x/y)dx = \int_{0}^{1-y} \frac{2(1-x-y)}{(1-y)^2} \, dx = \frac{1}{3}(1-y).$$

The conditional mean E(X/Y = y) is, of course, a function of the conditioning value y. This function is called regression function of X on Y. It is used as a predictor for X when Y is given.

EXAMPLE 5.14

Let (X, Y) have the joint density

$$f(x,y) = \begin{cases} 2, \ for \ x+y \le 1, \ x \ge 0, \ y \ge 0 \\ 0, \ elsewhere \end{cases}$$

The marginal density of Y is

$$f_Y(y) = \int_{-\infty}^{\infty} f(x,y)\,dx = \int_0^{1-y} 2\,dx = 2(1-y),\ 0<y<1$$

and the conditional density of X given $Y = y$ is

$$f(x/y) = \frac{f(x,y)}{f_Y(y)} = \frac{2}{2(1-y)}\ \textit{for } 0<x<1-y$$

The conditional mean of X given $Y = y$ is then

$$E(X/y) = \frac{1}{1-y} \int_0^{1-y} x\,dx = \frac{1-y}{2}$$

Integrating this with respect to the distribution of Y yields (unconditional) mean of X:

$$E(X) = E_Y\left[E(X/Y)\right] = \int_0^1 \frac{1-y}{2}\cdot 2(1-y)\,dy = \frac{1}{3}.$$

NOTE:

(i) $E(X/X = c) = c$

(ii) $E\left(\sum_{i=1}^k X_i \,/\, X\right) = \sum_{i=1}^k E(X_i \,/\, X)$

(iii) $E(XY/X) = XE(Y \,/\, X)$

(iv) $E(F(Y)/X) = E(F(Y))$

5.8.4   Linear Regression Function $E(Y \,/\, X)$ and $E(X \,/\, Y)$

5.8.4.1 Theorem.

If X and Y are random variables such that
$E(X) = \mu_x,\ E(Y) = \mu_y,\ Var(X) = \sigma_x^2 > 0,\ Var(Y) = \sigma_y^2 > 0$ and if
$E(Y \,/\, X)$ and $E(X \,/\, Y)$ are both linear, then

$$\begin{cases} E(Y/X) = \mu_y + \rho\dfrac{\sigma_y}{\sigma_x}(x - \mu_x) \\[4mm] E(X/Y) = \mu_x + \rho\dfrac{\sigma_x}{\sigma_y}(y - \mu_y) \end{cases}$$

Where ρ is the correlation between X and Y.

PROOF

Let $E(Y/X) = Ax + B$. Multiplying this equation by $g(x)$ the marginal pdf of X and integrating from $-\infty$ tp $+\infty$, we have

$$\int_{-\infty}^{\infty} E(Y/x)\, g(x)dx = \int_{-\infty}^{\infty} (Ax + B)\, g(x)dx$$

i.e.

$$E(Y) = AE(X) + B - - - - - - - - - - (1) \text{ i.e}$$

$$\mu_y = A\mu_x + B$$

Next multiplying $(Ax + B)$ by $xg(x)$ and integrating we have

$$\int_{-\infty}^{\infty} E(Y/x)\, xg(x)dx = \int_{-\infty}^{\infty} (Ax^2 + Bx)\, g(x)dx$$

i.e

$$E\big(XE(Y/x)\big) = AE(x^2) + BE(x)$$

i.e

$$EE(XY / x) = AE(x^2) + BE(x)$$

i.e

$$E(XY) = AE(x^2) + BE(x) - - - -(2)$$

Solving equations (1) and equations (2) simultaneously yields the required results after finding A and B

5.8.4.2 Some consequences

(i) $\rho = 0 \Rightarrow$ Independence i.e. $E(Y/x) = \mu_y$ and $E(y) = \mu_x$

(ii)   Sign of slope is determined by ρ since $\sigma_{y,x} > 0$

(iii)   The two regression functions have a common sign of slope

(iv)   The product of slopes $= \left(\rho\dfrac{\sigma_y}{\sigma_x}\right)\left(\rho\dfrac{\sigma_x}{\sigma_y}\right) = \rho^2$

(v)   Since slope is either $\rho\dfrac{\sigma_x}{\sigma_y}$ or $\dfrac{\sigma_y}{\sigma_x}$, then the value of slope enables us to calculate σ_x or σ_y given the other, having calculated ρ

(vi)   If $E(Y/x)$ and $E(X/y)$ are both linear then $(E(x), E(Y))$ is the point of intersection

5.9. Expected Value of a Function of Two Random Variables

A function of two random variables also defines a random variable:

$Z = H(X,Y)$, and it is desirable to be able to compute $E(Z)$ from the joint distribution of (X,Y) without going through the intermediate step of determining the distribution of Z. Since (X,Y) has a continuous distribution in the plane with density function $f(x,y)$, the formula is

$$E[H(X,Y)] = \iint H(x,y)\, f(x,y)\,dx\,dy,$$

where the double integral extends over the whole x-y plane, and is evaluated in the usual way as an iterated integral (either with respect to the given co-ordinates, or in terms of some other convenient coordinates).

The sum of two random variables is an instance of a function of two random variables

$Z = X + Y$. It has already been pointed out earlier that the expected value of a sum is the sum of the expected values, and this can now be seen as follows for the continuous case:

$$E(X+Y) = \iint (x+y)\, f(x,y)\,dx\,dy$$
$$= \iint x\, f(x,y)\,dx\,dy + \iint y\, f(x,y)\,dx\,dy$$
$$= E(X) + E(Y)$$

where all integrals extend from $-\infty$ to $+\infty$. Thus, the additivity of the expected value is essentially the additivity of the double integral in terms of which $E(\bullet)$ is evaluated. One step in the above reasoning exploits the fact that X is also a special case of a function of (X, Y). For this function, the equivalence of the two ways of evaluating the mean is seen as follows:

$$E(X) = \iint x\, f(x,y)\,dy\,dx = \int x \left\{ \int f(x,y)\,dy \right\} dx$$
$$= \int x\, f_X(x)\,dx = E(X).$$

The product xy is another common instance of a function of two variables – and one might naively expect that the average product is the product of the averages; but this is not generally so, although it can happen. One class of bivariate distributions for which it is true is that in which the marginals are independent. In such a case, the average of the product of any function of the other is the product of the averages. That is, if X and Y are independent.

$$E[g(X)\, h(Y)] = E[g(X)]\, E[h(Y)].$$

(In particular, this would mean that $E(X, Y) = E(X) E(Y)$ whenever X and Y are independent). This factorization of expectation follows easily from the factorization of the joint density into the product of the two marginals:

$$E[g(x)h(Y)] = \iint g(x)h(y)f(x, y)dx\,dy$$

(X and Y independent)

$$
\begin{aligned}
&= \iint g(x)h(y)f_X(x)f_Y(y)dxdy \\
&= \int g(x)f_X(x)\left\{\int h(y)f_Y(y)dy\right\}dx \\
&= E[g(X)]\ E[h(Y)]
\end{aligned}
$$

5.10. Bivariate Transformations (Jacobians)

In two dimensions, a transformation

$$
\begin{cases}
U = g(x, y), \\
V = h(x, y),
\end{cases}
$$

maps a region R of points in the xy-plane into a region S of points in the UV – plane. It is assumed that g and h are continuously differentiable. The quantity that plays the role of the derivative is now the Jacobian of the transformation

$$
J(x, y) = \frac{\partial(u, v)}{\partial(x, y)} =
\begin{vmatrix}
\dfrac{\partial u}{\partial x} & \dfrac{\partial u}{\partial y} \\[2ex]
\dfrac{\partial v}{\partial x} & \dfrac{\partial v}{\partial y}
\end{vmatrix}
$$

Suppose further that there is an inverse transformation

$$
\begin{cases}
x = G(u, v), \\
y = H(u, v),
\end{cases}
$$

which takes each point (u, v) of S into a unique point (x, y) in R such that

$$
\begin{cases}
g\{G(u, v), H(u, v) = U \\
h\{G(u, v), H(u, v) = V
\end{cases}
$$

these being identities in (u, v). Then, if the Jacobian of the transformation is non-vanishing in R, the Jacobian of the inverse transformation is defined in S as the reciprocal of the Jacobian of the direct transformation:

$$\frac{\partial(x,y)}{\partial(u,v)} = \left(\frac{\partial(u,v)}{\partial(x,y)}\right)^{-1},$$

and is the factor needed for conversion of area elements. Thus, the change of variables in a double integral is accomplished as follows:

$$\iint_R f(x,y)dx\,dy = \iint_S f(G(u,v),\ H(u,v)) \left|\frac{\partial(x,y)}{\partial(u,v)}\right| du\,dv,$$

where S is the image of the region R under the transformation.

If (x, y) is a possible value of the random vector (X, Y), the transformation being considered defines the new random vector (U, V) with

$$\begin{cases} U = g(X,Y) \\ V = h(X,Y) \end{cases}.$$

This transformation induces a probability distribution in the UV − plane as follows: If S is a set in the UV-plane and R is the set of all points in the XY − plane that have "images" in S under the transformation then
P[U, V) in S] = P[(X, Y) in R].
If the distribution of (X, Y) is of the continuous type the probability of S can be expressed as an integral over R:

$$P[(U,V) in S] = \iint_S f_{X,Y}(x,y)dx\,dy$$

$$= \iint_S f_{X,Y}(G(u,v),\ H(u,\ v)) \left|\frac{\partial(x,y)}{\partial(u,v)}\right| du\,dv$$

Where $f_{X,Y}(x,y)$ is the joint density function of (X, Y). Since this relation holds for each event S in the UV − plane, the density of (U, V) is the integrand function of the UV − integral:

$$f_{UV}(u,v) = f_{XY}(G(u,v),H(u,v)) \left|\frac{\partial(x,y)}{\partial(u,v)}\right| .$$

That is, to obtain the density of (U, V), solve the transformation equations for x and y in terms of U and V, substitute for x and y in the joint density of X and Y, and multiply by the absolute value of the Jacobian of (x, y) with respect to (U, V); and then, similarly the density of (X, Y) as obtained from that of (U,V) is

$$f_{X,Y}(x,y) = f_{U,V}(g(x,y), h(x,y)) \left| \frac{\partial(u,v)}{\partial(x,y)} \right|.$$

EXAMPLE 5.15

(Product of independent random variables)

Let X and Y be two independent random variables such that Y has a uniform distribution in the interval (0, 1) whilst X has probability density function (pdf.) given as

$$f_X(x) = \begin{cases} xe^{-x}, & x > 0 \\ 0, & otherwise \end{cases}$$

If Z is a new random variable defined as
Z = X Y (the product of X and Y), what is the density function of Z?

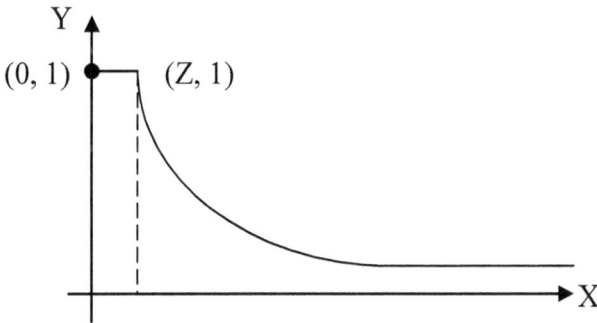

SOLUTION:
Since X and Y are independent, the joint density function

$$f_{XY}(x,y) = f_X(x). f_Y(y)$$
$$\text{But } f_X(x) = xe^{-x}, x > 0$$

and $f_Y(y) = 1, \ 0 < y < 1$ [since y is uniform in (0, 1)]

Hence

$$f_{XY}(x, y) = f_X(x) . f_Y(y) = x \, e^{-x} . 1$$

Given that $Z = X \cdot Y$, the distribution function of Z is given as

$$F_Z(z) = P(Z \le z) = P(XY \le z) = P\left(Y \le Z \cdot \frac{1}{X}\right)$$

$$= \int_Z^\infty \int_0^{\frac{Z}{X}} 1 \cdot X \, e^{-X} \, dy \, dx + \int_0^Z \int_0^1 X \cdot e^{-X} \, dy \, dx$$

$$= \int_Z^\infty Z \cdot e^{-X} \, dx + \int_0^Z X \cdot e^{-X} \, dx$$

$$= Z \cdot \left[-e^{-x} \right]_Z^\infty + \left[X \cdot \left(-e^{-X} \right) \right]_0^Z + \int_0^Z e^{-X} \, dx$$

$$= Z \, e^{-Z} - Z \, e^{-Z} + \left[-e^{-X} \right]_0^Z = 1 - e^{-Z}$$

$$\therefore f_Z(z) = \frac{d}{dZ} F_Z(z) = e^{-Z}$$

Alternative Solution.

Let Z = XY. Introduce a new variable U=X.

This implies that in the new coordinate system, X = U and $Y = \frac{Z}{U}$ with limits:

$$x > 0 \ \Rightarrow \ u > 0: \ \ y < 1 \Rightarrow \frac{Z}{U} < 1 \Rightarrow Z < U: \ y > 0 \ \Rightarrow Z > 0$$

The Jacobian of transformation is

$$\left\| \frac{\partial(x, y)}{\partial(u, z)} \right\| = \begin{vmatrix} \dfrac{\partial x}{\partial u} & \dfrac{\partial x}{\partial z} \\ \dfrac{\partial y}{\partial u} & \dfrac{\partial y}{\partial z} \end{vmatrix} = \begin{vmatrix} 1 & 0 \\ -\dfrac{z}{u^2} & \dfrac{1}{u} \end{vmatrix} = \frac{1}{u}, \qquad u > 0$$

$$\therefore f_{Z,U}(z, u) = f_{X,Y}\left(u, \frac{z}{u}\right) . \frac{1}{u} = u e^{-u} \cdot 1 . \frac{1}{u} = e^{-u}$$

$$f_Z(z) \int\limits_{z}^{\infty} e^{-u} \, du = [-e^{-u}]_z^{\infty} = e^{-z}$$

EXAMPLE 5.16

(Quotient of independent random variables)
Let X and Y be two independent random variables such that

$$f_X(x) = \begin{cases} 2x, \; 0<x<1 \\ 0, \; elsewhere \end{cases} \quad and$$

$$f_Y(y) = \begin{cases} 2y, \; 0<y<1 \\ 0, \; elsewhere \end{cases}$$

If Z is a new random variable defined as $Z = \dfrac{X}{Y}$ (the quotient of X and Y), what is the density function of Z?

SOLUTION:

Since $Z = \dfrac{Y}{X} \, , \; Z > 0 \, ,$

The distribution function of Z is given as

$$F_Z(z) = P[Z \le z] = P\left[\frac{Y}{X} \le Z\right] = P[Y \le Z \cdot X]$$

<u>Case 1</u> $Z \le 1$

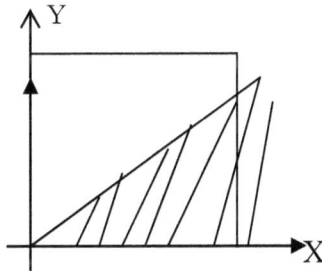

$$F_Z(z) \begin{bmatrix} for \; independent \\ X \; and \; Y \end{bmatrix} = 4 \int_0^1 \int_0^{zx} xy \; dy \; dx$$

$$= 4 \int_0^1 \left(x \left[\frac{y^2}{2} \right]_0^{zx} \right) dx$$

$$= 4 \int_0^1 x \cdot \frac{z^2 x^2}{2} dx = 2 z^2 \int_0^1 x^3 dx = 2z^2 \left[\frac{x4}{4} \right]_0^1 = \frac{z^2}{2}$$

$$\therefore f_Z(z) = \frac{d}{dz} F_Z(z) = \frac{1}{2} \cdot 2z = z$$

Case 2 Z > 1

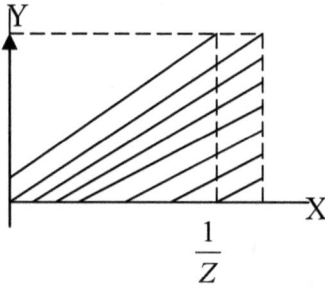

$$F_Z(z) = 4 \int_0^{\frac{1}{z}} \int_0^{zx} xy \, dy \, dx + 4 \int_0^1 \int_{\frac{1}{z}}^1 xy \, dx \, dy$$

$$= 4 \int_0^{\frac{1}{z}} \left(x \left[\frac{y^2}{2} \right]_0^{zx} \right) dx + 4 \int_0^1 \left(y \left[\frac{x^2}{2} \right]_{\frac{1}{z}}^1 \right)$$

$$= 4 \int_0^{\frac{1}{z}} \frac{z^2 x^3}{2} \, dx + 4 \int_0^1 y \left(\frac{1}{2} - \frac{1}{2z^2} \right) dy$$

$$= 2 z^2 \int_0^{\frac{1}{z}} x^3 \, dx + 2 \left(1 - \frac{1}{z^2} \right) \int_0^1 y \, dy$$

$$= 2 z^2 \left[\frac{x^4}{4} \right]_0^{\frac{1}{z}} + 2 \left(1 - \frac{1}{z^2} \right) \left[\frac{y^2}{2} \right]_0^1$$

$$= 2z^2 \cdot \frac{1}{4z^2} + \left(1 - \frac{1}{z^2}\right) = 1 - \frac{1}{2z^2}$$

$$\therefore f_Z(z) = \frac{d}{dz} F_Z(z) = -\frac{1}{2} \cdot \frac{-2}{z^3} = \frac{1}{z^3}$$

$$\therefore f_Z(z) = \begin{cases} z \quad, & if \quad 0 < z < 1 \\ \dfrac{1}{z^3}, & if \quad z > 1 \end{cases}$$

Alternative Solution

Let $Z = \dfrac{Y}{X}$, $\quad 0 < z < \infty \quad$ and introduce a new random variable.

$U = X$, $0 < u < 1$ so that in the new co-ordinate system

$X = U \quad$ and $Y = U Z$.

The Jacobian of transformation is

$$\left| \frac{\partial(x,\, y)}{\partial(u,\, z)} \right| = \begin{vmatrix} \dfrac{\partial x}{\partial u} & \dfrac{\partial x}{\partial z} \\ \dfrac{\partial y}{\partial u} & \dfrac{\partial y}{\partial z} \end{vmatrix} = \begin{vmatrix} 1 & 0 \\ z & u \end{vmatrix} = u$$

Since X and Y are independent,

$$f_{Z,U}(z,u) = f_{X,Y}(u,\, uz) \cdot u = 4 \cdot zu \cdot u \cdot u = 4\, zu^3$$

The limits of integration:

$$y < 1 \Leftrightarrow uz < 1 \quad \Rightarrow u < \frac{1}{z}$$

i.e when z > 1, the limits of u are $\left(0, \dfrac{1}{z}\right)$ and when z ≤ 1 the limits of u

are (0, 1).

<u>Case 1</u>: z≤ 1:

$$f_Z\left(z\right) = \int_0^1 f_{Z},u\left(z,u\right)du = \int_0^1 4\,zu^3 \ du = 4z\left[\frac{u^4}{4}\right]_0^1 = z$$

<u>Case 2</u>: z > 1:

$$f_Z\left(z\right) = \int_0^{\frac{1}{z}} f_{Z,U}\left(z,u\right)du = \int_0^{\frac{1}{z}} 4\,zu^3 \ du = 4z\left[\frac{u^4}{4}\right] = \frac{1}{z^3}$$

$$\therefore f_Z(z) = \begin{cases} z, & if \ \ 0 < z < 1 \\ \dfrac{1}{z^3}, & if \ \ \ z > 1 \end{cases}$$

5.11. Multivariate Transformations
Given the outcome $(X_1, \ldots X_2)$ one may wish to define,

$$\begin{cases} U_1 = g_1(x_1, \cdots\cdots, x_n) \\ \quad\vdots \\ \quad\vdots \\ U_2 = g_n(x_1, \cdots\cdots, x_n) \end{cases} \qquad \text{for given functions}$$

$(g_1, \cdots\cdots, g_n)$. For inverses and integral transformations, the crucial

quantity is again the Jacobian of the transformation:

$$\frac{\partial(u_1, \cdots\cdots u_n)}{\partial(x_1, \cdots\cdots x_n)} = \begin{vmatrix} \dfrac{\partial u_1}{\partial x_1} & \cdots & \cdots & \cdots & \dfrac{\partial u_1}{\partial x_n} \\ \vdots & \vdots & \vdots & \vdots & \vdots \\ \vdots & \vdots & \vdots & \vdots & \vdots \\ \dfrac{\partial u_n}{\partial x_1} & \cdots & \cdots & \cdots & \dfrac{\partial u_n}{\partial x_n} \end{vmatrix}$$

If this does not vanish in a Region R, and there is an inverse transformation defined on the image of R:

$$\begin{cases} x_1 = G(u_1, \cdots\cdots\cdots, u_n) \\ \quad\vdots \\ \quad\vdots \\ x_n = G(u_1 \cdots\cdots\cdots u_n) \end{cases}, \qquad \text{whose Jacobian is the reciprocal of}$$

that of the forward transformation, then the transformation from $\tilde{X}$ to $\tilde{U}$ in a multiple integral over R is accomplished by subtracting $\tilde{X}$ from the inverse transformation and transforming the volume element as follows: $dx_1 \ldots\ldots\ldots\ldots dx_n \rightarrow \left|\dfrac{\partial(x_1, \ldots\ldots\ldots x_n)}{\partial(u_1, \ldots\ldots\ldots u_n)}\right| du_1 \ldots\ldots\ldots\ldots du_n,$

the integral in $\tilde{U}$ then extending over the image of R. If $(x_1, \ldots\ldots\ldots\ldots\ldots\ldots, x_n)$, is a possible value of the random vector $(X_1, X_2, \ldots X_n)$, then the transformation from $\tilde{X}$ to $\tilde{U}$ defines a random

vector $\tilde{U}$ with components $\begin{cases} U_1 = g_1(X_1 \ldots\ldots\ldots\ldots, X_n) \\ \quad\vdots \\ \quad\vdots \\ \quad\vdots \\ U_n = g_n(X_1, \ldots\ldots\ldots\ldots, X_n) \end{cases}$

If S denotes a set of points $(u_1, \ldots\ldots\ldots\ldots\ldots u_n)$ and R the set of all points $(x_1, \ldots\ldots\ldots x_n)$ having images in S, then
$P[(U_1, \ldots U_n)] = P[(g_1(X_1, \ldots, X_n) \ldots g_n(X_1, \ldots\ldots X_n) in\ S] = P[(X_1, \ldots\ldots X_n)\ in\ R]$. And in the continuous case, this last probability can be expressed as an integral of the joint density function of X's:

$$P(S) = \int_R \ldots\ldots\ldots \int f(x_1 \ldots\ldots\ldots, x_n) dx_1 \ldots\ldots\ldots\ldots dx_n$$
$$= \int_S \ldots \int f(G_1(u_1, \ldots, u_n), \ldots G_n(u_1, \ldots u_n)) \left|\frac{\partial(x_1, \ldots\ldots x_n)}{\partial(u_1, \ldots\ldots u_n)}\right| du_1, \ldots\ldots du_n$$

The integrand of this $\tilde{U}$-integral is then the joint density functions of the U's. So far, the transformed variable has been assumed to be of the same dimension as the X's, but it is possible that one may be interested in a smaller number of the U's than X's:

$$\begin{cases} U_1 = g_1(X_1, \ldots \ldots \ldots , X_n) \\ \vdots \\ \vdots \\ \vdots \\ U_m = g_n(X_1, \ldots \ldots \ldots , X_n) \end{cases}, \quad \text{where } m \leq n. \quad \text{(If m were greater}$$

than n, the U's would be overdetermined, and if not incompatible, then not all necessary). One could supply 'n-m' additional relations and then determine the distribution of the mU's of interest as a marginal distribution of that of nU's. However, the transformation to $(U_1, \ldots \ldots \ldots, U_n)$ and the distribution in the $\widetilde{X}$ space determine a probability distribution directly in the m-dimensional $\widetilde{U}$ space, in the usual way:

$$P[(U_1, \ldots \ldots \ldots U_n) \text{ in } T] =$$

$$\int_R \ldots \ldots \ldots \int f(x_1, \ldots \ldots \ldots x_n) \, dx_1, \ldots \ldots \ldots dx_n.$$

Where R is the set of $(x_1, \ldots \ldots \ldots \ldots x_n)$ with image points in T. Taking T to be a semi-infinite rectangle and differentiating yields, as usual, the density function of $(U_1, \ldots \ldots \ldots \ldots U_m)$.

EXAMPLE 5.16

Let $X_1, \ldots X_n$ be successive interarrival times in a Poisson process-independent random variables with the exponential density $e^{-x}, x > 0$. the times to the successive arrivals measured from time zero, the reference point of X_1, are the sums:

$$\begin{cases} Y_1 = X_1 \\ Y_2 = X_1 + X_2 \\ \vdots \quad \vdots \\ Y_n = X_1 + X2 + \cdots + X_n \end{cases}$$

The inverse of this transformation of X's to Y's is given by
$$X_1 = Y_1$$
$$X_2 = Y_2 - Y_1$$

$$\begin{aligned} & \vdots \qquad \vdots \\ & X_n = Y_n - Y_{n-1} \end{aligned}$$

and the Jacobian of the transformation (either way) is 1. The joint density of the Y's is, therefore,

$$f_Y(y) = f_Y(y_1, y_2 - y_1, \cdots, y_n - y_{n-1})$$
$$= \exp[-y_1 - (y_2 - y_1) - \cdots, (y_n - y_{n-1})] = e^{-y_n},$$

a formula that holds for $0 < y_1 < y_2 < \cdots < y_n$. The marginal density of $y_n = \sum_{i=1}^{n} x_i \sum_{i=1}^{n} x_i$ can then be obtained by integrating out $y_1, \cdots, y_{n-1}$

$$f_{Y_n}(y_n) = e^{-y_n} \int_0^{y_n} \int_0^{y_{n-1}} \cdots \cdots \int_0^{y_1} dy_1 dy_2 \cdots \cdots dy_{n-1}.$$
$$= \frac{y_n^{n-1}}{(n-1)!} e^{-y_n}, \quad for \ 0 < y_n.$$

5.12. DIAGNOSTIC TEST
D 5.11.1 Find the value of the constant c such that the density function

$$f(x) = \begin{cases} 0, & if \ x < 0 \\ cx, & if \ 0 \leq x < 2 \\ 6c - 2cx, & if \ 2 \leq x < 3 \\ 0, & if \ x \geq 0 \end{cases}$$

Hence compute the mean, $E(X)$, and the median value $\tilde{m}$.

SOLUTION:

F(x) is a probability density function if $\int_{-\infty}^{\infty} f(x) dx = 1$.

$$ie \quad \int_{-\infty}^{\infty} f(x) dx = \int_0^2 cx dx + \int_2^3 (6c - 2cx) dx = 1$$

implies that $3c = 1$

or $\qquad c = \frac{1}{3}$

$$E(X) = \int_{-\infty}^{\infty} x f(x) dx = \int_0^2 \frac{x^2}{3} dx + \int_2^3 x \left(2 - \frac{2x}{3}\right) dx$$
$$= \frac{5}{3}.$$

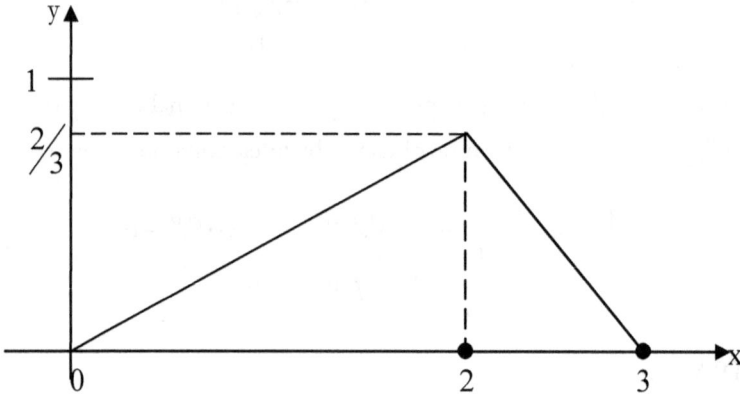

The median $\tilde{m}$ is obtained from the equation

$$\int_{-\infty}^{\tilde{m}} f(x)dx = \frac{1}{2}$$

Since $\int_{-\infty}^{2} f(x)dx = \int_{0}^{2} \frac{x}{3} dx = \frac{2}{3}$, it means that the median lies in the interval

(0, 2), and therefore,

$$\frac{1}{2} = \int_{0}^{\tilde{m}} \frac{x}{3} dx = \frac{\tilde{m}^2}{6}$$

$$\Rightarrow \tilde{m} = \sqrt{3} = \text{median.}$$

D5.11.2

Let X and Y be independent random variables, and define a new variable

$$Z = \frac{Y - X}{X}$$

If the distribution functions are defined as

$$F_X(x) = \begin{cases} 1 - (1+x)e^{-x}, & x \geq 0 \\ 0, & otherwise \end{cases}$$

$$F_Y(y) = \begin{cases} 1 - e^{-y}, & y \geq 0 \\ 0, & otherwise \end{cases}$$

(a) Obtain the density function of Z.
(b) Compute the median value of Z.

SOLUTION:

$$f_X(x) = \frac{d}{dx} F_X(x) = xe^{-x}, \qquad x \geq 0$$

$$f_Y(y) = \frac{d}{dy} F_Y(y) = e^{-y}, \qquad y \geq 0$$

$$Z = \frac{Y-X}{X}, \qquad -1 < Z < \infty, \qquad \left(\frac{Y-X}{X} = \frac{Y}{X} - 1\right)$$

(a) $\quad F_Z(z) = P(Z \leq z) = P\left(\frac{Y-X}{X} \leq z\right) = P(Y \leq (1+Z) \cdot X)$

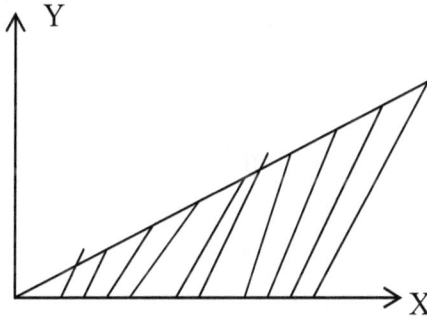

Since X and Y are independent,

$$= \int_0^\infty \int_0^{(1+z)x} f_{X,Y}(x,y)\, dy\, dx = \int_0^\infty \int_0^{(1+z)x} xe^{-x}.e^{-y} dy\, dx$$

$$= \int_0^\infty \left(xe^{-x}[-e^{-y}]^{(1+z)x}\right) dx = [-e^{-x}.x]_0^\infty + \int_0^\infty e^{-x}\, dx$$

$$= [-e^{-x}.x]_0^\infty + \int_0^\infty e^{-x}\,dx - \left[-\frac{1}{2+z}.e^{-(2+z)x}.x\right]_0^\infty$$
$$-\frac{1}{2+z}\int_0^\infty e^{-(2+z)x}dx$$

$$= 0 + 1 - 0 - \frac{1}{2+z}\left[\frac{-1}{2+z}\,e^{-(2+z)\cdot x}\right]_0^\infty$$

$$= 1 - \frac{1}{(2+z)^2}, \qquad z > -1$$

The density function

$$f_z(z) = \frac{d}{dz}\,F_z(z) = \frac{2}{(2+z)^3}, \qquad z > -1$$

(b) The median is computed from equation

$$F_z(z) = \frac{1}{2}$$

i.e. $\qquad 1 - \dfrac{1}{(2+z)^2} = \dfrac{1}{2} \quad iff \quad (2+x)^2 = 2$

$$\Rightarrow X + 2 = \pm\sqrt{2} \ or \ X = -2 \pm \sqrt{2}$$

$\therefore$ the median value $= \sqrt{2} - 2$

D 5.11.3 Let X and Y be independent random variables with density functions given as
$$f_X(x) = 1, 0 < X < 1 \ and \ f_Y(y) = e^{-y}, y > 0$$

(a)   Obtain the frequency density function of the new random variable Z defined as (i) Z = Y − X (ii) Z = X + Y
(b)   Find the probability that Y is greater than X (i.e. P(Y > X)).

SOLUTION (a) (i)

$$f_X(x) = 1, 0 < X < 1, \quad f_Y(y) = e^{-y}, y > 0$$
$$Z = Y - X, \quad -1 < Z < \infty.$$

$$F_Z(z) = P(Z < z) = P(Y - X \le z) = P(Y \le X + Z)$$

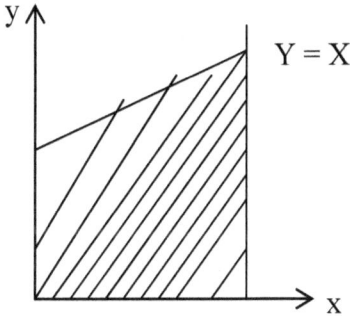

Y=X+Z

Z ≥ 0

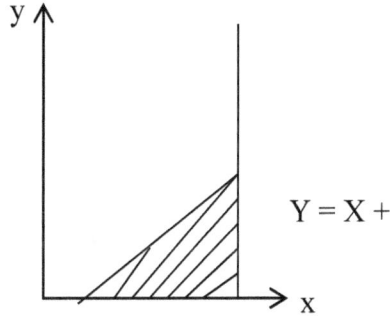

Y=X+Z

$-1 \le Z < 0$

Case 1: z≥ 0

$$F_Z(z) \int_0^1 \int_0^{x+z} e^{-y} \cdot 1 \, dy \, dx = [\text{since X and Y are independent}]$$

$$= \int_0^1 \left(\left[-e^{-y} \right]_0^{x+z} \right) dx = \int_0^1 \left(1 - e^{-x-x} \right) dx$$

$$= 1 - e^{-Z} \left[-e - x \right]_0^1$$

$$= 1 - e^{-Z} \left(-e^{-1} + 1 \right) = 1 - e^{-Z} \left(1 - \frac{1}{e} \right)$$

$$\therefore f_Z(z) = \frac{d}{dZ} F_Z(z) = e^{-Z} \left(1 - \frac{1}{e} \right)$$

CASE 2: $-1 < Z \le 0$

$$F_Z(z) = \int_{-z}^{x+z} e^{-y} \cdot 1 \, dy \, dx = \int_{-z}^1 (1 - e^{-x-z}) \, dx$$

$$= (1+z) - e^{-z} \left[-e^{-x} \right]_{-z}^1$$

$$= (1+z) - e^{-z} \left(-e^{-1} + e^z \right) = z + e^{-(1+z)}$$

$$\therefore f_z(z) = \frac{d}{dz} F_Z(z) = 1 - e^{-(1+z)}$$

$$\therefore f_Z(z) = \begin{cases} 1 - e^{-(1+Z)}, & for \ -1 < z < 0 \\ e^{-z}\left(1 - \frac{1}{e}\right), & for \ z \geq 0 \end{cases} ,$$

Alternative Solution:

$Z = Y - X$,   for $-1 < z < \infty$.

Introduce a new random variable $U = X$, $0 < u < 1$ so that in the new coordinate system

$X = U$ $, 0 < X < 1 \Rightarrow 0 < u < 1$

$Y = Z + u$, $Y > 0$ $\Rightarrow U > -Z$

The Jacobian of, the transformation is

$$\left\| \frac{\partial(x, y)}{\partial(u, z)} \right\| = \begin{vmatrix} 1 & 0 \\ 1 & 1 \end{vmatrix} = 1$$

$F_{Z,U}(z,u) = f_{X,Y}(u, z+u) \cdot 1 = e^{-(Z+u)}$

<u>Case 1</u>: $0 < u < 1$,   for $z \geq 0$

$$f_Z(z) = \int_0^1 e^{-(z+u)} \, du = e^{-z}[-e^{-u}]_0^1 = e^{-z}\left(1 - \frac{1}{e}\right)$$

<u>Case 2</u>: $-z < u < 1$ for $-1 < z < 0$

$$f_Z(z) = \int_0^1 e^{-(Z+u)} \, du = e^{-Z}\left[-e^{-u}\right]_{-Z}^1$$

$$= e^{-Z}\left(e^{-1} + e^Z\right) = 1 - e^{-(1+Z)}$$

(b) $P(Y > X) = P(Y - X > 0) = P(Z > 0)$

$$= 1 - F_Z(0) = 1 - e^{-(1+0)} = 1 - \frac{1}{e}$$

Alternatively:
P(Y > X)

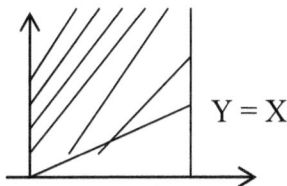

$$= \int_0^1 \int_x^{\infty} f_{X,Y}(x,y)dydx = \int_0^1 \int_x^1 e^{-y} \cdot 1 dydx = \int_0^1 ([-e^{-y}]_x^{\infty}) dx$$

$$= \int_0^1 e^{-x} dx = [-e^{-x}]_0^1 = 1 - \frac{1}{e}$$

D 5.11.4 The random variable X has distribution function (cdf) given as

$$F_X(x) \begin{cases} 0, & \text{if } x < 1 \\ \dfrac{x-1}{2}, & \text{if } 1 \le x < 3 \\ 1, & \text{if } x \ge 3 \end{cases}$$

(a) Obtain the probability density function (pdf) of X.
(b) Compute the following probabilities

P(X ≤ 2), P(X < 2), P(2 < X ≤ 3)

(i)   Using the distribution function $F_X(x)$ defined above.
(ii)   Using the probability density function (pdf).

SOLUTION:
The density function

(a) $f_X(x) = \dfrac{d}{dx} F_X(x) = \begin{cases} \dfrac{1}{2}, & 1 < x < 3 \\ 0, & otherwise \end{cases}$

(c)   (i) $P(X \le 2) = F_X(2) = \dfrac{2-1}{2} = \dfrac{1}{2}$

$P(X < 2) = F_X(2) = \dfrac{2-1}{2} = \dfrac{1}{2}$

$$P(2<X<3) = F_X(3) - F_X(2) = \left(\frac{3-1}{2}\right) - \left(\frac{2-1}{2}\right) = \frac{1}{2}$$

(ii) $P(X \le 2) = \int_{-\infty}^{2} f(x)dx = \int_{1}^{2} \frac{1}{2} dx = \left[\frac{x}{2}\right]_{1}^{2} = \frac{1}{2}$

$$P(X<2) = \int_{-\infty}^{2} f(x)dx = \int_{1}^{2} \frac{1}{2} dx = \left[\frac{x}{2}\right]_{1}^{2} = \frac{1}{2}$$

$$P(2<X<3) = \int_{2}^{3} f(x)dx = \int_{2}^{3} \frac{1}{2} dx = \left[\frac{x}{2}\right]_{2}^{3} = \frac{1}{2}$$

D 5.11.5 A random variable X has density function given as

$$f_X(x) = \begin{cases} \frac{1}{6}, & -2 < x < 4 \\ 0, & otherwise \end{cases}$$

(a) Compute the mean E(X) and the variance V(X) of X.
(b) Obtain the distribution function $F_X(x)$ of X.
(c) Determine the median of X.

SOLUTION:

(a) $E(X) = \int_{-\infty}^{\infty} x f(x)dx = \int_{-2}^{4} \frac{x}{6} dx = \left[\frac{x^2}{12}\right]_{-2}^{4} = 1$

$$E(X^2) = \int_{-\infty}^{\infty} x^2 f(x)dx = \int_{-2}^{4} x^2 \cdot \frac{1}{6} dx = \left[\frac{x3}{18}\right]_{-2}^{4} = 4$$

$$V(X) = E(X^2) - [E(X)]^2 = 4 - 1 = 3$$

Alternatively: It can be noted that X is uniformly distributed in the interval
[– 2, 4]. Hence using theorems 5.6.3.1 and 5.6.3.2 respectively, we have

$$E(X) = \frac{a+b}{2} = \frac{-2+4}{2} = 1, \quad and$$

$$V(X) = \frac{(b-a)^2}{12} = \frac{(4+2)^2}{12} = 3$$

(b) <u>Case 1</u>

$x < -2$

$$F_X(x) = \int_{-\infty}^{x} f(u)\,du = 0$$

<u>Case 2</u>

$-2 \le x < 4$

$$F_X(x) = \int_{-2}^{x} f(u)\,du = \int_{-2}^{x} \frac{1}{6}\,du = \frac{(x+2)}{6}$$

<u>Case 3</u>

$X \ge 4$: $F(\infty) - F(4) = 1 - 1 = 0$

$\Rightarrow F_X(x) = 1$

Hence $F_X(x) = \begin{cases} 0 & , x < -2 \\ \dfrac{(x+2)}{6}, & -2 \le x < 4 \\ 1 & , x \ge 4 \end{cases}$

(See also section 5.1.7.3)

(c) The median $\tilde{m}$ is such that

$\int_{-2}^{\tilde{m}} f(x)\,dx = \frac{1}{2},$ ie. $\left[\frac{x}{6}\right]_{-2}^{\tilde{m}} = \frac{1}{2},$ or $\left(\frac{\tilde{m}}{6} + \frac{2}{6}\right) = \frac{3}{6}, giving\ \tilde{m}=1.$

Note: Students of Mathematics and Statistics could find it more convenient solving the above integrals, using Mathematical programming software packages such as MathLab or Mathematheca.

EXERCISE 5

5.1 The distribution function of the random variable X is given as

$$F_X(x) = \begin{cases} 0 & , if\ x < 2 \\ \dfrac{x-2}{3}, & if\ 2 \le x < 5 \\ 1 & , if\ x \ge 5 \end{cases}$$

Determine

(a)   The mean of X, E(X)

(b)   The standard deviation of X, $\sqrt{V(X)}$

5.2 A random variable X is uniformly distributed in the interval (− 1, 1). Determine

(a) E(X) (b) E(X²) (c) E(X³) (d) $\sqrt{V(X)}$

5.3 The distribution function for the random variable X is given as

$$F_X(x) = \begin{cases} 0 & , \text{if } x < 3 \\ \dfrac{1}{4} & , \text{if } 3 \le x < 5 \\ 1 & , x \ge 5 \end{cases}$$

Determine

(a) E(X) and (b) $\sqrt{V(X)}$

5.4   A random variable X has density function given as
$$f_X(x) = \begin{cases} 2(x-1), & \text{if } 1 < x < 2 \\ 0, & \text{otherwise} \end{cases}$$
Determine
(a) E(X) (b) V(X)
(c) the distribution function for X
(d) P(1.1 < X ≤ 1.2)

5.5   Suppose that the random variable X is uniformly distributed over

[− a, 3a]. Find the variance of X.

5.6   Suppose that the random variable X is uniformly distributed on − 1 < x < 1.

(a)   Determine the probability that $X > \dfrac{3}{4}$, given that $|X| > \dfrac{1}{2}$

(b)   Obtain the distribution function (cdf) of X.

5.7   A random variable X has density.

$$f(x) = \begin{cases} \dfrac{x}{2} \; , & \text{if } 0 < x < 2 \\ 0 \; , & \text{elsewhere} \end{cases}$$

(a)   Determine P(X > 1)
(b)   Calculate the mean and variance of X
(c)   Determine the c.d.f. of X
(d)   Determine the c.d.f. of Y = X²

5.8   Let X and Y be independent observations on a uniform distribution on
[0 , 1]. Let Z = X Y, and W = Y, and obtain the density of Z = X Y as

a marginal density of (ZW).

5.9   The random variable X has its mean equal to its variance equal to 2, and has distribution function.

$$F_X(x) = \begin{cases} 1 - (1+x)e^{-x} \; , & \text{if } x > 0 \\ 0 \; , & \text{if } x \le 0 \end{cases}$$

If X_1 and X_2 are two independent observations of X, obtain the

frequency density function of a new random variable $Z = \dfrac{X_1}{X_2}$.

5.10   Let X and Y be two random variables (not necessarily independent) with density functions.

$$f_X(x) = \begin{cases} 1, & \text{if } 0 < x < 1 \\ & \quad\quad\quad\quad\quad\quad \text{and} \\ 0 \; , & \text{otherwise} \end{cases}$$

$$f_y(y) = \begin{cases} 1, & \text{if } 0 < y < 1 \\ 0 \; , & \text{otherwise} \end{cases}$$

respectively. Define a new random variable $Z = \dfrac{Y}{X}$ and obtain the

density function of Z.

5.11 Let X and Y be independent random variables with density functions.

$$f_X(x) = \begin{cases} 2e^{-2X} & , \ x > 0 \\ 0 & , \ otherwise \end{cases}$$

$$f_y(y) = \begin{cases} e^{-y} & , \ y > 0 \\ 0 & , \ otherwise \end{cases}$$

respectively. If a new random variable Z is defined as

$$Z = \frac{Y + 2X}{X}$$

(a) Obtain the density function of Z.
(b) Determine the median of Z.

5.12 Let X and Y be two independent random variables having similar density function.

$f(t) = e^{-t}$, $t > 0$.

Suppose Z is a random variable defined as

$$Z = \frac{X - Y}{X + Y}, \ -1 < Z < 1.$$

Obtain the density function of Z.

5.13 The random variable X has distribution function (cdf) given as

$$F_X(x) \begin{cases} 1 - e^{-x} & , \ for \ x \geq 0 \\ 0 & , \ x < 0 \end{cases}$$

Obtain the distribution function and the median of the random variable $Y = \sqrt{X}$.

5.14 Let X and Y be independent random variables having similar density functions given as

$$f(t) = \begin{cases} 2t & \text{for} \quad 0 < t < 1 \\ 0 & \text{otherwise} \end{cases}$$

Obtain the frequency density function of the random variable $Z = XY$

5.15 Two independent random variables X and Y have frequency density functions

$$f_X(x) = \frac{1}{\sqrt{2\pi}} e^{-\frac{x^2}{2}} \quad -\infty < x < \infty \quad \text{and} \quad f_Y(y) =$$

$$\begin{cases} e^{-y}, & y > 0 \\ 0, & \text{otherwise} \end{cases} \quad \text{respectively.}$$

Obtain the frequency density function for the random variable

$$Z = \frac{X}{\sqrt{Y}}$$

5.16 The independent random variables X and Y have frequency density functions

$$f_X(x) = \begin{cases} a\, e^{-ax} & x > 0 \\ 0 & \text{otherwise} \end{cases} \quad \text{and} \quad f_Y(y) =$$

$$\begin{cases} b\, e^{-by} & y > 0 \\ 0 & \text{otherwise} \end{cases} \quad \text{respectively}$$

Obtain the frequency density function of the random variable $Z = X - Y$

5.17 Let X and Y be independent random variables having similar frequency density functions:

$$f(t) = \begin{cases} e^{-t} & t > 0 \\ 0 & \text{otherwise} \end{cases} \qquad \text{Obtain}$$

(a)   The frequency density function $f_Z(z)$ and
(b)   The cumulative density function $f_Z(z)$ respectively, of the
random variable $Z = \dfrac{X}{X + Y}$

5.18 (a) Let (X, Y) be the distribution defined by the density function

$$f(x,y) = \begin{cases} e^{-x-y} & \text{for} \quad x > 0, \quad y > 0 \\ 0 & \text{otherwise} \end{cases}$$

(a) Determine the marginal probability density functions $f_X(x)$ and $f_Y(y)$ respectively.

(b)The joint cumulative density function $f_{XY}(x, y)$

5.19 (a) The cumulative distribution function of the random variable X is given below:

$$F(x) = \begin{cases} 0 & if \quad -\infty < x < 0 \\ x^2 & if \quad \quad 0 \le x < 1 \\ 1 & if \quad \quad \quad x \ge 1 \end{cases}$$

Determine the density function of the distribution

(b)   The random variable Y has density function

$$f(y) = \begin{cases} 2e^{-2y} & \text{for } y > 0 \\ 0 & \text{otherwise} \end{cases}$$

Determine the distribution function $F_Y(y)$ of Y and calculate the probability that $Y > 1$
{i.e. $P(Y > 1)$}

5.20 (a) Let X be a random variable with $E(X) = \mu$ and $V(X) = \sigma^2$. Suppose that $Y = H(X)$,
    prove that, if H is at least twice differentiable at $X = \mu$, then
$$E(Y) \approx H(\mu) + \frac{H''(\mu)}{2!}\sigma^2 \text{ and } V(Y) \approx \left[H'(\mu)\right]^2 \sigma^2$$

(b)   Under certain conditions the profit realized on a capital investment is given by the formula: $P = 2(1 - 0.005T)^{1.2}$ million Leones, where T is the period in years over which the capital was invested.

Suppose that T is a continuous random variable with the following density function

$$f(t) = \begin{cases} 3000t^{-4} & t \geq 10 \\ 0 & \text{otherwise} \end{cases}$$

Use the expression in (a) above to obtain an appropriate value for $E(P)$.

5.21 X and Y are two independent random variables, and define
$$Z = \frac{Y - X}{X}.$$

Suppose the cumulative distribution functions for X and Y are given as follows:

$$F_X(x) = \begin{cases} 1 - (1 + x)e^{-x} & x \geq 0 \\ 0 & \text{otherwise} \end{cases}$$

and

$$F_Y(y) = \begin{cases} 1 - e^{-y} & y \geq 0 \\ 0 & \text{otherwise} \end{cases}$$

Obtain the probability density function of Z

5.22 (a) Let X_1 and X_2 have joint probability density function given by:

$$f(x_1, x_2) = \begin{cases} 2 & 0 < x_1 < x_2 < 1 \\ 0 & \text{elsewhere} \end{cases}$$

Obtain (i) $f(x_1)$ (ii) $f(x_2)$ (iii) $f(x_1/x_2)$
(iv) $f(x_2/x_1)$

(b)   Determine whether $f(x_1/x_2)$ and $f(x_2/x_1)$ in (a) above are true probability functions.

5.23 The continuous random variable x has distribution function $F_X(x)$ given as

$$F_X(x) = \begin{cases} 1 - (1 + x)e^{-x} & \text{for } x \geq 0 \\ 0 & \text{otherwise} \end{cases}$$

Also $E(X) = V(X) = 2$. If X_1 and X_2 are two independent observations of X, obtain an
expression for the frequency density function $f_Z(z)$ of the
random variable $Z = \dfrac{X_1}{X_2}$.

5.24 Suppose that the two dimensional continuous random variables (X, Y) have joint
probability density function given by

$$f(x,y) = \begin{cases} x^2 + \frac{xy}{3}, & 0 \le x \le 1,\ 0 \le y \le 2 \\ 0 & \text{otherwise} \end{cases}$$

Obtain (i) $E(X)$ (ii) $E(Y)$ (iii) $f(x/y)$ (iv) $f(y/x)$
(v) $E(Y/X)$ (vi) $E(X/Y)$
Hence show that $E_X[E(Y/X)] = E(Y)$ and $E_Y[E(X/Y)] = E(X)$

5.25 (a) If (X,Y) is a two dimensional discrete random variable, prove that the expectation of
the conditional expectation i.e. $E_X[E(Y/X)] = E(Y)$.

(b) Suppose that shipments involving a varying number of parts arrive each day. If X is the number of items in the shipment, the probability distribution of the random variable X is given as follows:

x	8	9	10	11	12	13
$P(X = x)$	0.05	0.10	0.10	0.20	0.35	0.20

The probability that any particular part is defective is the same for all parts and equal to 0.10. If Y is the number of defective parts arriving each day, what is the expected value of Y?

5.26 The continuous random variable X has frequency density function given as:

$$f(x) = \begin{cases} k(2 - x), & 1 \leq x \leq 4 \\ 0 & \text{otherwise} \end{cases}$$

Find (i) the value of k (ii) $E(x)$ (iii) $V(x)$

(iv) $P(x > 1.5)$ (v) $P(x < 3)$ (vi) $P(1.2 < x < 2.4)$

CHAPTER 6
SOME IMPORTANT CONTINUOUS RANDOM VARIABLES

6.1. The Normal (or Gaussian) Distribution

The Normal distribution is the most useful continuous probability distribution in the entire field of Statistics. Its graph is called the normal (or Gaussian) curve and is bell-shaped as shown in Figure 6.1 below.

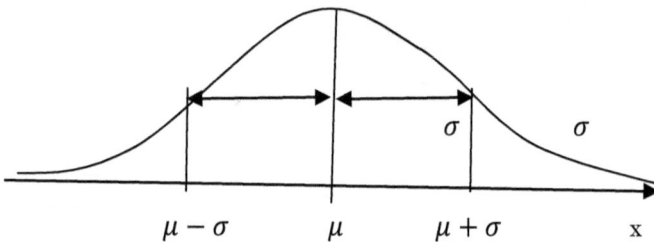

Figure 6.1

This shape describes so many sets of data that occur in nature, industry, and research. For example, errors made in measuring physical and economic phenomena often are distributed normally. In addition, many probability distributions (such as the Binomial) often can be approximated by the normal curve.

A random variable X having the bell-shaped distribution of Figure 6.1 above, is called a normal random variable.
The mathematical equation for the probability distribution of the continuous normal variable depends upon the two parameters μ (its mean) and σ (its standard deviation). Hence we shall denote the density function of X by f(x: μ, σ).

6.1.1   Definition (Normal or Gaussian Distribution)

The random variable X, assuming all real values $-\infty < x < \infty$, has a Normal (or Gaussian) distribution if its probability density function (pdf) is of the form

$$f(x:\mu, \sigma) = \frac{1}{\sqrt{2\pi}\sigma} \cdot \exp\left[-\frac{1}{2}\left(\frac{x-\mu}{\sigma}\right)^2\right],$$

The parameters μ and σ must satisfy the conditions ($-\infty < \mu < \infty, \sigma > 0$). Since we shall have many occasions to refer to the above distribution we shall use the following notation: X has a normal distribution $N(\mu, \sigma^2)$ or

$X \sim N(\mu, \sigma^2)$ if and only if its probability density function is $f(x: \mu, \sigma)$ given above, or $f(x)$ in short.

Once μ and σ are specified, the normal curve is completely determined (see Figure 6.1 above).

The following situations can now be illustrated:

(i)Two normal curves having the same standard deviation but different means. The two curves are identical in form but are centred at different positions along the horizontal axis. (See Figure 6.2 below)

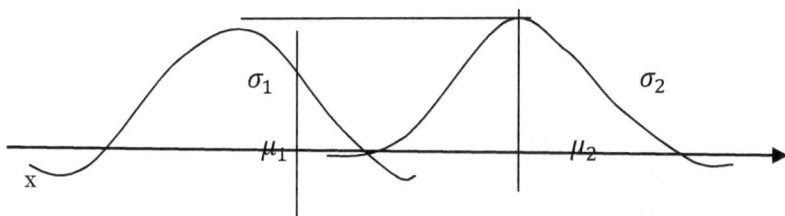

Figure 6.2: Standard deviation: $\sigma_1 = \sigma_2 = \sigma$

Mean $\mu_2 > \mu_1$

(ii)Two normal curves with the same mean but different standard deviations. (See Figure 6.3 below).

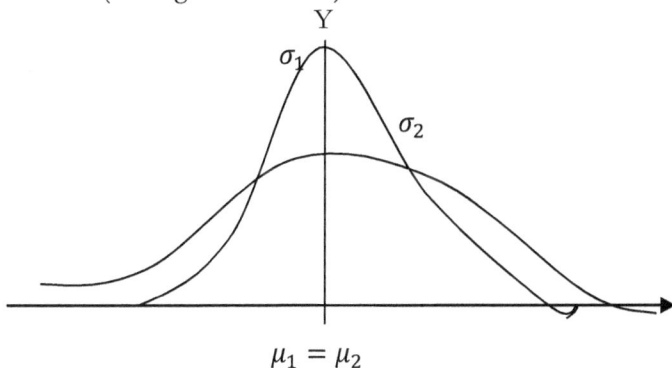

Figure 6.3 : Standard deviation: $\sigma_1 < \sigma_2$

Mean: $\mu_1 = \mu_2$

This time we see that the two curves are centered at exactly the same position on the horizontal axis, but the curve with the larger standard deviation is lower (in height) and spreads out further. Since the area under a probability curve is equal to 1, therefore the more variable the set of observations, the lower (in height) and the wider the spread of the corresponding curve will be.

(iii)Two normal curves having different means and different standard deviations. (See Figure 6.4 below)

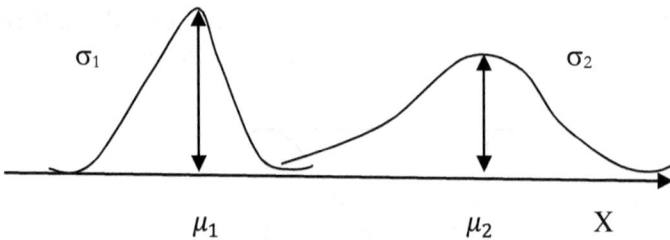

Figure 6.4 : Standard deviation: $\sigma_1 < \sigma_2$

Mean: $\mu_1 < \mu_2$

The curves are centred at different positions on the horizontal axis and their shapes reflect the two different values of σ.

6.1.2   Properties of the Normal Distribution

(i)The mode, which is the point on the horizontal axis where the curve is a maximum, occurs at $x = \mu$.

(ii)The curve is symmetric about a vertical axis through the mean μ.

(iii)The normal curve approaches the horizontal axis asymptotically as we proceed in either direction away from the mean.

(iv)Since $f(x: \mu, \sigma)$ is a legitimate pdf, we have, $f(x, \mu, \sigma) \geq 0$ and

(v) $\int_{-\infty}^{\infty} f(x)\, dx = 1$. [Note: $f(x)$ is the short form of $f(x: \mu, \sigma)$.]

That is the total area under curve and above the horizontal axis is equal to 1.

Proof (v):

$$f(x: \mu, \sigma) = \frac{1}{\sqrt{2\pi}\,\sigma} \exp\left[-\frac{1}{2}\left(\frac{x-\mu}{\sigma}\right)^2 \right],$$

Let $z = (x-\mu)/\sigma$, we may write

$$\int_{-\infty}^{\infty} f(x)\, dx \quad \text{as} \quad \frac{1}{\sqrt{2\pi}} \int_{-\infty}^{\infty} e^{-z^2/2}\, dz = I$$

The trick used to evaluate this integral is to consider, instead of I, the square of this integral, namely, I^2. Thus

$$I^2 = \frac{1}{2\pi} \int_{-\infty}^{\infty} e^{-z^2/2}\, dz \int_{-\infty}^{\infty} e^{-t^2/2}\, dt = \frac{1}{2\pi} \int_{-\infty}^{\infty} \int_{-\infty}^{\infty} \exp -(z^2 + t^2)\, dt\, dz$$

The double integral can now be evaluated by introducing polar coordinates.

Let $z = r\cos\theta$, $t = r\sin\theta$

The Jacobian of transformation $|J| = \left|\frac{\partial(z,t)}{\partial(r,\theta)}\right| = |r| = r$ for $r \neq 0$.

so that the element of area $dz\, dt = |J|\, dr\, d\theta = r\, dr\, d\theta$ does not vanish. As z and t vary between $-\infty$ and $+\infty$, r varies between 0 and ∞, while θ varies between 0 and 2π. Thus

$$I^2 = \frac{1}{2\pi} \int_0^{2\pi} \int_0^{\infty} r e^{-r^2/2}\, dr\, d\theta = \frac{1}{2\pi} \int_0^{2\pi} [-e^{r^2/2}]_0^{\infty}\, d\theta$$

$$= \frac{1}{2\pi} \int_0^{2\pi} d\theta = 1. \ \text{Hence } I = 1.$$

6.1.3   Expected Value (or Mean) of X

$$E(x) = \int_{-\infty}^{\infty} x f(x) dx$$

Since the density function

$$f(x : \mu, \sigma) = \frac{1}{\sqrt{2\pi}\,\sigma} \exp\left[-\frac{1}{2}\left(\frac{x-\mu}{\sigma}\right)^2\right]$$

$-\infty < x < \infty,$ $-\infty < \mu < \infty,$ $\sigma^2 > 0.$

We have

$$E(X) = \frac{1}{\sqrt{2\pi}\,\sigma} \int_{-\infty}^{\infty} x \cdot \exp\left[-\frac{1}{2}\left(\frac{x-\mu}{\sigma}\right)^2\right] dx$$

Let $z = \frac{X-\mu}{\sigma}$, then $dx = \sigma dz$.   Hence,

$$E(X) = \frac{1}{\sqrt{2\pi}} \int_{-\infty}^{\infty} (\sigma z + \mu) e^{-z^2/2} \, dz$$

$$= \frac{\sigma}{\sqrt{2\pi}} \int_{-\infty}^{\infty} z e^{-z^2/2} \, dz + \frac{\mu}{\sqrt{2\pi}} \int_{-\infty}^{\infty} e^{-z^2/2} \, dz$$

The first of the above integrals equals zero since the integrand, say $g(z)$, has the property that $g(z) = -g(-z)$, and hence g is an odd function. The second integral (without the factor μ) represents the total area under the normal pdf and hence equals unity. Thus E(X) = μ.

6.1.4   Variance of X

Consider

$$E(X^2) = \frac{1}{\sqrt{2\pi}\,\sigma} \int_{-\infty}^{\infty} x^2 \exp\left[-\frac{1}{2}\left(\frac{x-\mu}{\sigma}\right)^2\right] dx$$

Again letting $z = \dfrac{x - \mu}{\sigma}$, we obtain

$$E\left(X^2\right) = \frac{1}{\sqrt{2\pi}} \int_{-\infty}^{\infty} \left(\sigma z + \mu\right)^2 e^{-z^2/2} \, dz$$

$$= \frac{1}{\sqrt{2\pi}} \int_{-\infty}^{\infty} \sigma^2 z^2 e^{-z^2/2} \, dz + 2\mu\sigma \frac{1}{\sqrt{2\pi}} \int_{-\infty}^{\infty} z e^{-z^2/2} \, dz$$

$$+ \mu^2 \frac{1}{\sqrt{2\pi}} \int_{-\infty}^{\infty} e^{-z^2/2} \, dz$$

The second integral again equals zero by the argument of 6.1.3 above. The last integral (without the factor μ^2) equals unity. To evaluate

$$\frac{1}{\sqrt{2\pi}} \int_{-\infty}^{\infty} z^2 e^{-z^2/2} \, dz \, ,$$

We integrate by parts letting $z\, e^{-z^2/2} = dv$ and $z = u$. Hence $v = -e^{-z^2/2}$ while $dz = du$.

We obtain

$$\frac{1}{\sqrt{2\pi}} \int_{-\infty}^{\infty} z^2 e^{-z^2/2} \, dz = \frac{-z e^{-z^2/2}}{\sqrt{2\pi}} \Bigg]_{-\infty}^{\infty} + \frac{1}{\sqrt{2\pi}} \int_{-\infty}^{\infty} e^{-z^2/2} \, dz$$

$$= \quad 0 \quad + \quad 1$$

Therefore, $E(X^2 = \sigma^2 + \mu^2)$, and hence $V(X) = E(X^2) - [E(X)]^2 = \sigma^2$

6.1.5 Theorem

If X has the distribution $N(\mu, \sigma^2)$ and if $Y = aX + b$,
Then Y has the distribution $N(a\mu + b, \ a^2\sigma^2)$, a and b being constants.

EXAMPLE 6.1

Suppose the random variable X has distribution given as $X \sim N(\mu, \sigma) = N(7, 2)$.

Obtain the distribution of $Y = 3X - 10$.

SOLUTION:

$$Y = 3X - 10$$
$$E(Y) = E(3X - 10) = 3E(X) - E(10)$$
$$= 3(7) - 10 = 11$$

$$V(Y) = V(3X - 10) = 3^2 \, V(X) - 60\text{cov} \, (X, 10) + V(10)$$
$$= 9(2^2) \quad - \quad 0 \qquad\qquad + \quad 0$$
$$= 36$$
$$\therefore Y \sim N(\mu, \sigma^2) = N(11, 36)$$

EXAMPLE 6.2

The independent random variables X and Y have distributions. $X \sim N (\mu, \sigma) = N (3, 4)$, and $Y \sim N (\mu, \sigma) = N(2, 3)$ respectively. Let $Z = X - Y$ and obtain the distribution of Z.

SOLUTION:

$$E(Z) = E(X - Y) = E(X) - E(Y) = 3 - 2 = + 1$$
$$V(Z) = V(X - Y) = V(X) + V(y) - 2 \text{ cov } (X, Y)$$
$$= 4^2 + 3^2 - 0 = 25$$

Note: Cov $(X, Y) = 0$ since X and Y are independent.

Hence $Z \sim N (\mu, \sigma) = N(1, 5)$
Or $Z \sim N (\mu, \sigma^2) = N(1, 25)$

6.1.6   The Standard Normal Distribution

Consider again the random variable X having normal distribution N(μ, σ²). Suppose now μ = 0 and σ = 1, so that the new random variable is distributed as Z ~ N (0, 1), we say that Z has a standard normal distribution.

The probability density function (pdf) of Z may be written as

$$f(Z) = \frac{1}{\sqrt{2\pi}} e^{-\frac{1}{2}z^2} \quad, -\infty < z < \infty$$

The importance of the standard normal distribution is due to the fact that it is tabulated. Whenever X has distribution N(μ, σ²), we can always obtain the standardized form by simply taking a linear function of X, - that is, by a linear transformation involving a shift of origin and a change of scale. This transformation makes any random variable into a standardized variable with mean 0 and variance 1.

6.1.6.1   Theorem

If X has distribution N (μ, σ²) and if $z = \frac{(X - \mu)}{\sigma}$, then z has a standard normal distribution N(0, 1).

The function $e^{-\frac{1}{2}z^2}$ is nonnegative; it goes rapidly to 0 as $|z| \to \infty$, so that the area under its graph is finite. It can be shown that

$$\int_{-\infty}^{\infty} e^{-\frac{1}{2}z^2} \, dz = \sqrt{2\pi}$$

so that a density function is obtained upon division by $\sqrt{2\pi}$. This density f(z), for a standard normal distribution is defined above.

The function $e^{-\frac{1}{2}z^2}$ is symmetric about z = 0, having the same value at − z as at + z, and has a maximum at z = 0. The usual method of calculus involving the second derivative shows that there are points of inflection at z = ± 1. The graph is shown in Figure 6.5 below.

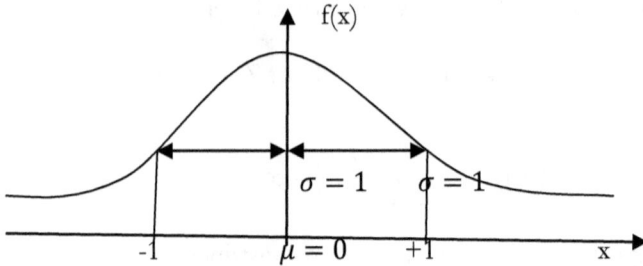

Figure 6.5

Let Z be a random variable having the standard normal distribution. Because the distribution is symmetric about $Z = 0$, its mean value is this centre of symmetry: $E(Z) = 0$. The variance is

$$V(Z) = E(Z^2) - [E(Z)]^2 = \frac{1}{\sqrt{2\pi}} \int_{-\infty}^{\infty} z^2 e^{-z^2/2} dz - 0 = 1$$

which can be verified by an integration by parts. That is, the standard deviation is 1, the distance from the mean to the points of inflection.

6.1.6.2   Theorem

If X has distribution $N(\mu, \sigma^2)$ and if we do a transformation $Z = \dfrac{X - \mu}{\sigma}$, then Z has a standard normal distribution.

The general procedure for transforming any normal variable into a standard normal variable is as follows:
(a) obtain the mean of the random variable.
(b) obtain the variance of the random variable, and hence the standard deviation.
(c) carry out the transformation by subtracting the mean from the random variable and dividing the result by the standard deviation.
(d) The final result is a standard normal variable.

EXAMPLE 6.3

Let X_i, $i=1,2,3,....n$; be normal distributed random variables each having mean μ and variance σ^2. We wish to transform
$$\overline{X} = \frac{X_1 + X_2 + \cdots + X_n}{n} \text{ into a standard normal variable.}$$

Procedure:

(a) $E(\overline{X}) = E\left\{\dfrac{X_1+X_2+X_3+\ldots\ldots\ldots+X_n}{n}\right\} = \dfrac{E(X_1)+E(X_2)\ldots\ldots+E(X_n)}{n}$

$$= \frac{n\mu}{\mu} = \mu$$

(b) $V(\overline{X}) = V\left\{\dfrac{X_1+X_2+X_3+\ldots\ldots\ldots+X_n}{n}\right\} =$

$$\frac{V(X_1) + V(X_2) \ldots \ldots \ldots + V(X_n)}{n^2} + 2\sum_{i<j} cov(X_i, X_j)$$

$$= \frac{1}{n^2}\underbrace{\left\{\sigma^2 + \sigma^2 + \ldots + \sigma^2\right\}}_{n \text{ times}} = \frac{n}{n^2}\sigma^2 = \sigma_n$$

(c) Hence the transformation: $Z = (\overline{X} - \mu)/\frac{\sigma}{\sqrt{n}} \sim N(0,1)$ makes Z a standard normal variable with mean value equal to 0(zero) and standard deviation equal to 1(one).

Similarly, the transformation: $Z = \frac{X-\mu}{\sigma} \sim N(0,1)$ makes Z a standard normal variable with mean 0(zero) and standard deviation 1(one).

6.1.7   Standard Normal Tables
Suppose that Z is a standard normal variable, ie. Z has distribution given as
$Z \sim N(0,1)$. The integral over the density function of Z gives the cumulative distribution function (cdf.). This cdf. of Z is denoted by $\emptyset(z)$, and represents the area under the curve up to some cut-off point $Z=z$:- It is the probability that the random variable Z is less than or equal to some assigned value z (as depicted in Figure 6.6 below).
$$\emptyset(z) = P(Z \leq z) = \int_{-\infty}^{z} \frac{1}{\sqrt{2\pi}} e^{-r^2/2} dr$$

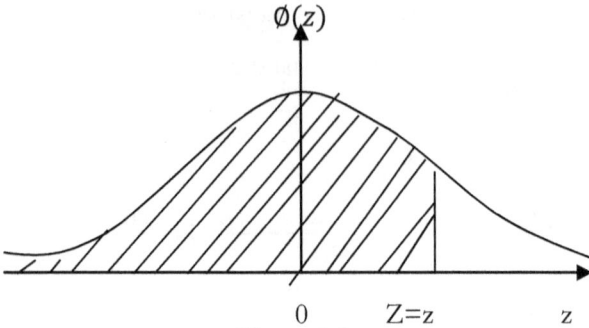

Figure 6.6

If the random variable is now bounded by two ordinates, say a and b (as shown in Figure 6.7 below), the area under the curve bounded by the two ordinates Z=a and Z=b equals the probability that the random variable Z assumes value between a and b.

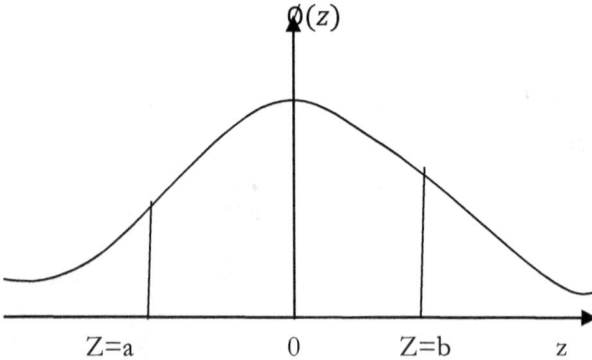

Figure 6.7

That is, $P(a \leq Z \leq b) = \frac{1}{\sqrt{2\pi}} \int_a^b e^{-z^2/2} \, dz$. Integrals of the type presented in section 6.1.7 above cannot be evaluated by ordinary means. (The difficulty stems from the fact that we cannot apply the Fundamental Theorem of calculus since we cannot find a function whose derivative equals $e^{-z^2/2}$). However, methods of numerical integration (using Computer programming packages such as MathLab and Mathematica) can be used to evaluate integrals of that form, and in fact a table of values for $P(Z \leq z) = \emptyset(z)$. Probabilities of intervals or simple combination of intervals can be computed by referring to such tables. We simply use Theorem 6.1.6.1 to note that if X has distribution $N(\mu, \sigma^2)$, then $Z = \frac{X-\mu}{\sigma}$ has distribution N(0,1). Hence $P(a \leq X \leq b)$, after standardization of X, becomes

$$P(a \leq X \leq b) = P(\frac{a-\mu}{\sigma} \leq \frac{X-\mu}{\sigma} \leq \frac{b-\mu}{\sigma})$$

$$= P\left(\frac{a-\mu}{\sigma} \leq Z \leq \frac{b-\mu}{\sigma}\right) = \emptyset\left(\frac{b-\mu}{\sigma}\right) - \emptyset\left(\frac{a-\mu}{\sigma}\right)$$

It is evident from the definition of $\emptyset$ (section 6.1.7) that :

$$P(Z > z) = 1 - P(Z \leq z) = 1 - \emptyset(z) \text{ and } \emptyset(-z) = 1 - \emptyset(z).$$

The latter relationship is particularly useful since in most tables the function $\emptyset$ is tabulated only for positive values of z.

EXAMPLE 6.4

Suppose that X has distribution $N(\mu, \sigma^2) = N(3,4)$. We wish to find a number C such that $P(X > C) = 2P(X \leq C)$.

SOLUTION:

To standardize X, we have
$Z = \frac{(X-\mu)}{\sigma} = \frac{(X-3)}{2} \sim N(0,1)$. Hence $P(X > C) = P\left(\frac{X-3}{2} > \frac{C-3}{2}\right)$

$$= P(z > \frac{C-3}{2} = 1 - \emptyset(\frac{C-3}{2}).$$

Also $2P(X \leq C) = 2P\left(\frac{X-3}{2} \leq \frac{C-3}{2}\right) = 2\emptyset\left(\frac{C-3}{2}\right)$. The condition

$P(X > C) = 2P(X \leq C)$ implies that
$1 - \emptyset\left(\frac{C-3}{2}\right) = 2\emptyset\left(\frac{C-3}{2}\right)$ ie. $\emptyset\left(\frac{C-3}{2}\right) = \frac{1}{3}$

Hence, from the Normal distribution we find that $\frac{C-3}{2} =$ -0.43, yielding C=2.14

EXAMPLE 6.5

Let X be a standard normal distributed random variable (ie. $X \sim N(0,1)$). Compute the probability P(X> -0.6).

SOLUTION:
Since X is already a standard normal variable, we need not standardize X.

$$P(X > -0.6) = P(Z > -0.6) = 1 - \emptyset(-0.6) = 1 - [1 - \emptyset(0.6)]$$
$$= \emptyset(0.6) = 0.7257$$

EXAMPLE 6.6

A random variable X has distribution $X \sim N(\mu, \sigma) = N(5,8)$. Find the value of v such that :

(i) $\quad P\{|X - 5| \le v = 0.95\}$ (ii) $P\{|X - 5| > v = 0.95\}$

SOLUTION:

(I) $\quad P\{|X - 5| \le v\} = P\{-v < X - 5 < v\} = 0.95$
$$= P\{-v + 5 < X < v + 5\} = 0.95$$

Standardization of X yields, $P\left\{\frac{-v}{8} < \frac{X-5}{8} < \frac{v}{8}\right\} = 0.95$ $ie. [\emptyset\left(\frac{v}{8}\right) - \emptyset\left(\frac{-v}{8}\right)] = 0.95$

$$ie. 2\emptyset\left(\frac{v}{8}\right) - 1 = 0.95 \; giving \; \emptyset\left(\frac{v}{8}\right) = 0.95 \; or \; v = 1.96 \times 8$$
$$= 15.68$$

(ii)$P\{|X - 5|\} > v = 0.95 = P\{X - 5 < -v\} + P\{X - 5 > v\} =$ 0.95

$$= P\{X < -v + 5\} + P\{X > v + 5\} = 0.95$$

$$= P\left\{\frac{X-5}{8} < \frac{-v}{8}\right\} + P\left\{\frac{X-5}{8} > \frac{v}{8}\right\} = 0.95$$

$$= \emptyset\left(\frac{-v}{8}\right) + 1 - \emptyset\left(\frac{v}{8}\right) = 2\left\{1 - \emptyset\left(\frac{v}{8}\right)\right\} = 0.95$$

$$\therefore \emptyset\left(\frac{v}{8}\right) = 0.525, \ giving \ v = 0.520$$

6.1.8   Central Limit Theorem

6.1.8.1 The Distribution of the sum of independent random variables.

In this section we will try to derive the distribution of sums of independent random variables. This will generally be quite complicated, but for large samples the situation is simplified by an important fact: The sum of a large number of independent observations on a random variable is approximately normally distributed. Let $X_1, X_2,$ $X_3.................,X_n$ be independent random variables, all having common mean μ and common *variance* σ^2.

(i)   The expected value of their sum ie.
$$E(X_1 + X_2 + X_3 + \cdots \ldots \ldots + X_n)$$
$$= E\left(\sum_{i=i}^{n} X_i\right) = \sum_{i=1}^{n} E(X_i)$$
$$= n\mu$$

(ii)   The variance of their sum, ie.
$$V(X_1 + X_2 + X_3 + \cdots \ldots \ldots + X_n)$$
$$= V\left(\sum_{i=i}^{n} X_i\right) = \sum_{i=1}^{n} V(X_i)$$
$$= n\sigma^2$$

where the X's are independent.
If S_n denotes the sum: $S_n = X_1 + X_2 + X_3 + \cdots \ldots \ldots + X_n = \sum_1^n X_i$, then the mean and variance of S_n are equal to $n\mu$ *and* $n\sigma^2$ respectively as stated above: ie. $S_n \sim N[n\mu, n\sigma^2]$. Standardization of S_n will, therefore yield

$$Z = \left(\frac{S_n - E(S_n)}{\sqrt{V(S_n)}}\right) = \left(\frac{S_n - n\mu}{\sqrt{n\sigma^2}}\right) \sim N(0,1).$$

The following important result given without proof, states that the limiting distribution of the standardized sum is $N(0,1)$, standard normal distribution with mean 0(zero) and standard deviation 1(one).

6.1.8.2 Theorem

If, for each integer n, the observations $X_1, X_2, X_3, \ldots\ldots\ldots, X_n$ are independent with common distribution having mean μ *and finite variance* σ^2, then the sum $S_n = X_1 + X_2 + X_3 + \ldots\ldots\ldots + X_n$ is asymptotically normal, in the following sense:

$$\lim_{n\to\infty} P\left(\frac{S_n - n\mu}{\sqrt{n\sigma^2}} \le z\right) = \emptyset(z), \ where \ \emptyset(z) \text{ is a Standard}$$

Normal

Cumulative Distribution Function. The Central Limit Theorem which is stated above, can be given in an equivalent form that is more directly useful.

6.1.8.3 Practical Version of Theorem 6.1.8.2

Let $X_1, X_2, X_3, \ldots\ldots\ldots X_n$ be independent, having a common distribution with mean μ and finite variance σ^2. Their sum $S_n = X_1 + X_2 + \cdots \ldots X_n$ is distributed approximately as $N(n\mu, n\sigma^2)$, and probabilities can be approximated as

$$P(S_n \le y) \cong \emptyset\left(\frac{y - n\mu}{\sqrt{n\sigma^2}}\right), \text{ and the values of } \emptyset \text{ are read from a standard}$$
normal table.

The larger the value of n, of course, the better the approximation, but no general rule can be given as to how large n should be. It would depend on (i) how precise an answer is desired, and (ii) how nearly normal the common distribution is .

If the common distribution is itself normal, then S_n is exactly normal. For most situations, an n-value greater than or equal to 10(ten) would be sufficiently large for a reasonable approximation.

EXAMPLE 6.7

The random variables $X_1, X_2, X_3, \ldots \ldots \ldots X_{1200}$ are independent and uniformly distributed in the interval (-10, 10). If $Y = X_1 + X_2 + \ldots \ldots + X_{1200}$, compute $P(|Y| > 200)$

SOLUTION:

Since X is uniformly distributed in the interval (a,b)=(-10,10), we have

$$E(X) = \frac{a+b}{2} = \frac{-10+10}{2} = 0, \text{ and } V(X) = \frac{(b-a)^2}{12} = \frac{400}{12}$$

For $Y = X_1 + X_2 + \cdots + X_{1200} = \sum_{i=1}^{1200} X_i$
$E(Y) = 1200E(X) = 1200 \times 0 = 0$, and $V(Y) = 1200V(X) = 1200 \times \frac{400}{12} = 40,000$ (when the X's are independent) using the Central Limit Theorem.

Hence $Y \sim N(0; 40,000)$.

$P\{|Y| > 200\} = P\{Y < -200\} + P\{Y > 200\}$. Standardisation of Y yields,

$$P\left\{ \frac{Y - E(Y)}{\sqrt{V(Y)}} < \frac{-200 - E(Y)}{\sqrt{V(Y)}} \right\} + P\left\{ \frac{Y - E(Y)}{\sqrt{V(Y)}} > \frac{200 - E(Y)}{\sqrt{V(Y)}} \right\}$$
$$= 1 - \emptyset\left(\frac{200}{\sqrt{40,000}} \right) + 1 - \emptyset(\frac{200}{\sqrt{40,000}}) = 2 - 2\emptyset(1) = 2 - 1.6826$$
$$= 0.3174$$

6.1.9   Normal Approximation To Binomial Theorem

The Binomial formula for the probability of k successes in n independent trials of a Binomial experiment with parameter p, is given as:

$P(n, k) = \binom{n}{k} p^k q^{n-k}$, where q=1-p and the number of trials n is fixed. If n is small, the individual terms of the Binomial distribution are relatively easy to compute. However, if n is relatively large, these computations become rather cumbersome. Let us consider the sum $Y_n = X_1 + X_2 + \cdots \ldots X_n$ of n independent Bernoulli trials, each with probability of success p. The sum Y_n being the number of successes among the n trials. For large n, the mean and variance of Y_n are:

$E(Y_n) = np$ and $V(Y_n) = np(1 - p) = npq$ respectively.

The Central Limit Theorem asserts that the sum Y_n is approximately normal with $Y_n \sim N(np, npq)$. Standardizing Y_n such that the random variable Z is defined as

$Z = \frac{Y_n - E(Y_n)}{\sqrt{V(Y_n)}} = \frac{Y_n - np}{\sqrt{npq}}$, then for large n, Z has approximately standard normal distribution [ie. $Z \sim N(0,1)$.] in the sense that $\lim_{n \to \infty} P(Z \le z) = \emptyset(z)$. This approximation is valid for values of n greater than 10 provided p is close to $\frac{1}{2}$. If p is close to 0 or 1, n should be somewhat larger to ensure a good approximation. Equivalently, the mean
$\bar{Y}_n = \frac{Y_n}{n}$, which is the relative frequency or sample proportion of the successes is approximately normal: ie. $\bar{Y}_n \sim N\left(p, \frac{pq}{n}\right)$.   Again, standardizing $\bar{Y}_n$, yields

$\left(\frac{\bar{Y}_n - p}{\sqrt{\frac{pq}{n}}}\right) \sim N(0,1)$ as a standard normal distribution. Thus, to work in

terms of the number of successes (an integer), we use the relation for Y_n
. To work in terms of the proportion of successes (a number between 0 and 1), we use the relation $\bar{Y}_n$.

6.1.9.1   Continuity Correction

In approximating the discrete Binomial distribution function by the continuous normal cumulative distribution function, the degree of success may be improved by making what is called a continuity correction. The value of the normal cumulative distribution function at $k = \frac{1}{2}$ is closer to the desired probability. Thus
$P[Y_n \le k] = \emptyset\left[\frac{k + 0.5 - np}{\sqrt{npq}}\right]$.   When n is very large, the correction is not significant.

EXAMPLE 6.8

Compute the probability of obtaining
(a)   40 heads (b) at most 40 heads (c) at least 40 heads in 100 tosses
of a fair coin.

SOLUTION: For a fair coin, p=0.5 and p(1-p)=pq=0.25. Given
n=100 tosses, np=50 and npq=25.
(a)P[40 heads in 100 tosses]$=P[Y_n \leq 40] - P[Y_n \leq 39]$

$$= \emptyset \left[\frac{(40+0.5)-50}{\sqrt{25}} \right] - \emptyset \left[\frac{39.5-50}{\sqrt{25}} \right]$$

$$= \emptyset(2.1) - \emptyset(1.9) = 0.0108$$

(b)P[at most 40 heads in 100 tosses of a coin]

$$= P[Y_n \leq 40] = P \left[\frac{(40+0.5)-50}{\sqrt{25}} \right] = 1 - \emptyset(1.9) = 0.0287$$

(c)P[at least 40 heads in 100 tosses of a coin]

$$= P[Y_n \geq 40] = 1 - P[Y_n \leq 40] = 0.9713$$

6.1.10   Normal Approximation to the Poisson Distribution

The Poisson Distribution table only extends to mean value m=15. This
is because if m>15, the Poisson distribution tends to a Normal
distribution. This is not surprising when it is realized that a Poisson
variable Y with mean m, an integer, can be expressed as:
$Y = X_1 + X_2 + X_3 + \cdots \ldots \ldots \ldots +X_m$, where the X's are independent
Poisson variables each with mean equal to 1.   The Central Limit
Theorem would then say that

$P[Y \leq y] \approx \emptyset(\frac{Y-m}{\sqrt{m}})$, where it is known that for a Poisson distribution
the mean and variance are the same. For the random variable X above,

with $E(X) = V(X) = 1$, we have $E(Y) = \sum_{i=1}^{m} E(X) = m$,
and VY=i=1mVY=m.

EXAMPLE 6.9

Suppose for a Poisson variable Y, $E(Y) = V(Y) = 15$. We wish to compute $P[Y \leq 12]$.

SOLUTION:

$$P[Y \leq 12] \approx \emptyset \left[\frac{(12 + 0.5) - 15}{\sqrt{15}} \right] = \emptyset \left[\frac{-2.5}{\sqrt{15}} \right] = 0.26$$

6.1.11 Chebyshev's and Related Inequalities

The Chebyshev Inequality is a useful theoretical tool as well as a relation that connects the variance of a distribution with the intuitive notion of dispersion in the distribution.
If we know the probability distribution of a random variable X (either the pdf. in the continuous case or the point probabilities in the discrete case), we may then compute E(X) and V(X), if these exist. However, the converse is not true. That is, from a knowledge of E(X) and V(X), we cannot construct the probability distribution of X and hence cannot compute quantities such as
$P[|X - E(X)| \leq C]$, *where C is a constant.* Nonetheless, it turns out that although we cannot evaluate such probabilities (from knowledge of E(X) and V(X)) we can give a very useful upper (or lower) bound to such probabilities. The result is in what is known as Chebyshev's inequality.

6.1.11.1 Markov's Inequality

Theorem: If X is a r.v. which takes only non-negative values, then for any

$$a \geq 0, P(X \geq a) \leq \frac{E(X)}{a}$$

Proof:
Define the discrete r.v. Y as

$$Y = \begin{cases} 0 \ if \ X < a \\ \\ a \ if \ X \geq a \end{cases}$$

Then $P(Y = 0) = P(X < a)$ since Y=0 and X<a are equivalent.
$P(Y = a) = P(X \geq a)$ since Y=a and $X \geq a$ are equivalent events.
Therefore

$$E(Y) = 0P(Y = 0) + aP(Y = a)$$
$$= aP(X \geq a)$$

Now, $X \geq Y$ and hence $E(X) \geq E(Y)$. Thus $E(X) \geq E(Y) = aP(X \geq a)$. This implies that $P(X \geq a) \leq \frac{E(X)}{a}$

6.1.11.2   Corollary: Chebyshev's Inequality

Let X be a random variable with $E(X) = \mu$ and let c be any real number.
Then, if $E(X - c)$ is finite and ε is any number, we have

$$P[|X - c| \geq \varepsilon] \leq \frac{1}{\varepsilon^2} E(X - c)^2$$

Proof :

First we use Markov's Inequality for the proof.

Since $(X - c)^2$ is a non-negative r.v. we can apply Markov's Inequality to it with $a = \epsilon^2$ to obtain $P\{(X - c)^2 \geq \epsilon^2\} \leq \frac{E[(X-c)^2]}{\epsilon^2}$.
However since $(X - c)^2 \geq \epsilon^2 \Leftrightarrow |X - c| \geq \epsilon$ we have $P(|X - c| \geq \epsilon \leq EX - c2\epsilon2$

Alternatively,

(We shall prove this for the continuous case. In the discrete case the argument is very similar with integrals replaced by sums. However, some care must be taken with endpoints of intervals): Consider,
$P[|X - c| \geq \varepsilon] = \int f(x)dx$, the integral taken over $(x: |x - c| \geq \varepsilon$.
(The domain of the integral indicates that we are integrating between $-\infty$ and $c - \varepsilon$

and between $c + \varepsilon$ *and* $+ \infty$). Now $|x - c| \geq \varepsilon$ is equivalent to $\frac{(x-c)^2}{\varepsilon^2} \geq 1$. Hence, the above integral $\leq \int_R \frac{(x-c)^2}{\varepsilon^2} f(x) dx$, *where* $R = \{x: |x - c| \geq \varepsilon\}$. This integral is, in turn $\leq \int_{-\infty}^{\infty} \frac{(x-c)^2}{\varepsilon^2} f(x) dx$, ie.

$\int f(x) dx$, *for* $\{x: |x - c| \geq \varepsilon\} \leq \int_R \frac{(x-c)^2}{\varepsilon^2} f(x) dx \leq \int_{-\infty}^{\infty} \frac{(x-c)^2}{\varepsilon^2} f(x) dx$

$$\leq \frac{1}{\varepsilon^2} E(x - c)^2.$$

The above inequality can be expressed in other useful forms.
(a)By considering the complementary event we obtain

$$P[|X - c| < \varepsilon] \geq 1 - \frac{1}{\varepsilon^2} E(X - c)^2$$

(b) Choosing $c = \mu$, *we obtain* $P[|X - \mu| \geq \varepsilon] \leq \frac{E[X - \mu]^2}{\varepsilon^2} = \frac{V(X)}{\varepsilon^2} = \frac{\sigma^2}{\varepsilon^2}$

(c) Choosing $c = \mu$ *and* $\varepsilon = k\sigma$, *where* $\sigma^2 = V(X) > 0$ we obtain,

$$P[|X - \mu| \geq k\sigma] \leq \frac{1}{k^2}$$

This last form (c) is particularly indicative of how the variance measures the degree of concentration of probability near $E(X) = \mu$.

EXAMPLE 6.10

Suppose that X assumes the values -1 and +I, each with probability $\frac{1}{2}$. Then $E(X) = 0$ and $V(X) = 1$, and $P[|X - E(X)| \geq \sigma_x] = 1 - P[-1 < X < 1] = 1$.
But 1 is precisely the Chebyshev bound for this probability, and so the inequality becomes an equality.
As we note from section 6.1.11.2 (c) above, if $V(X)$ is small, most of the probability distribution of X is concentrated near $E(X)$. This may be expressed more precisely in the following Theorem:

6.1.11.3 Theorem

Suppose that $V(X) = 0$. Then, $P[X - \mu] = 1$, $where$ $\mu = E(X)$. (Informally, $X = \mu$ with probability 1.

Proof.

From 6.1.11.2 (b) we find that $P[|X - \mu| \geq \varepsilon] = 0$ for any $\varepsilon > 0$. Hence,
$P[|X - \mu| < \varepsilon] = 1$ for any $\varepsilon > 0$. Since ε may be chosen arbitrarily small, the Theorem is established.

6.1.11.4 NOTE:

(a) Theorem 6.1.11.3 shows that zero variance does imply that all the probability is concentrated at a single point namely, at $E(X)$.
(b) If $E(X) = 0$, $then$ $V(X) = E(X)^2$, and hence in this case $E(X)^2 = 0$, implies the same conclusion.
(c) It is in the above sense that we say that a random variable X is degenerate: it assumes only one value with probability 1.

6.1.12 Law of Large Numbers (Bernoulli's Form)

Let E be an experiment and let A be an event associated with E. Consider n independent repetitions of E. Let n_A be the number of times event A occurs among the n repetitions, and let $f_A = \frac{n_A}{n}$. Let P(A)=p (which is assumed to be the same for all repetitions), and $P(\bar{A}) = 1 - p$ be the probability that A does not occur.
Then for every positive number, ε, we have
$$P[|f_A - p| \geq \varepsilon] \leq \frac{p(1 - p)}{n\varepsilon^2}$$
or equivalently,
$$P[|f_A - p| < \varepsilon] \geq 1 - \frac{p(1-p)}{n\varepsilon^2}.$$

The Law of Large Numbers would assert that $\lim_{n \to \infty} f_A = \frac{n_A}{n} = p$, in some sense.

Proof:

Let n_A be the number of times the event A occurs. This is a Binomially distributed random variable. Then $E(n_A) = np$ and

$V(n_A) = npq = np(1 - p)$. Now applying Chebyshev's inequality to the random variable f_A, we obtain

$$P[|f_A - p| < k \sqrt{\frac{p(1-p)}{n}} \geq 1 - \frac{1}{k^2}, \text{ for some constant } k.$$

Let $\varepsilon = k\sqrt{p(1-p)/n}$.

Then $k^2 = \frac{n\varepsilon^2}{[p(1-p)]}$, and thus $P[|f_A - p| < \varepsilon] \geq 1 - \frac{p(1-p)}{n\varepsilon^2}$.

The above result immediately implies that:

$\lim_{n \to \infty} [|f_A - p| < \varepsilon] = 1, for \ all \ \varepsilon > 0.$ It is in this sense that we say that the relative frequency f_A converges in probability to $P(A) = p$.

6.1.12.1 NOTE:

(a) Another form of the Law of Large Numbers is obtained when we ask ourselves the following question: How many times should we perform the experiment in order to have a probability of at least 0.95, say, that the relative frequency differs from $p = P(A)$ by less than 0.01 say? That is, for $\varepsilon = 0.01$ we wish to choose n so that

$$1 - \frac{P(1-P)}{[n(0.01)^2]} = 0.95.$$ Solving for n we obtain

$$n = \frac{p(1-p)}{(0.01)^2(0.05)}.$$ Replacing the specific values of 0.05 and 0.01 by

$\delta \ and \ \varepsilon \ respectively,$ we have $P[|f_A - p| < \varepsilon] \geq 1 - \delta, whenever$
$n \geq \frac{p(1-p)}{\varepsilon^2 \delta}$. It should be stressed that taking $n \geq p(1-p)/\varepsilon^2\delta$ does not guarantee anything about $|f_A - p|$. It only makes it probable that $|f_A - p|$ will be very small.

EXAMPLE 6.11

(a)How many times would we have to toss a fair die in order to be at least 95 per cent sure that the relative frequency of having a six come up is within 0.01 of the theoretical probability $\frac{1}{6}$?

SOLUTION:
Here $p = \frac{1}{6}, \ 1 - p = \frac{5}{6}, \varepsilon = 0.01, \ and \ \delta = 0.05. \ Hence$

$n \geq \frac{p(1-p)}{\varepsilon^2 \delta} = \frac{\left(\frac{1}{6}\right)\left(\frac{5}{6}\right)}{(0.01)^2(0.05)} = 27777.78.$ Recall that f_A is a random variable and

not just an observed value.

(b) In many problems we do not know the value $p = P(A)$ and hence would be unable to use the above bound on n. In that case we can use the fact that p(1-p) assumes its maximum value when $P = \frac{1}{2}$ and this maximum value equals $\frac{1}{4}$.

Hence we would be on the safe side if we stated that for $n \geq \frac{1}{4\varepsilon^2 \delta}$

We have $P[|f_A - p| < \varepsilon] \geq 1 - \delta.$

EXAMPLE 6.12

Items are produced in such a manner that the probability of an item being defective is p (assumed unknown). A large number of items, say, n, are classified as defective or non-defective. How large should n be so that we may be 99 per cent sure that the relative frequency of defectives differs from p by less than 0.05?

SOLUTION:
Since we do not know the value of p, we assume maximum value of $p(1 - p) = \frac{1}{4}.$

Thus with $\varepsilon = 0.05, \delta = 0.01, \ we \ have \ n \geq \frac{1}{4\varepsilon^2 \delta} = \frac{1}{4(0.05)^2(0.01)} = 10.0$

6.1.13 Weak Law of Large Numbers

Suppose that $X_1, X_2, X_3, \ldots\ldots,X_n$ are identically distributed, independent random variables with finite mean and variance.

Let $E(X) = \mu$ and $V(X) = \sigma^2$. Define $\bar{X} = \frac{1}{n}(X_1 + X_2 + X_3 + \cdots \ldots +$ Xn.

Now $\bar{X}$ is a function of $X_1, X_2, X_3, \ldots\ldots,X_n$, and hence is a

random variable, with $E(\bar{X}) = \mu$ and $V(\bar{X}) = \frac{\sigma^2}{n}$. Then for $\epsilon > 0$

$$P(|\bar{X} - \mu| \geq \epsilon) \to 0 \text{ as } n \to \infty$$

Comments: Note that $P(|\bar{X} - \mu| \geq \epsilon) = 1 - P(|\bar{X} - \mu| < \epsilon)$.
Therefore

$$P(|\bar{X} - \mu| \geq \epsilon) \to 0 \text{ as } n \to \infty \Leftrightarrow P(|\bar{X} - \mu| < \epsilon) \to 1 \text{ as } n \to \infty$$

Proof:
By Chebyshev's inequality, if Y is a r.v. with mean m and variance v, then for any real value k>0

$$P(|Y - m| \geq k) \leq \frac{v}{k^2}$$

Now let $Y = \bar{X} = \frac{1}{n}\sum_{i=1}^{n} X_i$ and let $k = \epsilon$. Then $E(\bar{X}) = \mu$ and

$$Var(\bar{X}) = \frac{\sigma^2}{n}$$

Then from Chebyshev's inequality,

$$P(|\bar{X} - \mu| \geq \epsilon) \leq \frac{\sigma^2}{n\epsilon^2} \Rightarrow 0 \leq P(|\bar{X} - \mu| \geq \epsilon) \leq \frac{\sigma^2}{n\epsilon^2}$$

i.e. $0 \leq \lim_{n \to \infty} P(|\bar{X} - \mu| \geq \epsilon) \leq \lim_{n \to \infty} \frac{\sigma^2}{n\epsilon^2} = 0$

this implies $P(|\bar{X} - \mu| \geq \epsilon) = 0 \text{ as } n \to \infty$ by the sandwich theorem of Analysis
i.e.

$$P(|\bar{X} - \mu| < \epsilon) \to 1 \text{ as } n \to \infty$$

Equivalently, the weak law says that

$$P(|\bar{X} - \mu| < \epsilon) \geq 1 - \frac{\sigma^2}{n\epsilon^2}$$

Proof:

$$0 \leq P(|\bar{X} - \mu| \geq \epsilon) \leq \frac{\sigma^2}{n\epsilon^2} \implies$$

$$-P(|\bar{X} - \mu| \geq \epsilon) \geq \frac{\sigma^2}{n\epsilon^2} \implies$$

$$1 - P(|\bar{X} - \mu| \geq \epsilon) \geq 1 - \frac{\sigma^2}{n\epsilon^2} \implies$$

$$P(|\bar{X} - \mu| < \epsilon) \geq 1 - \frac{\sigma^2}{n\epsilon^2}$$

As $n \to \infty$, the right hand side of the above inequality approaches 1(one). It is in this sense that the arithmetic mean converges to E(X).

EXAMPLE 6.13

A large number of electronic tubes are tested. Let T_i be the time to failure of the i-th tube. Suppose, furthermore, that all the tubes come from the same stockpile and that all may be assumed to be exponentially distributed with the same parameter α.

Hence $E(T_i) = \frac{1}{\alpha}$. Let $\bar{T} = \frac{1}{n}(T_1 + T_2 + T_3 + \cdots \ldots + T_n)$.

The above form of the Law of Large Numbers says that if n is quite large, it would be very probable that the value obtained for the arithmetic mean of a large number of failure times would be close to $\frac{1}{\alpha}$.

6.1.14 Bivariate Normal Distribution

The Bivariate Normal Distribution is one of the most important continuous two-dimensional random variables. It is a direct generalization of the one-dimensional normal distribution and is defined as follows:

6.1.14.1 Definition (Bivariate Normal Distribution)

Let (X,Y) be a two dimensional continuous random variable assuming all values in the Euclidean plane. We say that (X,Y) has a bivariate normal distribution if its joint pdf. is given by the following expression:

$$f(x,y) = \frac{1}{2\pi\sigma_x\sigma_y\sqrt{1-\rho^2}} exp\left\{-\frac{1}{2(1-\rho^2)}\left[\left(\frac{X-\mu_x}{\sigma_x}\right)^2 -\frac{2\rho(X-\mu_x)(Y-\mu_y)}{\sigma_x\sigma_y} + \left(\frac{Y-\mu_y}{\sigma_y}\right)^2\right]\right\},$$

$$-\infty < x < \infty, -\infty < y < \infty$$

The above pdf. depends on 5(five) parameters. For f to define a legitimate pdf. [That is, $f(x,y) \geq 0$ and $\int_{-\infty}^{\infty}\int_{-\infty}^{\infty} f(x,y)dxdy = 1$], we must place the following restrictions on the parameters:

$$-\infty < \mu_x < \infty, \qquad -\infty < \mu_y < \infty, \;\; \sigma_x > 0, \;\; \sigma_y > 0,$$
$$-1 < \rho < 1.$$

6.1.14.2 Theorem

Suppose that (X,Y) is a two-dimensional, continuous random variable having a normal distribution. If the pdf. of (X,Y) is given as defined in 6.1.14.1 above, then

(i)   The marginal distribution of X and Y are
$N(\mu_x, \sigma_x^2)$ and $N(\mu_y, \sigma_y^2)$ respectively.

(ii)   The parameter ρ appearing above is the correlation coefficient between X and Y.

(iii)   The conditional distribution of $X(given\ Y = y)$ and of $Y(given\ X = x)$ are respectively

$N[\mu_x + \rho\frac{\sigma_x}{\sigma_y}(y - \mu_y), \sigma_x^2(1 - \rho^2)]$ and $N[\mu_y + \rho\frac{\sigma_y}{\sigma_x}(x - $
$\mu x, \sigma y 2 1 - \rho 2]$

6.1.15 The Exponential Distribution (or Laplace Distribution)

6.1.15.1 Definition (Exponential Distribution)

Let X be a continuous random variable assuming all non-negative values. Then X is said to be an exponential distribution with parameter $\theta > 0$ if its probability density function (pdf) is given by

$$f(x) = \begin{cases} \theta e^{-\theta x}, & x > 0 \\ 0, & elsewhere \end{cases}$$

Since f(x) is a true pdf, straightforward integration shows that

$$\int_{-\infty}^{\infty} f(x)dx = \int_{-\infty}^{0} f(x)dx + \int_{0}^{\infty} f(x)dx$$

$$= 0 + \int_0^\infty \theta e^{-\theta x}\, dx = 1$$

The exponential distribution plays an important role in describing a large class of phenomena, particularly in the area of reliability theory which is not treated in this book. For now we shall investigate some of the properties of the exponential distribution.

6.1.15.2 The Cumulative Distribution Function (cdf) F

The cdf. (F) of the exponential distribution is given by

$$F(x) = P[X \leq x] = \int_0^x \theta e^{-\theta s}\, ds. \quad \text{Straightforward integration gives}$$

$$F(x) = \begin{cases} 1 - e^{-\theta x}, & x \geq 0 \\ 0, & elsewhere \end{cases}$$

Since $P[X \leq x] + P[X > x] = 1$, $this\ means\ P[X > x] = 1 - P[X \leq x] = e^{-\theta x}$.

6.1.15.3 The Expected Value, E(X)

The Expected value of the exponential random variable X is given as

$$E(X) = \int_{-\infty}^{\infty} xf(x)dx = \int_0^\infty x\theta e^{-\theta x}.\,dx$$

Integrating by parts and letting $\theta e^{-\theta x}dx = dv$, $x = u$, $we\ obtain$

$V = -e^{-\theta x}$, $du = dx$. Hence $E(X) = [-xe^{-\theta x}]_0^\infty +$ $\int_0^\infty e^{-\theta x}dx = \frac{1}{\theta}$

Thus the expected value equals the reciprocal of the parameter θ.

NOTE: *Suppose we relabel the parameter* $\theta = \frac{1}{\beta}$. *The pdf. will now be*

written as $f(x) = \begin{cases} \frac{1}{\beta}e^{-\left(\frac{1}{\beta}\right)x}, & x > 0 \\ 0, & elsewhere \end{cases}$

In this form, the parameter β is the expected value of X. ie. $E(X) = \beta$.

6.1.15.4 The Variance of X

$V(X) = E(X^2) - [E(X)]^2$. We find by integration that

$E(X^2) = \int_0^\infty x^2 e^{-\theta x}\, dx = \frac{2}{\theta^2}$ Hence, $V(X) = \frac{2}{\theta^2} - \left(\frac{1}{\theta}\right)^2 = \frac{1}{\theta^2}$

6.1.15.5 The Forgetfulness Property

The exponential distribution has the following forgetfulness property, analogous to that described (in section 4.7.4.3 of Chapter 4) for the Geometric distribution. For any
k,t >0,

$$P[X > k + t / X > k = \frac{P[X > k + t]}{P[> k]} = \frac{e^{-\theta(k+t)}}{e^{-\theta k}} = e^{-\theta t}$$

Hence
$P[X > k + t / X > k] = P[X > t]$.
Thus, we have shown that the exponential distribution also has the property of having
'no memory' as did the geometric distribution.

EXAMPLE 6.14

Suppose that X has an exponential distribution with parameter θ.
Compute the probability that X exceeds the expected value, ie.
$P[X > E(X)]$.

SOLUTION:

$$E(X) = \frac{1}{\theta}, \quad hence \ P[X > E(X)] = P[X > \frac{1}{\theta}]$$

$$= \int_{\frac{1}{\theta}}^{\infty} f(x)dx = e^{-\theta\left(\frac{1}{\theta}\right)} = e^{-1}$$

EXAMPLE 6.15

Suppose that the length of a phone call in minutes is an exponential random variable with mean 10. If someone arrives immediately ahead of you at a public telephone booth, find the probability that you will have to wait between 10 and 20 minutes

SOLUTION

Let X denote the length of the call made by the person in the booth. Then the required probability is

$$P(10 \leq X \leq 20) = \int_{10}^{20} \frac{1}{10} e^{-\frac{1}{10}x} dx = \left[-e^{\frac{1}{10}x} \right]_{10}^{20} = e^{-1} - e^{-2}$$
$$= 0.233$$

Comment:

(1)   The exponential distribution models fairly well the life-time distribution of several living things and man-made objects. For instance if the fluorescent tube manufactured by a factory has an exponential distribution with mean λ i.e. it's pdf is given by

$$f(x) = \frac{1}{\lambda} e^{-\frac{1}{\lambda}x}, x \geq 0$$

where x is life time of tube in years, then the probability that a tube fails in time T is

$$P(X \leq T) = \int_0^T \frac{1}{\lambda} e^{-\frac{1}{\lambda}x} dx$$

The probability that a tube survives for at least T years is

$$P(X \geq T) = \int_T^{\infty} \frac{1}{\lambda} e^{-\frac{1}{\lambda}x} dx = e^{-\frac{1}{\lambda}T}$$

(2)   If X_1 and X_2 are iid exponential random variables each having parameter λ then the random variable $T = X_1 + X_2$ has pdf

$$f(t) = \lambda^2 t e^{-\lambda t}, t \geq 0$$

The distribution above is known as the Gamma distribution with parameters 2 and λ

(3)   Given that $X \sim Exp(\frac{1}{\lambda})$ with pdf $f(x) = \frac{1}{\lambda}e^{-\frac{1}{\lambda}x}, x \geq 0$. Let $P(X \geq T) = \int_T^\infty \frac{1}{\lambda}e^{-\frac{1}{\lambda}x}dx = C$ then the density function $g(x) = \frac{1}{C}\frac{1}{\lambda}e^{-\frac{1}{\lambda}x}, x \geq T$ is known as the truncated exponential pdf.

6.2. The Gamma Distribution

Before discussing the Gamma distribution, we shall first introduce an important function (used in Probability Theory and in many areas of Mathematics) called the Gamma function.

6.2.1.   Definition (The Gamma Function)

The Gamma function is denoted by the symbol Γ, and is defined as follows:

$$\Gamma(t) = \int_0^\infty x^{t-1}e^{-x}\,dx, \quad defined\ for\ t > 0.$$

It can be shown that the above improper integral exists (converges) whenever t>0. If we integrate $\Gamma(t)$ by parts, letting $e^{-x}dx = dv\ and\ x^{t-1} = u,$ we

obtain $\Gamma(t) = \int_0^\infty x^{t-1}e^{-x}dx = \frac{x^t e^{-x}}{t}\Big|_0^\infty + \int_0^\infty \frac{x^t e^{-x}}{t}dx$

$$= 0 + \frac{1}{t}\Gamma(t+1)$$

giving $t\Gamma(t) = \Gamma(t+1)$.

....6.2.1

This relationship automatically gives the values of $\Gamma(t)$, on any interval $n < t \leq n + 1$ in terms of its values on the interval $0 < t \leq 1$. On the latter range, the point t=1 is easy to handle: $\Gamma(1) = \int_0^\infty e^{-x} dx = 1$.

One interesting property of the Gamma function is that it can be considered as a factorial function- its value at a possible integer is a factorial. To see how this comes about, if we use equation 6.2.1, we have:

$$\Gamma(2) = \Gamma(1) = 1 = 1!$$
$$\Gamma(3) = 2\Gamma(2) = 2.1 = 2!$$
$$\Gamma(4) = 3\Gamma(3) = 3.2.1 = 3!$$
$$\Gamma(5) = 4\Gamma(4) = 4.3.2.1 = 4!$$

And by simple induction $\qquad \Gamma(n) = (n-1)!$

It is also easy to verify that $\qquad \Gamma\left(\frac{1}{2}\right) = \int_0^\infty x^{-\frac{1}{2}} e^{-x} dx = \sqrt{\pi},\qquad$ if we make the change of variable $x = \frac{u^2}{2}$ in the above integral. The value of $\Gamma(t)$ at the points halfway between the integers is obtainable from the value $\Gamma\left(\frac{1}{2}\right) = \sqrt{\pi}$ given above. Again from the recurrence relation $\Gamma(t+1) = t\Gamma(t)$ there follows:

$$\Gamma\left(\frac{3}{2}\right) = \frac{1}{2}\Gamma\left(\frac{1}{2}\right) = \frac{1}{2}\sqrt{\pi}, \text{ and } \Gamma\left(\frac{5}{2}\right) = \frac{3}{2}\Gamma\left(\frac{3}{2}\right) = \frac{3}{2} \times \frac{1}{2}\sqrt{\pi}$$

and in general (by induction):

$$\Gamma\left(k + \frac{1}{2}\right) = (2k-1)(2k-3)\ldots\ldots\ldots 5.3.1.\frac{\sqrt{\pi}}{2^k}.$$

With the aid of the Gamma function we can now introduce the Gamma probability distribution.

The integration formula $\qquad \int_0^\infty x^{t-1} e^{-\lambda x} dx = \frac{\Gamma(t)}{\lambda^t}$, for $t, \lambda > 0$ gives the proper

multiplying constant to make the integration (which is a non-negative function) a density.

6.2.2. Definition (Probability Density Function)

Let X be a continuous random variable assuming only non-negative values. We say that X has a Gamma distribution if its probability density function (pdf.) is given by

$$f(x) = \begin{cases} \dfrac{\lambda^t \, x^{t-1} e^{-\lambda x}}{\Gamma(t)} & , \quad x > 0 \\ 0, \quad elsewhere \end{cases}$$

This distribution depends on two parameters, t and λ, of which we require t>0, $\lambda > 0$.

6.2.3 NOTE:

(a) If $t = 1$, $f(x) = \lambda e^{-\lambda x}$.   Hence the exponential distribution is a special case of the Gamma distribution.

(b) If the random variable X has a Gamma distribution given by the pdf defined in section 6.2..2 above, we have

$E(X) = \dfrac{t}{\lambda,}$ $,and$ $V(X) = \dfrac{t}{\lambda^2}.$ The simple proofs are left as assignments to the student

6.3. Beta Distribution

6.3.1.   The Beta Function

The Beta function, B(s, t), is a symmetric function of two parameters say, s and t, and is obtained as a product of two Gamma functions. This derivation of B(s, t) is along the same lines as the calculation of the integral of the normal density (see section 6.1.2 above), and involves the use of exponential functions of the type $\exp\left(-\dfrac{x^2}{2}\right)$

Thus, if Γ(s) and Γ(t) are two Gamma functions we may write, by letting

$$x = \frac{u^2}{2} \quad and \quad y = \frac{v^2}{2} ,$$

$$\Gamma(s)\,\Gamma(t) = \int_0^\infty \int_0^\infty \left(\frac{u^2}{2}\right)^{s-1} \left(\frac{v^2}{2}\right)^{t-1} \exp\left[-\frac{1}{2}(u^2 + v^2)\right] v \, du \, dv$$

using polar co-ordinates u = r cos θ and v = r sin θ would give,

$$\Gamma(s)\Gamma(t)$$
$$= \int_0^{\frac{\pi}{2}} \int_0^\infty \left(\frac{r^2}{2}\right)^{s+t-1} 2(\cos\theta)^{2s-1}(\sin\theta)^{2t-1} \exp\left(-\frac{1}{2}r^2\right) r \, dr \, d\theta$$

On letting $W = \dfrac{r^2}{2}$

$$\Gamma(s)\,\Gamma(t) = \int_0^\infty w^{s+t-1} e^{-w} dw \int_0^{\frac{\pi}{2}} 2(cos\theta)^{2s-1}(sin\theta)^{2t-1} d\theta$$

$$= \Gamma(s+t)\,B(s,t), \quad \text{so that}$$

$$B(s,t) = 2\int_0^{\frac{\pi}{2}} Cos^{2s-1}\theta Sin^{2t-1}\theta d\theta = \frac{\Gamma(s)\,\Gamma(t)}{\Gamma(s+t)}$$

$$= \int_0^1 x^{s-1}(1-x)^{t-1} dx = \int_0^1 (1-y)^{s-1} y^{t-1} dy = B(t,s)$$

This symmetric function of s and t is called the Beta function.

6.3.2. Definition (The Family of Beta Distributions)

Let X be a continuous random variable assuming only non-negative values. We say that X belongs to the family of Beta distributions if its pdf is given as

$$f(x) = \begin{cases} \dfrac{1}{B(r,s)} x^{r-1}(1-x)^{s-1} & ,for\ 0 < x < 1 \\ 0, & elsewhere \end{cases}$$

where r and s are positive parameters.

6.3.3 NOTE: The family of Beta distributions, belonging to the two-parameter exponential family, provides a versatile and useful set of models for distributions on the unit interval $0 < x < 1$. For $r > 1$ and $s > 1$ they are unimodal, with maximum density at the value $x = (r-1)/(r+s-2)$, having various degrees of peakedness depending on the size of $r + s$, and various degrees of skewness depending on the ratio of r to s. (For $r = s$ they are symmetric). For $r = s = 1$ the density is uniform on $(0, 1)$, for $r < 1$ it is infinite at $x = 0$, and for $s < 1$ it is infinite at $x = 1$. A related family of distributions, on the positive real axis $0 < x < \infty$, is suggested by a form of the Beta function obtained by making the change of variable $x = u/(1 + u)$ in the earlier integral formula; the result of the change is

$$B(r,s) = \int_0^\infty \frac{u^{r-1}}{(1+u)^{r+s}} du.$$

The integrand, divided by B(r, s), is a density function on $(0, \infty)$; it will be encountered shortly in the study of the ratio of chi-square variables.

6.4. Chi-Square Distribution

The chi-square distribution is a special, very important, case of the Gamma distribution and is obtained by letting $\lambda = \frac{1}{2}$ and $t = \frac{n}{2}$ in the pdf of the Gamma distribution defined in section 6.2.2 above, where n is a positive integer. A random variable X having the resulting pdf given by

$$f(x) = \begin{cases} \dfrac{1.x^{\frac{n-2}{2}} e^{-\frac{x}{2}}}{2^{\frac{n}{2}} \Gamma\left(\frac{n}{2}\right)}, & x > 0 \\ 0, & elsewhere \end{cases}$$

is said to have a chi-square distribution with n degrees of freedom (denoted by χ_n^2).

6.4.1.   Properties of the Chi-square Distribution

(a) The chi-square distribution has many important applications in statistical inference. Because of its importance, the chi-square distribution is tabulated for various values of the parameter n. Thus, we may find in the table that value, denoted by χ_α^2, satisfying the probability $(X \leq \chi_\alpha^2) = \alpha, 0 < \alpha < 1$. (see figure 6.8 below).

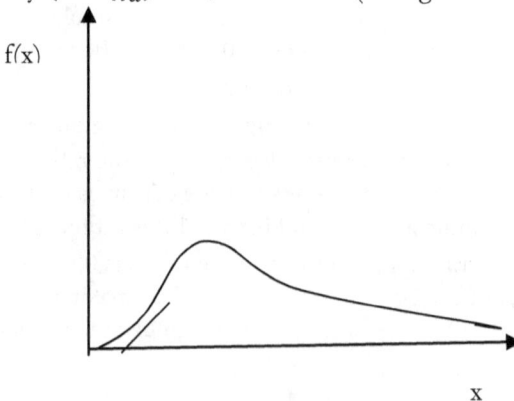

FIGURE

$$1 - \alpha$$

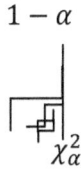

$$\chi_\alpha^2$$

(b)The square of the standardized observation $Z = \left(\frac{X-\mu}{\sigma}\right)$ in a random sample from a normal population with mean μ and variance σ^2 has a Chi-square distribution with
1 degree of freedom: i.e.

$$Z^2 = \left(\frac{X - \mu}{\sigma}\right)^2 \sim \chi_1^2$$

The sum of squares of these independent normal variables then has the Chi-square distribution with n degrees of freedom. The number of terms in the sum is called the number of degrees of freedom.
i.e.

$$\sum_{i=1}^{n} Z_i^2 = (Z_1^2 + Z_2^2 + \cdots + Z_n^2) \sim \chi_n^2$$

(c)The Chi-square distribution is often encountered in the large-sample theory as an asymptotic distribution of a test statistic. For instance, it is the asymptotic distribution of

$- 2 \log L$, where L is the likelihood statistic and it will provide the asymptotic distribution of a goodness-of-fit statistic and of certain contingency table statistics referred to in
statistical inference.

In small-sample theory, the Chi-square distribution is what is needed in studying the sample variance in the case of a normal population and various estimates of variance in the analysis of designed experiments involving normal models.
(d)The Chi-square distribution obeys an additive law, to be stated formally along with some important consequences:

If X and Y have independent chi-square distributions with k and m degrees of freedom, respectively, then their sum X + Y is chi-square with (k + m) degrees of freedom.

(e) If $X_1, \ldots X_n$ are independent and X_i is chi-square with k_i degrees of freedom

(i = 1, ... n), then $X_1 + X_2 + \ldots + X_n$ is chi-square with $k_1 + k_2 + \ldots + k_n$

degrees of freedom.

(f) If X and Y are independent, and if X is Chi-square with k degrees of freedom and

(X + Y) is chi-square with m degrees of freedom, then Y is chi-square with (m − k) degrees of freedom.

6.4.2. Theorem

Suppose that the random variable Y has distribution χ_n^2. Then for sufficiently large n the random variable $\sqrt{2Y}$ has approximately the distribution $N(\sqrt{2n-1}, 1)$, with

$$E(\sqrt{2Y}) \approx \sqrt{2n-1} \quad and \quad V(\sqrt{2Y}) \approx 1.$$

6.4.3 NOTE:

The above theorem may be used as follows: Suppose that we require $P(Y \le t)$, where Y has distribution χ_n^2 and n is so large that the above probability cannot be directly obtained from the table of the chi-square distribution. Using Theorem 6.4.1.1, we may write

$$P(Y \le t) = P[\sqrt{2Y} \le \sqrt{2t}]$$

$$= P\left[\frac{\sqrt{2Y} - \sqrt{2n-1}}{1} \le \frac{\sqrt{2t} - \sqrt{2n-1}}{1}\right] = \phi\left(\frac{\sqrt{2t} - \sqrt{2n-1}}{1}\right)$$

The value of ϕ may be obtained from the tables of the normal distribution.

EXAMPLE 6.15

Let X be a normally distributed random variable with mean equal 0 and standard deviation 1. If $Y = X^2$, then Y has a chi-square distribution with 1 degree of freedom. Obtain the probability density function of Y. [Hint: if $Z \sim \chi_n^2$ (chi-square with n degrees of freedom, then Z has pdf given below

$$f_Z(z) = \begin{cases} \dfrac{1}{\Gamma\left(\dfrac{n}{2}\right)2^{\frac{n}{2}}} Z^{\frac{n}{2}-1} e^{-\frac{Z}{2}}, & Z > 0 \\ 0 & , elsewhere \end{cases}$$

SOLUTION:

If $X \sim N(0,1)$ then the pdf of X is given by

$$f_X(x) = \frac{1}{\sqrt{2\pi}} e^{-\frac{x^2}{2}}, \quad -\infty < x < \infty$$

The distribution function of Y is defined as:

$$F_Y(y) = P(Y \le y) = P(X^2 \le y)$$
$$= P(-\sqrt{y} < x < \sqrt{y})$$
$$= F_x(\sqrt{y}) - F_x(-\sqrt{y})$$

The pdf of Y becomes

$$f_y(y) = \frac{d}{dy} F_1(y) = \frac{d}{dy}\left[F_x(\sqrt{y}) - F_x(-\sqrt{y})\right]$$
$$= \frac{d}{dx}\{F_x(\sqrt{y}) - F_x(-\sqrt{y})\}\frac{dx}{dy}$$
$$= f_x(\sqrt{y}).\frac{1}{2\sqrt{y}} + f_x(-\sqrt{y}).\frac{1}{2\sqrt{y}} = \frac{1}{2\sqrt{y}}\{f_x(\sqrt{y}) + f_x(-\sqrt{y})\}$$
$$= \frac{1}{2\sqrt{y}}.\frac{1}{\sqrt{2\pi}}\{e^{-\frac{1}{2}y} + e^{-\frac{1}{2}y}\} = \frac{1}{\sqrt{y}}.\frac{1}{\sqrt{2\pi}}e^{-\frac{1}{2}y} = \frac{1.y^{-\frac{1}{2}}}{2^{\frac{1}{2}}}e^{-\frac{1}{2}y}, y > 0$$

Comparing with

$$f_Z(z) = \frac{1}{2^{\frac{n}{2}}\Gamma\left(\frac{1}{2}\right)} Z^{\frac{n}{2}-1} e^{-\frac{z}{2}}, Z > 0$$

where Z is χ_n^2 (Chi-square with n degrees of freedom), we have for n = 1

$$f_Z = f(y) = \frac{1}{2^{\frac{1}{2}}\Gamma\left(\frac{1}{2}\right)} Z^{\frac{-1}{2}} e^{-\frac{z}{2}}, Z > 0$$

since $\Gamma\left(\frac{1}{2}\right) = \pi$

$\therefore Y = X^2$ has chi-square distribution with 1 degree of freedom.

EXAMPLE 6.16

Let X and Y be independent random variables such that X has a normal distribution with mean equal to 0 and standard deviation equal to 1, while Y has chi-square distribution with n degrees of freedom. If

$$Z = \frac{X}{\sqrt{\frac{Y}{n}}}$$

then Z has t – distribution with n degrees of freedom. Obtain the frequency density function (pdf) of Z.

Hint: $f_Y(y) = \begin{cases} \dfrac{1.y^{\frac{n-2}{2}} e^{-\frac{y}{2}}}{\Gamma\left(\frac{n}{2}\right)2^{\frac{n}{2}}} & , \quad y > 0 \\ 0, \; elsewhere \end{cases}$

SOLUTION:

X ~ N (0, 1) , so pdf of X is given by $f(x) = \frac{1}{\sqrt{2\pi}} e^{-\frac{x^2}{2}}, -\infty < x < \infty$

Since X and Y are independent,

$$f_{XY}(x,y) = f_X(x)f_Y(y) = \frac{1}{\sqrt{2\pi}}e^{-\frac{x^2}{2}} \times \frac{1.y^{\frac{n-2}{2}}}{\Gamma\left(\frac{n}{2}\right).2^{\frac{n}{2}}}e^{-\frac{y}{2}}, -\infty < x$$

$$< \infty, -\infty < y < \infty.$$

If $Z = \dfrac{X}{\sqrt{\frac{Y}{n}}}$ Let $U = Y$, then $X = Z\sqrt{\dfrac{U}{n}}$ and $Y = U$.

This transformation now gives the joint density of Z and U as:

$$f_{ZU}(z,u) = f_{XY}\left\{X\left(Z\sqrt{\frac{U}{n}}\right), Y(u)\right\}|J|$$

$$= \frac{1}{\sqrt{2\pi}}e^{-\frac{1}{2}z^2\frac{u}{n}} \times \frac{1.u^{n/2-1}}{\Gamma\left(\frac{n}{2}\right).2^{\frac{n}{2}}}e^{-\frac{u}{2}}.|J|$$

Where J is the Jacobian of transformation and

$$J = \begin{vmatrix} \frac{\delta x}{\delta z} & \frac{\delta x}{\delta u} \\ \frac{\delta y}{\delta z} & \frac{\delta y}{\delta u} \\ \frac{\delta y}{\delta z} & \frac{\delta y}{\delta u} \end{vmatrix} = \begin{vmatrix} \sqrt{\frac{u}{n}} & \frac{z.}{\sqrt{n}}\frac{1}{2\sqrt{u}} \\ 0 & 1 \end{vmatrix}, \quad \text{ie.} \quad |J| = \sqrt{\frac{U}{n}}. \qquad \text{Hence,}$$

$$f_{zu}(z,u) = \frac{1}{\sqrt{\pi n}}\frac{1}{\Gamma\left(\frac{n}{2}\right)2^{\frac{n+1}{2}}}e^{-\frac{u}{2}\left(\frac{z^2}{n}+1\right)}u^{\frac{n+1}{2}-1}$$

The marginal density function $f_Z(z)$ is then obtained as

$$f_Z(z) = \int_{-\infty}^{\infty}f(z,u)du = \int_0^{\infty}f(z,u)du, \quad 0 < u < \infty.$$

$$= \frac{1}{\sqrt{\pi n}}\frac{1}{\Gamma\left(\frac{n}{2}\right)2^{\frac{n+1}{2}}}\int_0^{\infty}e^{-\frac{u}{2}\left(\frac{z^2}{n}+1\right)}u^{\frac{n+1}{2}-1}du.$$

Assume now that $r = \dfrac{u}{2}\left(\dfrac{z^2}{n}+1\right)$ then

$$dr = \frac{1}{2}\left(\frac{z^2}{n}+1\right)du \quad \text{or} \quad du = dr.\frac{2}{\left(\frac{z^2}{n}+1\right)}, \quad \text{therefore,}$$

$$f_Z(z) = \frac{1}{\sqrt{\pi n}}.\frac{1}{\Gamma\left(\frac{n}{2}\right)}.\frac{1}{2^{\frac{n+1}{2}}}.2^{\frac{n+1}{2}}.\frac{1}{\left(\frac{z^2}{n}+1\right)^{\frac{n+1}{2}}}\int_0^{\infty}e^{-r}r^{\left(\frac{n+1}{2}\right)-1}.dr$$

Since $\Gamma(P) = \int_0^\infty x^{p-1}e^{-x}dx$ we have, $\int_0^\infty r^{\left(\frac{n+1}{2}\right)-1}e^{-r}dr = \Gamma\left(\frac{n+1}{2}\right)$

Hence, the density function of Z:

$$f_Z(z) = \frac{1}{\sqrt{\pi n}} \frac{\Gamma\left(\frac{n+1}{2}\right)}{\Gamma\left(\frac{n}{2}\right)} \cdot \frac{1}{\left(\frac{z^2}{n}+1\right)^{\frac{n+1}{2}}}$$

$$= \frac{1}{\sqrt{n}} \cdot \frac{\Gamma\left(\frac{n+1}{2}\right)}{\Gamma\left(\frac{1}{2}\right)\Gamma\left(\frac{n}{2}\right)} \cdot \frac{1}{\left(\frac{z^2}{n}+1\right)^{\frac{n+1}{2}}},$$

$$-\infty < z < \infty$$

6.5. F and T-Distribution

In the problem of comparing Normal populations with respect to their variances, as well as in a variety of other problems, it will be necessary to know the distribution of the ratio of two Chi-square random variables. Let X and Y denote independent Chi-square random variables with n_1 and n_2 degrees

of freedom respectively. If $Z = \frac{X/n_1}{Y/n_2}$ and $U = Y$; $0 < z < \infty, 0 < u < \infty$, then $X = \frac{n_1}{n_2}ZU$, and $Y = U$. The probability density function of Z(ie. $f_z(z)$) is obtained from the joint density $f_{zu}(z,u)$ of z and u : That is,

$$f_{ZU}(z,u) = f_{XY}\left\{X\left(zu.\frac{n_1}{n_2}\right), Y(u)\right\}|J| = f_X\left(zu\frac{n_1}{n_2}\right)f_Y(u)|J|, \text{ since X}$$
and Y are independent.

But

$$f_X(x) = \frac{1}{\Gamma\left(\frac{n_1}{2}\right)2^{\frac{n_1}{2}}}x^{\left(\frac{n_1}{2}-1\right)}e^{-\frac{x}{2}}, \quad x > 0,$$

and

$$f_Y(y) = \frac{1}{\Gamma\left(\frac{n_2}{2}\right)2^{\frac{n_2}{2}}}y^{\left(\frac{n_2}{2}-1\right)}e^{-\frac{y}{2}}, \quad y > 0 \text{ which gives}$$

$$f(z,u) =$$

$$\left\{\left(\frac{1}{\Gamma\left(\frac{n_1}{2}\right)2^{\frac{n_1}{2}}}\right)\left(zu\frac{n_1}{n_2}\right)^{\left(\frac{n_1}{2}-1\right)}e^{-\left(zu\frac{n_1}{n_2}\right)/2}\right\} \times \left(\frac{1}{\Gamma\left(\frac{n_2}{2}\right)2^{\frac{n_2}{2}}}u^{\left(\frac{n_2}{2}-1\right)}e^{-u/2}\right),$$

$$0 < u < \infty, \qquad 0 < z < \infty.$$

$$= \left\{\frac{1}{\Gamma\left(\frac{n_1}{2}\right)\Gamma\left(\frac{n_2}{2}\right)2^{\frac{n_1+n_2}{2}}}\left(zu\frac{n_1}{n_2}\right)^{\left(\frac{n_1}{2}-1\right)}u^{\left(\frac{n_2}{2}-1\right)} \times e^{-\frac{u}{2}\left(\frac{n_1}{n_2}z+1\right)}|J|\right\},$$

where the Jacobian of transformation ,J, is given as:

$$J = \begin{vmatrix} \frac{\partial x}{\partial z} & \frac{\partial x}{\partial u} \\ \frac{\partial y}{\partial z} & \frac{\partial y}{\partial u} \end{vmatrix} = \begin{vmatrix} u\frac{n_1}{n_2} & z\frac{n_1}{n_2} \\ 0 & 1 \end{vmatrix} = u\frac{n_1}{n_2}.$$ The marginal density function

of z, ie.

$f_Z(z)$ is given as: $f_Z(z) = \int_{-\infty}^{\infty} f(z,u)du = \int_{0}^{\infty} f(z,u)du, \quad 0 < u < \infty.$ Hence

$$f_Z(z) = \int_{0}^{\infty} \frac{\left(\frac{n_1}{n_2}\right)^{\left(\frac{n_1}{2}-1\right)}z^{\left(\frac{n_1}{2}-1\right)}}{\Gamma\left(\frac{n_1}{2}\right)\Gamma\left(\frac{n_2}{2}\right)2^{\frac{n_1+n_2}{2}}}u^{\left(\frac{n_1+n_2}{2}\right)-1}e^{-\frac{u}{2}\left(\frac{n_1 z}{n}+1\right)}du$$

Let $t = \frac{u}{2}\left(\frac{n_1}{n_2}z+1\right)$, *then* $dt = \frac{1}{2}\left(\frac{n_1}{n_2}z+1\right)du$, *or* $du =$ $2dt\frac{1}{\left(\frac{n_1}{n_2}z+1\right)}$. So that

$$f_Z(z) = \begin{cases} \int_0^\infty \dfrac{\left(\frac{n_1}{n_2}\right)^{\frac{n_1}{2}} z^{\left(\frac{n_1}{2}-1\right)}}{\Gamma\left(\frac{n_1}{2}\right)\Gamma\left(\frac{n_2}{2}\right)2^{\frac{(n_1+n_2)}{2}}} \left(\dfrac{2t}{\frac{n_1z+n_2}{n_2}}\right)^{\frac{n_1+n_2-2}{2}} . e^{-t} \dfrac{2}{\left(\frac{n_1z+n_2}{n_2}\right)} dt, \\ 0, \quad otherwise \end{cases}$$

$0 < z < \infty$,

That is,

$f_Z(z)$

$$= \begin{cases} \dfrac{\left(\frac{n_1}{n_2}\right)^{\frac{n_1}{2}} z^{\left(\frac{n_1}{2}-1\right)}}{\Gamma\left(\frac{n_1}{2}\right)\Gamma\left(\frac{n_2}{2}\right)2^{\frac{n_1+n_2}{2}}} \dfrac{1}{2} . 2.2^{\frac{n_1+n_2}{2}} \dfrac{\left(\frac{n_1z}{n_2}+1\right)}{\left(\frac{n_1z}{n_2}+1\right)^{\left(\frac{n_1+n_2}{2}-1\right)}} \int_0^\infty t^{\left(\frac{n_1+n_2}{2}-1\right)} e^{-t} dt \\ 0, \quad otherwise \end{cases}$$

Which gives:

$$f_Z(z) = \begin{cases} \dfrac{\left(\frac{n_1z}{n_2}\right)^{\left(\frac{n_1}{2}-1\right)} \Gamma\left(\frac{n_1+n_2}{2}\right)}{\Gamma\left(\frac{n_1}{2}\right)\Gamma\left(\frac{n_2}{2}\right)\left(\frac{n_1z}{n_2}+1\right)^{\frac{n_1+n_2}{2}}}, \quad 0 < z < \infty , \\ 0, \quad otherwise \end{cases}$$

that is,

$$f_Z(z) = \begin{cases} \dfrac{1}{B\left(\frac{n_1}{2},\frac{n_2}{2}\right)}\left(\dfrac{n_1z}{n_2}\right)^{\frac{n_1}{2}-1}\left(1+\dfrac{n_1z}{n_2}\right)^{-\left(\frac{n_1+n_2}{2}\right)}, \quad 0 < z < \infty \\ 0 \ , otherwise \end{cases}$$

is obtained as the probability density function of Z, which has the F distribution, defined as the of the ratio of two independent Chi-square variables, each divided by the corresponding number of degrees of freedom.

The t distribution with n_2 degrees of freedom can be defined as that of a random variable symmetrically distributed about 0(zero) whose square has the F distribution with 1(one) and n_2 degrees of freedom in numerator and denominator, respectively (see for Example 6.16 above). Let T denote such a random variable so that T² has the F density:

$$f_{T^2}(z) = \frac{\Gamma(1+n_2)/2}{\sqrt{n_2\pi}\,\Gamma\left(\frac{1}{2}\right)\Gamma\left(\frac{n_2}{2}\right)}\left(\frac{z}{n_2}\right)^{\frac{1}{2}-1}\left(1+\frac{z}{n_2}\right)^{-(1+n_2)/2}, \quad z > 0.$$

Then for z>0,

$$f_{|T|}(z) = \frac{d}{dz}P(|T| < z) = \frac{d}{dz}P(T^2 < z^2) = 2zf_{T^2}(z^2).$$

But since T is symmetrically distributed, its distribution is obtained from that of $|T|$ as follows:

$$f_T(\pm z) = \frac{1}{2}f_{|T|}(|z|) = |z|f_{T^2}(z^2)$$

$$= \frac{\Gamma(n_2 + 1)/2}{\sqrt{n_2\pi}\,\Gamma\left(\frac{n_2}{2}\right)}\left(1 + \frac{z^2}{n_2}\right)^{-(n_2+1)/2}.$$

This is the desired density of the t distribution. The symmetry of the t distribution about $z = 0$ implies that the mean value of T, if it exists, must be 0(zero). The integral defining the mean is clearly absolutely convergent for $n_2 > 1$, and so for those values of n_2, the expected value $E(T) = 0$. For $n_2 = 1$, the density of T reduces to what is called the Cauchy density, having no absolute moments of any integral order. For $n_2 = 2$, the integral defining $E(T^2)$ is absolutely convergent, and so for those values of n_2, the variance of T:$Var(T) = E(T^2) = \frac{n_2}{(n-2)}$. As $n \to \infty$, the density function of T approaches that of a standard normal variate. For

$$\left(1 + \frac{z^2}{n_2}\right)^{-(n_2+1)/2} = \left\{\left(1 + \frac{z^2}{n_2}\right)^{n_2/z^2}\right\}^{-z^2/2}\left(1 + \frac{z^2}{n_2}\right)^{-\frac{1}{2}} \to exp\left(\frac{z^2}{2}\right).$$

Moreover, the constant factor tends to $\frac{1}{\sqrt{2\pi}}$. This can be seen using Stirlings formula $\Gamma(p) \sim \sqrt{\frac{2\pi}{p}}\left(\frac{p}{e}\right)^p$, for large p, which implies that

$\dfrac{\Gamma(p+h)}{\Gamma(p)} \sim p^k$, *for large p*.   This approach to normality accounts for the fact that
in t- tables, there is often a sequence of entries for $n = \infty$, which are simply the corresponding points on a standard normal distribution.

EXERCISE 6

6.1 A random variable X has normal distribution given as:
  $N \sim (\mu, \sigma^2) = N \sim (5, 64)$.

(a) Compute (i) $P(X \leq 17)$, (ii) $P(X < 13)$, (iii) $P(X > 21)$, (iv) P(X
$\leq 1)$ and (v) $P(13 < X < 17)$.
(b) Find the value of u such that $P(X \leq u) = 0.95$
(c) Find the value of V such that $P(|X - 5| \leq v) = 0.95$

6.2 X and Y are independent normally distributed random variables with X
  $\sim (\mu, \sigma^2) = (3, 16)$ and $Y \sim N (\mu, \sigma^2) = (2, 9)$. If $Z = X - Y$, obtain
  $P(-1 < Z < 15)$.

6.3 X_1, X_2 and X_3 are three independent normally distributed random
  variables with
  $X_1 \sim (\mu, \sigma^2) = (2, 4)$, $X \sim (\mu, \sigma^2) = (7, 16)$, $X_3 \sim N (\mu, \sigma^2) = (6, 1)$.

  If $Y = 2X_1 + 2X_2 - X_3 - 7$, obtain
  (i) $P(5 < Y < 14)$ (ii) $P\{|X - 5| < 3\}$ (iii) $P\{|X - 5| > 3\}$.

6.4 A random variable X has normal distribution with $X \sim N (\mu, \sigma^2) = (1, 16)$. Compute $P(|X| > 3)$.

6.5 X and Y are two independent normal distributed random variables
  with $X \sim N (\mu, \sigma^2) = (30, 9)$ and $Y \sim (\mu, \sigma^2) = (20, 4)$. Compute P(Y > X).

6.6 Let Y and $\overline{X}$ denote the sum and mean respectively, of 25 independent observations of a normally distributed random variable with mean equal to 2 and variance equal to 225.
Obtain (a) the distribution of Y (b) the distribution of $\overline{X}$.

6.7 A fair coin is tossed 100 times. Compute the probability of obtaining at most 40 heads.

6.8 The random variable X has Binomial distribution:

$X \sim$ Bin $(n, p) = (150, p)$. Compute $P(10 < X < 65)$ if
(a) $P = 0.4$
(b) $P = 0.04$

6.9 We may be interested only in the magnitude of X, say $Y = | X |$. If X has distribution
N (0, 1), determine the pdf of Y, and evaluate E(Y) and V(Y).

6.10 Suppose that X has distribution N(0, 25). Evaluate $P(1 < X^2 < 4)$.

6.11 In calculating Binomial probabilities, for the following cases, which method is to be preferred (given the usual tables, but no calculator) – Poisson approximation, Normal approximation or direct calculation?

(a) $n = 500$, $p = 0.15$
(b) $n = 40$, $p = 0.05$
(c) $n = 6$, $p = 0.4$
(d) $n = 10$, $p = 0.15$

6.12 Consider the discrete distribution for X with $f(-a) = f(a) = \dfrac{1}{8}$ and

$f(0) = \dfrac{3}{4}$.

Compute $P\{| X | \geq 2\sigma\}$ and compare with the bound given by the

Chebyshev inequality for this probability.

6.13 Suppose $\overline{X}$ is the mean of n independent observations of a random variable which has mean m and standard deviation σ. Prove that for each $\varepsilon > 0$,

$$\lim_{n \to \infty} P\left\{\left|\overline{X} - m\right| \geq \varepsilon\right\} = 0$$

(Hint: use Chebysev's inequality).

6.14(a) Items are produced in such a manner that 2 percent turn out to be defective. A large number of such items, say n, are inspected and the relative frequency of defectives, say f_D, is recorded. How large should n be in order that the probability is at least 0.98 that f_D differs from 0.02 by less than 0.05?

(b) answer (a) above if 0.02, the probability of obtaining a defective item, is replaced by p which is assumed to be unknown.

6.15 Suppose that a sample of size n is obtained from a very large collection of bolts, 3 percent of which are defective. What is the probability that at most 5 percent of the chosen bolts are defective if
(a) n = 6? (b) n = 60? (c) n = 600?

6.16 Suppose that 30 electronic devices say D_1, D_2, ...D_{30} are used in the following manner. As soon as D_1 fails D_2 becomes operative. When D_2 fails D_3 becomes operative etc. Assume that the time to failure of D_i is an exponentially distributed random variable with parameter β = 0.1 hour^{-1}. Let T be the total time of operation of the 30 devices. What is the probability that t exceeds 350 hours?

6.17 Suppose that X_i, i = 1, 2 ...50 are independent random variables each having a Poisson distribution with parameter λ = 0.03.
Let S = X_1 + X_2 + ... + X_{50}.
(a) Using the Central Limit Theorem, evaluate $P(S \geq 3)$.
(b) Compare the answer in (a) with the exact value of this probability.

6.18 Compute the following:
(a) $\Gamma(6)$ (b) $\Gamma\left(\dfrac{11}{2}\right)$ (c) $B\left(2, \dfrac{3}{2}\right)$

6.19 Show that $\Gamma\left(\dfrac{11}{2}\right) = \sqrt{\pi}$. (Hint: Make the change of variable

$x = \dfrac{u^2}{2}$ in the integral: $\Gamma\left(\dfrac{1}{2}\right) = \displaystyle\int_0^\infty x^{-\frac{1}{2}} e^{-x}\, dx$

6.20 Given that X has a Gamma distribution (α, λ), evaluate $E(X)$ and $E(X^2)$ by recognizing the integrals involved as Gamma functions. From these obtain Var (X).

6.21 Let W have the density : $f_W(w) = \dfrac{1}{B(r,s)} \cdot \dfrac{w^{r-1}}{(1+w)^{r+s}}$, $w > 0$.

Calcúlate (i)$E(w)$ (ii) $E(w^2)$
By recognizing the integrals involved as Beta functions.

6.22 Calculate $E\left(\dfrac{1}{u}\right)$, where u has a chi-square distribution with k degrees of freedom.

6.23 Three random independent normally distributed random variables are such that

$R_1 \sim N(\mu, \sigma^2) = N(2,\ 16), R_2 \sim N(\mu,\ \sigma^2) = N(6,\ 9)$ and
$R_3 \sim N(\mu,\ \sigma^2) = N(5,\ 1)$.
Let $Y = R_1 + R_2 + R_3$. Compute $P(5 < Y < 14)$

(b) Two normally distributed random variables are such that
$X \sim N(\mu,\ \sigma^2) = N(3,\ 16)$
and $Y \sim N(\mu,\ \sigma^2) = N(2,\ 9)$. If $Z = X + Y$ and
$E(XY) = 8$. Compute $P(-1 <$
$Z < 12)$.

(c) The random variable $X \sim N(\mu,\ \sigma^2) = N(1,\ 16)$.
Compute
    (i) $P\{|X| < 6\}$ (ii) $P\{|X - 3| > 5\}$

6.24 A normally distributed random variable X is such that $X \sim N(\mu, \sigma2 = N(1,\ 16)$.
        Compute
(i) $P\{|X| < 3\}$ (ii) $P\{|X| > 3\}$
(iii) $P\left\{\left|\dfrac{X}{5} -\right| < 8\right\}$
(iv) $P\{|X - 5| > 8\}$

(b) Let $\bar{X}$ be the sample mean of 4 independent normally distributed random variables. If

$$E(\bar{X}) = \mu = 60 \text{ and } \sigma^2 = 196, \text{ compute } P(\bar{X} < 50)$$

6.25 Let $X_1, X_2, \cdots X_5$ and $Y_1, Y_2, \cdots Y_6$ be independent normally distributed random variables with mean 0 (zero) and variance 1 (one). Compute

(a) $$E\left(\sum_{i=1}^{5} X_i\right)^2$$

(b) $$P\left\{\sum_{i=1}^{5} X_i > \sum_{i=1}^{5} (1 + Y_i)\right\}$$

(c) $$\frac{\left\{\sum_{i=1}^{5} X_i - \sum_{i=1}^{5} Y_i\right\}^2}{11}$$

(d) $$0.5\sum_{i=1}^{5} (X_i - Y_i)^2$$

(e) $$\frac{X_i}{\sqrt{\sum_{i=1}^{6} Y_i^2}}$$

CHAPTER 7
GENERATING FUNCTIONS

Suppose that X is a random variable, that is, X is a function from the sample space to the real members. In computing various characteristics of the random variable X, such as the first moment $E(X)$, or the second moment. $E(X^2)$ or $V(X)$, we work directly with the probability distribution of X. [The probability distribution is given by a function: either the pdf in the continuous case, or the point probabilities $P(x_i) = P(X = x_i)$ in the discrete case. The latter may also be considered as a function assuming non zero values only if $X = x_i, i = 1,2,3 \dots X$] Possibly, we can introduce some other function and make our required computation in terms of this new function (just as above we associated with each number, some new member). Of those functions that will be considered, the moment generating function and the factorial moment generating function can be used to generate the moments of a distribution, whilst the probability generating function is particularly useful in certain combinatorial problems.

7.1. Moment Generating Function

7.1.1 Definition (Moment Generating Function)
For any r.v. X, the $k - th$ moment of X about a constant b is defined as
$$E(X - b)^k, k = 1,2, \dots$$
Note: (1) if b=0, we have the $k - th$ moment of X about X (about 0). In this case we have :
$E(X)$ when $k = 1$, $E(X^2)$ when $k = 2$
(2) When $b = E(X)$, we have the $k - th$ central moment of X that is $E(X - E(X))^k$. In this case the first central moment (when k=1) is $E(X - E(X)) = 0$
If k=2, the second central moment of X is $EX - E(X)^2$ which just the variance of X

Let X be a discrete random variable with probability mass function $P(x_i) = P(X = x_i), i = 1, 2, \dots\dots$. The function $M_x(t)$, called the moment generating function of X, is defined by

$$M_X(t) = E(e^{tx}) = \sum_{i=0}^{\infty} e^{tx_i} p(x_i)$$

If X is a continuous random variable with pdf f, we define the moment generating function by

$$M_x(t) = E\left(e^{tx}\right) = \int_{-\infty}^{+\infty} e^{tx} f(x)dx..$$

In either the discrete or the continuous case, when it exists, the expectation depends on the choice of t, and so defines a function of t. For t = 0 it always exists: $M_x(0) = 1$, but for other values of t it may or may not exist, depending on the distribution.

7.1.2   Properties of the Moment Generating Function.

7.1.2.1   Moments (Expected Value and Variance).

Let us again look at the definition of

$$M_X(t) = E(e^{tx}) = \int_{-\infty}^{\infty} e^{tx} dF(x)$$

We recall the Maclaurin series expansion of the function e^x:

$$e^x = 1 + x + \frac{x^2}{2!} + \frac{x^3}{3!} + \cdots + \frac{x^n}{n!} + \cdots$$

(This series converges for all values of x).

Thus

$$e^{tx} = 1 + tx + \frac{(tx)^2}{2!} + \frac{(tx)^3}{3!} + \cdots \ldots \ldots + \frac{(tx)^k}{k!} + \cdots \ldots = \sum_{0}^{\infty} \frac{(tx)^k}{k!}$$

Thus

$$M_x(t) = E\left(e^{tx}\right) = \int_{-\infty}^{\infty} \sum_{k=0}^{\infty} \frac{(tx)^k}{k!} \, dFx$$

$$= \sum_{k=0}^{\infty} \frac{t^k}{k} \int_{-\infty}^{\infty} x^k \, dF(x) = \sum_{k=0}^{\infty} \frac{t^k}{k!} E\left(X^k\right)$$

provided that the interchange of summation and the integral is permitted, and

$$E(X^k) = \int_{-\infty}^{\infty} X^k \, dF(x)$$

is simply the k–th moment of a distribution of X. If the interchange of summation and integral is permitted, it means that the k– th moment of a distribution is simply the coefficient of $\dfrac{t^k}{k!}$ in the power series expansion of the moment-generating function.

We shall recall that t is a constant so far as taking the expectation is concerned and we may write

$$M_x(t) = 1 + t\, E(X) + \frac{t^2\, E(X^2)}{2!} + \cdots\cdots + \frac{t^K\, E(X^k)}{k!} + \cdots$$

Since $M_x(t)$ is a function of the real variable t, we may consider taking the derivative of $M_x(t)$ with respect to t, that is, $\left[\dfrac{d}{dt} M_x(t)\right]$ or $M_X^1(t)$ for short. Again

we are faced with a mathematical difficulty. The derivative of a finite sum is always equal to the sum of the derivatives (assuming, of course, that all derivatives in question exist). However, for an infinite sum this is not always so. Certain conditions must be satisfied in order to justify this operation: We shall assume that these conditions hold and proceed. (In most problems we shall encounter, such an assumption is justified). Thus,

$$\left[\frac{d}{dt} M_x(t)\right] \text{ or } M_x^1(t) = E(X) + t\, E(X^2) + \frac{t^2\, E(X^3)}{2!} + \cdots + \frac{t^{k-1}\, E(X^k)}{(k-1)!} + \cdots$$

setting t = 0 we find that M' (0) = E(X). Thus the first derivative of the moment generating function (mgf) evaluated at t = 0 gives the expected value, E(X), of the random variable X. Taking the second derivative of $M_x(t)$, i.e.

$$M_x''(t) = E(X^2) + tE(X^3) + \cdots \ldots \ldots + \frac{t^{k-2}}{(k-2)!} E(X^k) + \cdots \ldots$$

and setting t = 0, we have
$$M''(0) = E(X^2)$$

Continuing in this manner, we obtain [assuming that $M^k(0)$ exists]

$$M^k(0) = E(X^k)$$

Thus, if the function is one whose power series is not well known, the coefficients (or moments) can be found by differentiation. This result may also be seen by differentiating $M_x(t)$, k, times:

$$\frac{d^k}{dt^k} = E(e^{tx}) = E(X^k e^{tk})$$

provided that the operations of differentiation and expectation (which is an integration) can be interchanged. Justification of these various formal manipulations will not be carried out here; but it can be shown that if $M_x(t)$ exists in a proper interval about t = 0, then X has moments of all orders, and they can be computed as described.

EXAMPLE 7.1

Consider a simple Bernoulli trial for which the random variable X has just two possible values, 1 (for success) with probability P and 0 (for failure) with probability q = 1 – P. Obtain the moment generating function $M_x(t)$ for this distribution of X and hence the moments.

SOLUTION :

$$M_x(t) = E(e^{tx}) = \sum_{i=0}^{\infty} e^{tx_i} . P(x_i)$$
$$= e^{t.0} P(X = 0) + e^{t.1} P(X = 1) = 1.(1-p) + e^t.p$$

Moments: Method 1 - (Using Power Series Expansion).

$M_x(t)$ = p et + (1 – p) = Moment generating function.
Expanding e^t as a power series gives,

$$p e^t + (1-p) = p\left(1 + \frac{t}{1!} + \frac{t^2}{2!} + \cdots \frac{t^k}{k!} + \cdots\right) + (1-p)$$

$$= 1 + p\left(\frac{t}{1!}\right) + p\left(\frac{t^2}{2!}\right) + \left(\frac{t^3}{3!}\right) + \cdots + p\left(\frac{t^k}{k!}\right) + \cdots$$

The coefficient of $\dfrac{t^k}{k!}$ are seen to be all equal to p (for k = 1, 2, 3, …),

giving $E(X^k) = p$.

Thus $E(X) = p$, $E(X^2)$, = p, $E(X^3) = p$. etc.

Method 2. (Taking Derivatives):

$M_x(t) = pe^t + (1 - p)$

$$\frac{d}{dt^1} M_x(t) = M'(t) = pe^t$$

$$\frac{d}{dt^2} M_x(t) = M''(t) = pe^t$$

Continuing in this way,

$$\frac{d}{dt^k} M_x(t) = M^k(t) = pe^t$$

At t = 0, it can be seen that

$$M'(0) = p = E(X), \ M''(0) = E(X^2) = p,$$
$$M^k(0) = E(X^k) = p.$$

Method 3: (Direct Computation).

The result can be obtained directly with very little effort, using the probability distribution of X given below:

X:	0	1
P(xi)	1 − p	p

$E(X) = 0. (1 - p) + 1. p = p$
$E(X^2) = 0^2. (1 - p) + 1^2 . p = p$
Hence $E(X) = E(X^2) = \ldots = E(X^k) = p$.

EXAMPLE 7.2

Consider the integral, $\int_0^\infty x^k e^{-x}dx, \quad for \; x > 0$

Obtain (i) the probability density function of X and hence the moment generating function $M_x(t)$ of X.
(ii) Derive the r –th moment
(a)   by using a power series expansion (if it is possible), and
(b)   By taking derivatives of $M_x(t)$.

SOLUTION:

(i) Successive integrations by parts show that for any nonnegative integer k,
$$\int_0^\infty x^k e^{-x} \, dx = k! \, , \, x>0.$$
This implies that
$$\frac{1}{k!} \int_0^\infty x^k e^{-x} \, dx = 1 \, , \text{ giving the pdf of X as,}$$

$$f(x) = \begin{cases} \dfrac{x^k e^{-x}}{k!} & for \; x > 0. \\ 0, & elsewhere \end{cases}$$

The moment generating function,
$$M_x(t) = E(e^{tx}) = \int_0^\infty e^{tx} f(x)dx = \int_0^\infty e^{tx} \frac{x^k e^{-x}}{k!} dx$$
$$= \frac{1}{k!} \int_0^\infty x^k e^{-(1-t)x} dx$$

Put u = (1 − t)x, gives

$$M_x(t) = \frac{1}{k!} \int_0^\infty \frac{u^k}{(1-t)^k} \bullet e^{-(U)} \bullet \frac{1}{(1-t)} du$$

$$= \frac{1}{K!} \int_0^\infty \frac{u^k e^{-u}}{(1-t)^{k+1}} \, du = \frac{1}{(1-t)^{k+1}}$$

$$\textit{for } t < 1.$$

Moments: Method 1- (Using Power Series).

$M_x(t) = \dfrac{1}{(1-t)^{k+1}}$ can be expanded as a Binomial expansion i.e.

$$M_x(t) = (1-t)^{-(k+1)} = 1 + (-k-1)(-t) + \frac{(-k-1)(-k-2)(-t)^2}{2!} +$$

$$+ \frac{(-k-1)(-k-2)(-k-3)}{3!}(-t)^3 + \cdots$$

The moment $E(X^k)$ is the coefficient of $\dfrac{t^K}{k!}$ in the above

expansion.
Hence $E(X) = (k+1)$, $E(X^2) = (k+1)(k+2)$.

Method 2 – (Taking Derivatives).

$$M_x(t) = \frac{1}{(1-t)^{k+1}}$$

$$M_x'(t) = \frac{d}{dt}\left[\frac{1}{(1-t)^{k+1}}\right] = (k+1)(1-t)^{-k-2}$$

$$M''(t) = \frac{d^2}{dt^2}\left[\frac{1}{(1-t)^{k+1}}\right] = -(k+1)(-k-2)(1-t)^{-k-3} \quad \text{and so}$$

on.

At t = 0
$$M_x'(0) = E(X) = k+1$$
$$M_x''(0) = E(X^2) = (k+1)(k+2)$$

7.1.2.2   Sum of Independent Random Variables.

If X and Y are independent random variables, the moment generating function of the
Sum (X + Y) is a particular simple combination of the Moment Generating Functions of the summands, namely their product.

Proof:
Let $M_x(t)$ and $M_y(t)$ be the moment generating functions of X and Y respectively. Then, the mgf of their sum Z = X + Y.

$$M_Z(t) = M_{X+Y}(t) = E\left(e^{t(X+Y)}\right)$$

$$= E(e^{tx}.e^{ty})$$

$$= E(e^{tx}).E(e^{ty})$$

$$= M_X(t).M_Y(t)$$

since X and Y are independent.

Thus the moment generating function of the sum of two independent random variables is the product of their moment generating functions. Finite induction extends this result to the sum of any finite number of independent random variables:

If $X_1, X_2, \dots , X_n$ are independent, then

$$M_{(X_1 + X_2 + \cdots X_n)}(t) = M_{\Sigma X_i}(t) = E\left(e^{t(X_1 + X_2 + \cdots + X_n)}\right)$$

since the $X_i^{'s} \left(i = 1, 2, \cdots n\right)$ are independent.

$$= M_{X_1}(t) \cdot M_{X_2}(t) \cdots M_{X_n}(t)$$

i.e $M_{\Sigma X_i}(t) = \prod_{i=1}^{n} M_{X_i}(t).$

If, however, the $X_i^{'s}\,(i=1,2,\cdots n)$ have the same distribution, say, with common moment generating function.

M_X (t), then

$$M_{\sum X_i}(t) = \prod_{i=1}^{n} M_x(t) = \left[M_x(t)\right]^n$$

EXAMPLE 7.3

Suppose that the random variable X has mgf $M_x(t)$. Let Y = aX + b where a and b are constants. Then,

$$M_Y(t) = M_{aX+b}(t) = E\left(e^{t(aX+b)}\right) = E\left(e^{t(aX)} \bullet e^{tb}\right)$$
$$= E\left(e^{at(x)}\right) \bullet E\left(e^{tb}\right)$$
$$= e^{tb}\, E\left(e^{at(x)}\right)$$
$$= e^{tb} \bullet M_X(at)$$

In words, to find the mgf of Y = a + b X, where a and b are constants, evaluate the mgf of X at '(at)' [instead of t] and multiply by e^{bt}.

EXAMPLE 7.4

Consider again EXAMPLE 7.1 above, where X_1, X_2, ... X_n are independent random variables, each with the distribution
P(X = 1) = p, and P(X = 0) = q, where q = 1 – p. The moment generating function of the common distribution is $M_X(t)$ = $E(e^{tx})$ = pe^t + q, and so the moment generating function of the sum is

$$M_{\Sigma X}(t) = \left[M_X(t)\right]^n = \left(pe^t + q\right)^n, \ where \ q=1-p.$$

EXAMPLE 7.5

Let X be a random variable with cumulative distribution function (cdf), $F_X(x)$, given by,

$$F_X(x) = \begin{cases} 0, \ for \ x < -2 \\ \dfrac{1}{2}, \ for \ -2 \le x < 2 \\ 1, \ for \ 2 \le x \end{cases}$$

Also let X_1, X_2, X_3, X_4 be independent observations of X and each having the same distribution as X. Define the random variable

$Y = 2 + (X_1 - X_2 + X_3 - X_4)/4.$
 Obtain:
(a)   the moment generating function of Y.
(b)   the distribution of Y.
(c) $E(Y^3)$.

SOLUTION:

(a) From the cdf of X, we now obtain the probability distribution of X as follows:

X	-2	2
$P(X = x)$	$\dfrac{1}{2}$	$\dfrac{1}{2}$

Thus

$$M_X(t) = E(e^{tx}) = \frac{1}{2} \bullet e^{-2t} + \frac{1}{2} \bullet e^{2t} = \frac{1}{2}\left(e^{2t} + e^{-2t}\right)$$

Hence, the mgf of Y is given by

$$M_Y^{(t)} = E(e^{ty}) = E\{e^{t[2+(X1-X2+X3-X4)/4]}\}$$
$$= E\left(e^{2t+t(X1-X2+X3-X4)/4}\right)$$

$= [$ for X_1, X_2, X_3, X_4 independent]

$$= E(e^{2t}).E\left(e^{\frac{t}{4}X_1}\right).E\left(e^{\frac{-t}{4}X_2}\right).E\left(e^{\frac{t}{4}X_3}\right).E\left(e^{\frac{-t}{4}X_4}\right)$$

$$= e^{2t} M_X\left(\frac{t}{4}\right) M_X\left(\frac{-t}{4}\right) M_X\left(\frac{t}{4}\right) M_X\left(\frac{-t}{4}\right)$$

$$= e^{2t} \cdot \frac{1}{2}\left(e^{\frac{t}{2}} + e^{\frac{-t}{2}}\right) \frac{1}{2}\left(e^{\frac{t}{2}} + e^{\frac{-t}{2}}\right) \frac{1}{2}\left(e^{\frac{t}{2}} + e^{\frac{-t}{2}}\right) \frac{1}{2}\left(e^{\frac{t}{2}} + e^{\frac{-t}{2}}\right)$$

$$M_Y(t) = \frac{1}{16} e^{2t}\left(e^{\frac{t}{2}} + e^{\frac{-t}{2}}\right)^4 = \left(\frac{1}{2}e^t + \frac{1}{2}\right)^4$$

(b) $\left(\dfrac{1}{2}e^t + \dfrac{1}{2}\right)$ is the moment generating function for a random

variable X which has a Bernoulli distribution:

X	0	1
P(X = x)	$\dfrac{1}{2}$	$\dfrac{1}{2}$

Therefore, $M_Y(t) = \left(\dfrac{1}{2}e^t + \dfrac{1}{2}\right)^4$ must be the moment generating

function for a random variable $Y = \sum_{i=1}^{\infty} X_i$ where the $X_i^{'s}$ are

independent. Hence, Y has a Binomial distribution given by,

$$Y \sim \text{Bin (n, P)} = \text{Bin}\left(4, \frac{1}{2}\right)$$

(c) $E(y^3) = \sum_{i=1}^{4} i^3 \binom{4}{i}\left(\frac{1}{2}\right)^4$

$$= \frac{1}{16}(0^3 \times 0 + 1^3 \times 4 + 2^3 \times 6 + 3^3 \times 4 + 4^3 \times 1)$$

$$= \frac{1}{16}(4 + 48 + 108 + 64) = \frac{1}{16} \bullet 224 = 14$$

EXAMPLE 7.6

Let X_1 and X_2 be two independent observations of the random variable
X with moment generating function given by

$$M_X(t) = \frac{1 + 2e^{4t}}{3e^t}$$

Obtain,

(a)   The distribution of X.

(b)   Derive the moment generating function of $(X_1 - X_2)$

(c)   Compute $E[(X_1 - X_2)^n]$, where n is a positive integer.

SOLUTION:

(a) $M_X(t) = \dfrac{1 + 2e^{4t}}{3e^t} = \dfrac{1}{3}e^{-t} + \dfrac{2}{3}e^{3t}$, this means that X has

distribution given below:

X	− 1	3
P(X = x)	$\dfrac{1}{3}$	$\dfrac{2}{3}$

(b) Since X_1 and X_2 are independent variables, we have the mgf,

$$M_{[X_1 - X_2]}(t) = M_{X_1}(t).\, M_{-X_2}(t) = E(e^{tX_1})E(e^{-tX_2})$$

$$= \left(\frac{1}{3}e^{-t} + \frac{2}{3}e^{3t} \right)\left(\frac{1}{3}e^{t} + \frac{2}{3}e^{-3t} \right)$$

$$= \frac{1}{9}e^{0} + \frac{2}{9}e^{-4t} + \frac{2}{9}e^{4t} + \frac{4}{9}e^{0}$$

$$M_{[X_1 - X_2]}(t) = \frac{2}{9}e^{-4t} + \frac{5}{9} + \frac{2}{9}e^{4t}$$

(c) $M_X(t) = \dfrac{2}{9}e^{-4t} + \dfrac{5}{9} + \dfrac{2}{9}e^{4t}$

$M'(t) = (-4)^1 \cdot \dfrac{2}{9}e^{-4t} + 4^1 \cdot \dfrac{2}{9}e^{4t}, \Rightarrow M'_X(0) = \dfrac{2}{9}\left[(-4)^1 + 4^1 \right]$

$$M_X''(t) = (-4)^2 \cdot \frac{2}{9} \cdot e^{-4t} + 4^2 \cdot \frac{2}{9} e^{4t}, \Rightarrow M_X''(0) = \frac{2}{9}\left[(-4)^2 + 4^2\right]$$

$$M_X^n(t) = (-4)^n \cdot \frac{2}{9} e^{-4t} + 4^n \cdot \frac{2}{9} e^{4t}, \Rightarrow M_X^n(0) = \frac{2}{9}\left[(-4)^n + 4^n\right]$$

Thus,

$$E(X_1 - X_n)^n = \begin{cases} 0, \text{ if } n \text{ is odd} \\ 4^{n+1}\Big/9, \text{ if } n \text{ is even} \end{cases}$$

7.1.3   Theorem: If $M_X(t)$ is the moment generating function of a r.v. X

whose moments exist,
$$Var(X) = M_X''(0) - (M_X'(0))^2$$

Proof:
$$M_X(t) = E(e^{tx}), \qquad M_X'(t) = E(Xe^{tX}), \qquad M_X'(0) = E(X)$$
$$M_X''(t) = E(X^2 e^{tX}), \qquad M_X''(0) = E(X^2)$$

Therefore,

$$Var(X) = M_X''(0) - (M_X'(0))^2$$

7.1.4. Moment generating of some familiar distributions

7.1.4.1   Moment generating function of a Binomial distribution $B(n, p)$
$$P(X = k) = \binom{n}{k} p^k (1 - p)^{n-k}, k = 0, 1, \dots n$$

$$M_X(t) = E(e^{tX}) = \sum_{k=0}^{n} \binom{n}{k} p^k (1 - p)^{n-k} e^{tk}$$

$$= \sum_{k=0}^{n} \binom{n}{k} (pet)^k (1 - p)^{n-k}$$
$$= (pe^t + 1 - p)^n$$

By the binomial theorem.
$$M_X'(t) = npe^t(1 - p + pe^t)^{n-1}$$

Therefore $E(X) = M_X'(0) = npe^0(1 - p + pe^0)^{n-1} = np$
You may show that $Var(X) = npq$

7.1.4.2.   Moment generating function of a Poison r.v.X with pmf $P(\lambda)$

$$P(X = k) = \frac{e^{-\lambda}\lambda^k}{k!}, k = 0,1,2,\ldots$$

$$M_X(t) = E(e^{tX}) = \sum_{k=0}^{\infty} e^{tk}\frac{e^{-\lambda}\lambda^k}{k!} = e^{-\lambda}\sum_{k=1}^{\infty}\frac{(\lambda e^t)^k}{k!}$$

Noting that $e^x = \sum_{k=0}^{\infty}\frac{x^k}{k!}$ And $x = \lambda e^t$, we have

$$M(t) = e^{-\lambda}e^{\lambda e^t} = e^{\lambda(e^t-1)}$$

Now $M_X'(t) = \lambda e^t e^{\lambda(e^t-1)}, E(X) = M_X'(0) = \lambda e^0 e^{\lambda(e^0-1)} = \lambda$

You may want to show that $Var(X) = \lambda$

Alternatively

$$ln[M_X(t)] = \lambda(e^t - 1)$$

Therefore

$$\frac{M_X'(t)}{M_X(t)} = \lambda e^t \text{ } --------------- (1) \text{ Therefore}$$

$$\frac{M_X'(0)}{M_X(0)} = \lambda^0 = \lambda \Rightarrow M_X'(0) = E(X) = \lambda$$

Since $M_X(0) = 1$

Next,

$$\frac{M_X(t)M_X''(t) - \left(M_X'(t)\right)^2}{\left(M_X(t)\right)^2}$$

By the quotient rule from (1)
Therefore

$$\frac{M_X(0)M_X''(0) - \left(M_X'(0)\right)^2}{\left(M_X(0)\right)^2} = M_X''(0) - \left(M_X'(0)\right)^2 = var(X) = \lambda$$

We have in the process also proved that if a r.v. with mean μ and variance σ^2 has m.g.f. $M_X(t)$, if we put $\Psi_X(t) = \ln[M_X(t)]$, the result is

$$\Psi_X'(0) = \mu$$
$$\Psi_X''(0) = \sigma^2$$

7.1.4.3. Moment generating function of a Geometric r.v.X with pmf $P(X = k) = p(1-p)^{k-1}, k = 1,2,...$

$$M_X(t) = E(e^{tX}) = \sum_{k=1}^{\infty} p(1-p)^{k-1} e^{tk}$$

$$= \sum_{k=1}^{\infty} \left(\frac{p}{1-p}\right) [(1-p)e^t]^k = \left(\frac{p}{1-p}\right) \sum [(1-p)e]^k$$

$$= \left(\frac{p}{1-p}\right) \left[\frac{(1-p)e^t}{1-(1-p)e^t}\right] = \frac{pe^t}{1-qe^t}$$

7.1.4.4. Moment generating function of a Uniform distribution over (α, β)

Suppose X is a continuous r.v. with a uniform distribution over (α, β) then

$$M_X(t) = E(e^{tx}) = \int_{\alpha}^{\beta} \frac{1}{(\beta - \alpha)} e^{tx} dx$$

$$= \frac{1}{(\beta - \alpha)} \int_{\alpha}^{\beta} e^{tx} dx = \frac{1}{(\beta - \alpha)} \left\{\frac{1}{t} [e^{tx}]_{\alpha}^{\beta}\right\} = \frac{1}{(\beta - \alpha)} \frac{1}{t} \left\{e^{t\beta} - e^{t\alpha}\right\}$$

What is $M_X(0)$? What is $M_X'(0)$?
Note: $M_X(0) = \lim_{n \to \infty} M_X(t)$
Now

$$M_X(t) = \frac{1}{t(\beta-\alpha)} \left\{\left(1 + t\beta + \frac{(t\beta)^2}{2!} + \cdots \frac{(t\beta)^n}{n!}\right) - \left(1 + t\alpha + \frac{(t\alpha)^2}{2!} + \cdots \frac{(t\alpha)^n}{n!}\right)\right\}$$

$$= \frac{1}{t(\beta - \alpha)} \left\{t(\beta - \alpha) + \frac{t^2}{2!}(\beta^2 - \alpha^2) + \cdots + \frac{t^n}{n!}(\beta^n - \alpha^n) + \cdots\right\}$$

$$= 1 + \frac{t(\beta^2 - \alpha^2)}{2(\beta - \alpha)} + \cdots + \frac{t^{n-1}(\beta^n - \alpha^n)}{n!(\beta - \alpha)}$$

Therefore

$$M_X(0) = 1$$
$$M_X'(0) = \frac{(\beta^2 - \alpha^2)}{2(\beta - \alpha)} = \frac{\alpha + \beta}{2}$$

Which agrees with known results.

7.1.4.5. Moment generating function of an exponential distribution with parameter λ

$$M_X(t) = \int_0^\infty e^{tx} \lambda e^{-\lambda x} dx$$

$$= \lambda \int_0^\infty e^{-x(\lambda-t)} dx$$

$$= \frac{\lambda}{\lambda - t} \left[-e^{-x(\lambda-t)} \right]_0^\infty$$

$$= \frac{\lambda}{\lambda - t}$$

Clearly, $M_X(0) = 0, M_X'(t) = \lambda(\lambda - t)^{-2} \Rightarrow M_X'(0) = \frac{1}{\lambda} = E(X)$

$$M_X''(t) = 2\lambda(\lambda - t)^{-3} \Rightarrow M_X''(0) = \frac{2}{\lambda^2} = E(X^2) \Rightarrow$$

$$Var(X) = M_X''(0) - \left(M_X'(0) \right)^2 = \frac{2}{\lambda^2} - \frac{1}{\lambda^2} = \frac{1}{\lambda^2},$$

Therefore

$$Var(X) = \frac{2}{\lambda^2}$$

7.1.4.6. Moment generating function of a standard normal distribution

$$f(x) = \frac{e^{-\frac{x^2}{2}}}{\sqrt{2\pi}}, -\infty < x < \infty$$

$$E(e^{tx}) = \int_{-\infty}^\infty \frac{1}{\sqrt{2\pi}} e^{tx} e^{-\frac{x^2}{2}} dx$$

$$= \int_{-\infty}^\infty \frac{1}{\sqrt{2\pi}} e^{\frac{-1}{2}(x^2-2tx)} dx$$

$$= e^{\frac{t^2}{2}} \int_{-\infty}^\infty \frac{1}{\sqrt{2\pi}} e^{\frac{-1}{2}(x-t)^2} dx$$

on completing the square

$$M_X(t) = e^{\frac{t^2}{2}}$$

Since the integrand of the integral part is the pdf of a normal N(t,1)
Clearly, $M_X(0) = e^0 = 1$

$$M_X(0) = 0, M'_X(t) = te^{\frac{t^2}{2}} \Rightarrow M'_X(0) = 0 = E(X)$$
$$M''_X(t) = e^{\frac{t^2}{2}} + t^2 e^{\frac{t^2}{2}} \Rightarrow M''_X(0) = 1 = E(X^2) \Rightarrow$$
$$Var(X) = M''_X(0) - \left(M'_X(0)\right)^2 = 1 - 0^2 = 1$$

7.1.4.5. Note: If X is standard normal r.v, the r.v. $Y = X^2$ is called a chi-square r.v. and it has a chi-square distribution with 1 degree of freedom i.e. χ_1^2

If X is standard normal r.v, and $Y = X^2$, find the mgf of $Y = X^2$

$$M_X(t) = E\left(e^{tx^2}\right)$$

$$= \int_{-\infty}^{\infty} \frac{1}{\sqrt{2\pi}} e^{tx^2} e^{-\frac{x^2}{2}} dx = \int_{-\infty}^{\infty} \frac{1}{\sqrt{2\pi}} e^{\frac{-1}{2}\left[(1-2t)^{\frac{1}{2}}x\right]^2} dx$$

Let $z = \left(\sqrt{1-2t}\right)x, dz = \left(\sqrt{1-2t}\right)dx \Rightarrow dx = \frac{dz}{(1-2t)^{\frac{1}{2}}}$

Hence, $M_X(t) = \int_{-\infty}^{\infty} \frac{1}{\sqrt{2\pi}} e^{-\frac{z^2}{2}} \frac{dz}{(1-2t)^{\frac{1}{2}}}$

$$= \frac{1}{(1-2t)^{\frac{1}{2}}} \int_{-\infty}^{\infty} \frac{1}{\sqrt{2\pi}} e^{-\frac{z^2}{2}} dz$$

$$= \frac{1}{(1-2t)^{\frac{1}{2}}}$$

Since $\int_{-\infty}^{\infty} \frac{1}{\sqrt{2\pi}} e^{-\frac{z^2}{2}} dz$ is a legitimate pdf i.e $N(0,1)$

7.1.4.6. Note: Let X be a continuous r.v. assuming only non-negative values. We say that X has a Gamma distribution with parameters say $\alpha > 0$ and $r > 0$ if its pdf is given by

$$f(x) = \begin{cases} \dfrac{\alpha}{\Gamma(r)} (\alpha x)^{r-1} e^{-\alpha x}, & x > 0 \\ \\ 0, & otherwise \end{cases}$$

Where Γ is the gamma function defined by

$$\Gamma(p) = \int_0^{\infty} x^{p-1} e^{-x} dx, \qquad p > 0$$

Let X have a Gamma distribution with parameter α and r. Then the m.g.f of X is

$$M_X(t) = \frac{\alpha}{\Gamma(r)} \int_0^\infty e^{tx} (\alpha x)^{r-1} e^{-\alpha x} dx$$

$$= \frac{\alpha^r}{\Gamma(r)} \int_0^\infty x^{r-1} e^{-x(\alpha-t)} dx$$

Now let $u = x(\alpha - t)$, $dx = \frac{1}{(\alpha-t)} du$ and

$$M_X(t) = \frac{\alpha^r}{(\alpha - t)\Gamma(r)} \int_0^\infty \left(\frac{u}{\alpha-t}\right)^{r-1} e^{-u} du$$

$$= \left(\frac{\alpha}{(\alpha-t)}\right)^r \frac{1}{\Gamma(r)} \int_0^\infty (u)^{r-1} e^{-u} du$$

$$= \left(\frac{\alpha}{(\alpha-t)}\right)^r \frac{1}{\Gamma(r)} \Gamma(r)$$

$$= \left(\frac{\alpha}{(\alpha-t)}\right)^r , \alpha > t$$

Note: If r=1, the Gamma distribution becomes the exponential distribution with parameter α and m.g.f is $\left(\frac{\alpha}{(\alpha-t)}\right)$

7.1.4.7. Note: if $\alpha = \frac{1}{2}$ and $r = \frac{n}{2}$ (n a positive integer), the Gamma distribution becomes the special distribution called the chis-square distribution with n degrees of freedom denoted by χ_n^2 and its m.g.f. is

$$M_X(t) = (1 - 2t)^{\frac{-n}{2}} \text{ or } \left(\frac{\frac{1}{2}}{\frac{1}{2}-t}\right)^{\frac{-n}{2}}$$

Generally, if $X_i^2 (i = 1,2,3, \dots n)$ are n iid random variable where each $X_i \sim N(0,1)$, then $\sum_{i=1}^n X_i^2$ is said to have a χ_n^2 distribution.

7.1.5. Properties of MGF's $M_X(t)$
1. $M_X(0) = 1$
Proof:

$$M_X(t) = E(e^{tx}) \Rightarrow$$
$$M_X(0) = E(e^{t0}) = E(1) = 1$$

2. $M_X^n(0) = \frac{d^n}{dt^n} M_X(t)||_{t=0} = E(X^n)$ (Already proved)

3. If X has mgf $M_X(t)$ and $Y = \alpha X + \beta$ (α is a real non – zero number) then
$M_Y(t) = e^{\beta t} M_X(\alpha t)$

Proof:
$$M_Y(t) = E(e^{tY})$$
$$= E\left(e^{t(\alpha X + \beta)}\right)$$
$$= E\left(e^{\beta t}.e^{\alpha tx}\right) = e^{\beta t} E(e^{\alpha tx}) = e^{\beta t} M_X(t\alpha)$$

If X has the normal distribution $N(\mu, \sigma^2)$, we may want to find the mgf of X

Let $Y = \frac{X - \mu}{\sigma}$ then $Y \sim N(0,1)$ and $M_Y(t) = e^{\frac{1}{2}t^2}$. Now $X = \mu + \sigma Y$ then

$$M_X(t) = e^{\mu t} M_Y(\sigma t) = e^{\mu t} e^{\frac{1}{2}\sigma^2 t^2} = e^{\left\{\mu t + \frac{1}{2}\sigma^2 t^2\right\}}$$

4. If X and Y are independent, then if Z=X+Y, the mgf of Z is
$M_Z(t) = M_X(t). M_Y(t)$ where
$M_X(t)$ and $M_Y(t)$ are the m.g.f. of X and Y respectively.
5. The Uniqueness property
Every random variable has its own unique m.g.f. that is if X and Y are two r.v's such that $M_X(t) = M_Y(t)$ for all t then X and Y must have the same distribution and conversely.

Some consequence of the Uniqueness Property
1.   If X has the normal distribution $N(\mu, \sigma^2)$ then the r.v. $Y = aX + b$ has the normal distribution $N(a\mu + b, a^2\sigma^2)$
Proof: (by the Uniqueness Property of m.g.f)
$$M_X(t) = e^{\left\{\mu t + \frac{1}{2}\sigma^2 t^2\right\}} \implies$$
$$M_Y(t) = e^{bt} M_X(at) = e^{bt} e^{\left\{a\mu t + \frac{1}{2}t^2(a^2\sigma^2)\right\}}$$
$$= e^{\left\{(a\mu + b) + \frac{1}{2}(a^2\sigma^2 t^2)\right\}}$$
Which the m.g.f. of a r.v which is $N(a\mu + b, a^2\sigma^2)$. Therefore by the uniqueness property $Y \sim N(a\mu + b, a^2\sigma^2)$.

6. The Reproductive properties
    (a) of the normal distribution
   Let $X_1, X_2, \dots, X_n$ be n independent r.v.s such that $X_i \sim N(\mu_i, \sigma_i^2), i = 1, 2, \dots n$.
   Let $Z_1 = \sum_{i=1}^n X_i$ and $Z_2 = \sum_{i=1}^n a_i X_i$ where a_i are non-zero real numbers, then

Then

(i) Z_1 has the Normal distribution $N(\sum_{i=1}^{n}\mu_i, \sum_{i=1}^{n}\sigma_i^2)$

(ii) Z_2 has the Normal distribution $N(\sum_{i=1}^{n}a_i\mu_i, \sum_{i=1}^{n}a_i^2\sigma_i^2)$

Proof:

Let $M_{X_i}(t)$ be the mgf of $X_i, i = 1,2,3, \ldots n$

Also $M_{Z_i}(t)$ be the mgf of $Z_i, i = (1,2)$

Then

$$M_{Z_i}(t) = \prod_{i=1}^{n} M_{X_i}(t)$$

$$= \prod_{i=1}^{n} e^{\{\mu_i t + \frac{1}{2}\sigma_i^2 t^2\}}$$

$$= e^{\left\{\left(\sum_{i=1}^{n}\mu_i\right) + \frac{1}{2}\left(\sum_{i=1}^{n}\sigma_i^2\right)\right\}}$$

Which is the mgf of $N(\sum_{i=1}^{n}\mu_i, \sum_{i=1}^{n}\sigma_i^2)$

By the uniqueness property of mgf's Z_1 has the normal distribution $N(\sum_{i=1}^{n}\mu_i, \sum_{i=1}^{n}\sigma_i^2)$

Establish the distribution of Z_2 too similarly.

7.2. The Factorial Moment Generating Function

The factorial moment generating function $F_X^{(t)}$ is closely related to the moment generating function, and as the name implies it generates factorial moments. The function is defined as a logarithmic function of the moment generating function in the sense that:

$$F_X(t) = E(t^X) = E\left[e^{X(\log t)}\right] = M_X(\log\;)$$

Both functions can briefly be summarized in table 7.1 below.

Moment Generating Function	Factorial Moment Generating Function
$M_X(t) = E(e^{tx})$	$F_X(t) = E(t^x)$

Table 7.1

Note the absence of the exponential function, in the definition of the factorial generating function: Thus $F_X(t) = M_X(\log t)$.

7.2.1   Properties of Factorial Moment Generating Function

7.2.1.1   Moments (Expected Value and Variance)

In section 7.1.2.1 above, we observe that the $k - th$ moment of the moment generating function is obtained by taking the $k - th$ derivative of $M_X(t)$ and evaluating the result at
$t = 0$, i.e.

$$M_X^K(0) = \frac{d^k}{dt^k} M_X(t) = E(X^k).$$

The substitution for the value of t is, however, different for the factorial moment generating function defined as $F_X(t) = M_X(\log t) = E(t)$.

To obtain the moments, since $\log 1 = 0$, it is the point $t = 1$ that is of interest, and that produces the factorial moments from the derivatives. Thus,

$$F_X'(t) = \frac{d}{dt} F_X(t) = E\left(X t^{X-1}\right)$$

$$F_X''(t) = \frac{d^2}{dt^2} F_X(t) = E\left[X(X-1)t^{X-2}\right],$$

$$F_X'''(t) = \frac{d^3}{dt^3} F_X(t) = E\left[X(X-1)(X-2)t^{X-3}\right] \quad \text{and so on, and}$$

hence at t=1
$$F_X'(1) = E(X),\ F_X''(1) = E\left[X(X-1)\right],$$
$$F_X'''(1) = E\left[X(X-1)(X-2)\right],$$

and so on. In general then (assuming that the above moments exist, and that the steps of differentiation and averaging can be interchanged), we obtain the $k - th$ derivative as

$$F_X^k(t) = \frac{d^k}{dt^k} F_X(t) = E\left[X(X-1)(X-2)\cdots(X-k+1)t^{X-K+2}\right]$$

which yields the $k -$ moment (at $t = 1$) as

$$F_X^k(1) = E[X(X-1)(X-2)\cdots\cdots(X-k+1)] = E(X^k)$$

This is what is called the k – th factorial moment.
Like the moment generating function, the first factorial moment $F_X'(1)$, is the same as the first moment E(X). The second factorial moment,

$F_X''(1) = E[X(X-1)] = E(X^2) - E(X)$, is however, a combination of the first two moments: so that the second moment E(X²) is obtained by adding E(X) to $F_X''(1)$.

i.e.

$$F_X''(1) + E(X) = E[X(X-1)] + E(X) = E(X^2) - E(X) + E(X) = E(X^2).$$

Hence the variance of X, is obtained as

$$V(X) = F_X''(1) + E(X) - [E(X)]^2.$$

7.2.1.2   Sum of Independent Random Variables

Like the moment generating function, the factorial moment generating function has the property that if $X_1, X_2, \ldots X_n$ are independent random variables, each with factorial moment generating function $F_{Xi}(t)$, i = 1, 2, ...n, the factorial moment generating function of the sum of these independent random variables is just the product of the factorial moment generating function of the summands:

$$F_{X_1+X_2}(t) = E\left[t^{X_1+X_2}\right] = E\left[t^{X_1} \cdot t^{X_2}\right] = E\left(t^{X_1}\right) E\left(t^{X_2}\right) = F_{X_1}(t) \cdot F_{X_2}(t)$$

In general,

$$F_{\sum X_i}(t) = E\left(t^{\sum X_i}\right) = E\left(\prod_{i=1}^{n} t^{X_i}\right) = \prod_{i=1}^{n} E\left(t^{X_i}\right) = \prod_{i=1}^{n} F_{X_i}(t)$$

7.2.1.3 NOTE: Sometimes it is difficult to judge whether to use the moment generating function or the factorial moment generating function, or to calculate desired moments directly. The clue to this is the nature of the probability density function: Sometimes it combines easier with one function than the other, and therefore makes the computation(s) easier. This is left to the best judgement of the reader.

EXAMPLE 7.7

Let $X_1, \ldots, X_n$ be independent random variables, each with the Bernoulli distribution $P(X = 1) = p$, $P(X = 0) = q = 1 - p$. Obtain the mean and variance of the sum

$$Z = \sum_{i=1}^{n} X_i.$$

SOLUTION:

Using the Factorial Moment Generating function:
$$F_X(t) = E(t^X) = \sum t^X \cdot P(X=x) = t^1 \cdot p + t^0 \cdot (1-p) = pt + (1-p) = pt + q,$$
where $p + q = 1$

$$F_Z(t) = F_{\sum X_i}(t) = E(t^{\sum X_i}) = \prod_{i=1}^{n} E(t^{X_i}) = \prod_{i=1}^{n} F_{X_i}(t) =$$

[since the X's have identical distribution]$= (pt + q)^n$

$$F'_{\sum X_i}(t) = \frac{d}{dt}(pt+q)^n = np(pt+q)^{n-1}$$

$$F''_{\sum X_i}(t) = \frac{d^2}{dt^2}(pt+q)^n = n(n-1)p^2(pt+q)^{n-2}$$

The factorial moments are obtained by evaluating the above derivatives at
t = 1. Thus

$$F'_{\sum X}(1) = np = E(Z)$$
$$F''_{\sum X}(1) = n(n-1)p^2 = E[Z(Z-1)] = E(Z^2) - E(Z).$$
Hence $E(Z^2) = F''_{\sum X_i}(1) + E(Z) = n(n-1)p^2 + np,$ and

$$Var(Z) = E(Z^2) - [E(Z)]^2$$
$$= n(n-1)p^2 + np - [np]^2$$
$$= npq = np(1-p)$$

EXAMPLE 7.8

Let $W \sim$ Bin (n, p) denote that W is a Binomially distributed random variable having mean equal to np and variance equal to $np(1-p)$. Suppose that W and Y are independent random variables such that $W \sim$ Bin (n, p) and
$Y \sim$ Bin (m, p). Show that their sum $(W + Y) \sim$ Bin $(m + n, p)$.

SOLUTION:

For the distribution of the sum $Z = X_1 + X_2 + \ldots + X_n = \sum X_i$ in Example 7.4 above, the Factorial moment generating function $F_{\sum X_i}$,

was found to be

$(pt + q)^n$. Thus $X \sim Bin\,(n, p) \Rightarrow F_X\left(t^{\sum X_i}\right) = E\left(t^{\sum X_i}\right)$

$$= E\left(\prod_{i=1}^{n} t^{X_i}\right) = \begin{bmatrix} X^{\prime s}\ are \\ independent \end{bmatrix} = \prod_{i=1}^{n} E\left(t^{X_i}\right) = \prod_{i=1}^{n} F_{X_i}(t) = (pt + q)^n,$$

Here it is assumed that each of the X's has a Bernoulli distribution with $P(X = 1) = p$ and $P(X = 0) = 1 - p = q$, where $p + q = 1$; and
$X_1 + X_2 + \ldots + X_n = \sum X_i = Z$.

Thus

$$F_{X+Y}(t) = E\left(t^{X+Y}\right) = F_X(t) \cdot F_Y(t) = \left[Since\ X\ and\ Y\ are\ independent\right]$$
$$= (pt + q)^n \cdot (pt + q)^m$$
$$= (pt + q)^{m+n},$$

and this is the factorial moment generating function for a variable that has a Binomial distribution: Bin (m + n , p).
Thus, (X + Y) ~ Bin (m + n , p).

EXAMPLE 7.9

Let X be a random variable having a Poisson distribution with mean equal to λ,
i.e. X ~ P_o (λ). Suppose that X and Y are independent random variables each having a Poisson distribution, so that X ~ P_o (λ) and Y ~ P_o (λ).

(a) Show that their sum (X + Y) ~ P_o (2λ)
(b) Obtain the conditional distribution of X = x given that X + Y = s.
i.e P(X = x / X + Y = s).

SOLUTION:

(a) using the factorial moment generating function, we have:

$$F_X(t) = E\left(t^X\right) = e^{-\lambda} \sum_{X=0}^{\infty} t^X \frac{\lambda^x}{x!} = e^{-\lambda} \cdot e^{\lambda t} = e^{\lambda(t-1)}.$$

$$F_{X+Y}(t) = E\left(t^{X+Y}\right) = E\left(t^X \cdot t^Y\right) = \begin{bmatrix} X\ and\ Y\ are \\ independent \end{bmatrix} = E\left(t^X\right). E\left(t^Y\right)$$

$$F_X(t) \cdot F_Y(t) = e^{\lambda(t-1)} e^{\lambda(t-1)} = e^{2\lambda(t-1)}, \text{ which is the factorial}$$

moment generating function for a random variable that has a Poisson distribution:
P_O (2λ). Thus, (X + Y) ~ P_O (2λ).

(b)
$$P(X = x / X + Y = s) = \frac{P(X = x \cap X + Y = s)}{P(X + Y = s)} = \frac{P(X = x \cap Y = s - x)}{P(X + Y = s)}$$

$$= \frac{e^{-\lambda} \dfrac{\lambda^x}{x!} \cdot e^{-\lambda} \dfrac{\lambda^{s-x}}{(s-x)!}}{e^{-2\lambda} \cdot \dfrac{(2\lambda)^S}{s!}} = \frac{s!}{x!\,(s-x)!} \cdot 2^{-s} = \binom{s}{x} 2^{-s}, \quad \text{which is a}$$

Binomial

distribution: $Bin\left(s, \dfrac{1}{2}\right)$ i.e. $P\left(X = x / X + Y = s\right) \sim Bin\left(s, \dfrac{1}{2}\right)$.

7.3. Probability Generating Function

Suppose X is a random variable assuming possible values that are nonnegative integers: 0,1,2,3......The factorial moment generating function of the distribution of X is then given as

$F_X(t) = E(t^x) = \sum_{k=0}^{\infty} t^k P(X = k)$, which is a power series in t. It is the Maclaurin expansion of $F_X(t)$ about t=0, the coefficients of t^k in the expansion being the probabilities P(X=k), k=0,1,2,3,........ Recall that in section 7.2.1.1 above, the expansion of $F_X(t)$ about t=1 yielded the factorial moments E(X), E[X(X-1)] etc. So whereas expanding $F_X(t)$ about t=1 yields the factorial moments as coefficients, expanding about t=0 yields the probabilities.

EXAMPLE 7.10

The discrete random variable X, which assumes nonnegative integer values 0,1,2.....only, has probability distribution p(x) given by:

$$p(x) = \begin{cases} \dfrac{1}{2}, & for\ x = 0 \\ \left(\dfrac{1}{3}\right)^x, & for\ x = 1,2\ ... \end{cases}$$

(a) Obtain the probability generating function for X.
(b) Compute the mean E(X) and the variance V(X).

SOLUTION:

(a) The probability generating function of X:

$$F_X(t) = E(t^X) = \sum_{k=0}^{\infty} t^k \, p(X = k)$$

$$= \frac{1}{2}t^0 + \sum_{k=1}^{\infty} t^k \left(\frac{1}{3}\right)^k = \frac{1}{2} + \sum_{i=1}^{\infty} \left(\frac{t}{3}\right)^k$$

$$= \frac{1}{2} + \frac{t/3}{1 - t/3} = \frac{1}{2} + \frac{t}{3 - t} \quad , \qquad for \; t < 3.$$

(b) $F'_X(t) = \frac{3}{(3-t)^2}$, $\quad F''_X(t) = \frac{6}{(3-t)^3}$

$F'_X(t = 1) = E(X) = \frac{3}{4}$, $\quad F''_X(1) = E[X(X - 1)] = E(X^2) - E(X)$.

$$\Rightarrow E(X^2) = F''_X(1) + E(X) = \frac{6}{8} + \frac{3}{4} = \frac{3}{2}, and \; V(X)$$

$$= E(X^2) - [E(X)]^2 = \frac{3}{2} - \frac{9}{16} = \frac{15}{16}$$

EXAMPLE 7.11

X and Y are independent random variables having Poisson distributions with parameters λ and μ respectively. Obtain the probability generating function for
X (or Y) and use it to determine the distribution of the random variable Z=X+Y.

SOLUTION:

X and Y are independent Poisson distributed random variables with parameters λ and μ respectively: Thus $P(X = \lambda) = \frac{e^{-\lambda}\lambda^k}{k!}$ $,k = 0,1,2,3, \dots \dots \dots$ Also,

$P(Y = \mu) = \frac{e^{-\mu}\mu^r}{r!}$ $,r = 0,1,2, \dots \dots$ The probability generating function of X:

$$F_X(t) = E(t^X) = \sum_{k=0}^{\infty} t^k \frac{e^{-\lambda}\lambda^k}{k!} = \sum_{k=0}^{\infty} (\lambda t)^k \frac{e^{-\lambda}}{k!} = e^{-\lambda} \sum_{k=0}^{\infty} \frac{(\lambda t)^k}{k!}$$

$$= e^{-\lambda}e^{\lambda t} = e^{\lambda(t-1)}$$

Since X and Y have identical distribution, implies $F_Y(t) = e^{\mu(t-1)}$

Hence for $Z = X+Y$, $F_Z(t) = F_{X+Y}(t) = E(t^Z) = E(t^{X+Y}) = E(t^x)E(t^y)$

since X and Y are independent $= F_X(t).F_Y(t) = e^{\lambda(t-1)}e^{\mu(t-1)}$

$$= e^{-(\lambda+\mu)}e^{(\lambda+\mu)t} = \sum_{s=0}^{\infty} e^{-(\lambda+\mu)} \frac{[(\lambda+\mu)t]^s}{s!} = \sum_{s=0}^{\infty} e^{-(\lambda+\mu)} \frac{(\lambda+\mu)^s}{s!} t^s$$
$$= \sum_{s=0}^{\infty} t^s P(Z = s)$$

So that, $P(Z = s) = e^{-(\lambda+\mu)} \frac{(\lambda+\mu)^s}{s!}$, which is the coefficient of t^s in the Maclaurin series expansion of $F_Z(t)$ about t=0.

EXAMPLE 7.12

Let X denote the number of points on a fair die. The distribution of X has the probability
generating function $\quad F_X(t) = E(t^X) = \sum_{k=0}^{\infty} t^k P(X = k)$

$$\frac{1}{6}t + \frac{1}{6}t^2 + \frac{1}{6}t^3 + \frac{1}{6}t^4 + \frac{1}{6}t^5 + \frac{1}{6}t^6$$
$$= \frac{1}{6}\{t + t^2 + t^3 + t^4 + t^5 + t^6\}$$
$$=$$

$\frac{t(1-t^6)}{6(1-t)}$ *, for* $|t| < 1$.

If this die is tossed 3 (three) times, with results X_1, X_2, X_3, and if these are independent

random variables, then the total number of points shown is a random variable Y whose

factorial moment generating function is the cube of that for the outcome of a single toss:

$$F_Y(t) = E(t^Y) = E(t^{X_1+X_2+X_3}) = [X_1, X_2, X_3, \text{ are independent}$$

$$= E(t^{X_1})E(t^{X_2})E(t^{X_3})$$

$$= \left[\frac{t(1-t^6)}{6(1-t)}\right]^3 = \frac{t^3}{216}\sum_{k=0}^{3}\binom{3}{k}(-t^6)^k \sum_{t=0}^{\infty}\binom{-3}{j}(-t)^j \text{ , the last}$$

sum being the extension of the Binomial expansion to the case of the negative integer exponents, where

$$\binom{-n}{k} = \frac{(-n)(-n-1)........(-n-k+1)}{k!!}$$. This expansion can be shown to converge to

the right value when $|t| < 1$. Writing the expansion for $E(t^Y)$ as a double sum, one has

$$E(t^Y) = \frac{1}{216}\sum_{k=0}^{3}\sum_{j=0}^{\infty}(-1)^{j+k}\binom{3}{k}\binom{-3}{j}t^{6k+j+3}$$.

From this, one can read, for instance, the probability that the total number of points is 7(seven), as the coefficient of t^7. This power only occurs for k=0 and j=4:

$$P(Y = 7) = \frac{1}{216}\binom{3}{0}\binom{-3}{4}\frac{(-3)(-4)(-5)(-6)}{216 \times 24} = \frac{15}{216}$$

7.4. The Characteristic Function

The moment generating function $M_X(t)$ of a distribution may or may not exist. There is another function closely related to the moment generating function which is often used in its place. It is called the characteristic function denoted here by C_X, and is defined by

$C_X(t) = E(e^{itx})$, where $i = \sqrt{-1}$, the imaginary unit. Because of de Moivre formula $e^{itx} = cosxt + isinxt$, the definition can also be written as

$$Cx(t) = E(CostX) + iE(SintX)$$

This will be defined for all real t because the Sine and Cosine functions are bounded in magnitude by 1(unity). For example,

$E|CostX| = \int_{-\infty}^{\infty}|costX|dF(x) \le \int_{-\infty}^{\infty}dF(x) = 1$. For theoretical reasons, there is
Considerable advantage in using $C_X(t)$ instead of $M_X(t)$. [For one thing, $C_X(t)$ always exists for all values of t].

That $M_X(t)$ and $C_X(t)$ [when $C_X(t)$ exists] are related can be demonstrated using complex variable theory. Moreover, it can be shown that differentiating under the
   "E"- the expectation symbol-is permitted k times when X has k-finite moments, so that
$E(X^k) = \frac{1}{i^k} C_X^k(0)$. If moments up to a certain order, say, r exists, it is possible to express $C_X(t)$ as a Maclaurin series with a remainder –even though the complete series expansion would not exist.
Although the characteristic function does generate moments, its principal use is as a tool in deriving distributions. For this purpose it is necessary to know several facts about characteristic functions. It is beyond the scope of this treatment to present proofs of these facts, but their statements are not hard to comprehend.

7.4.1   Theorem

If $E(|X|^k)$ exists, so does $E(|X|^j) \, for \, j = 0,1,2, \ldots \ldots k-1,$ and
$C_X(t) = E(e^{itx}) =$

$$1 + E(X)(it) + E(X^2)\frac{(it)^2}{2!} + \cdots \ldots \ldots E(X^k)\frac{(it)^k}{k!} + 0(t^k),$$

where $0(h)$ denotes a function such that $\frac{0(h)}{h} \to 0 \; as \; h \to 0.$ The above Theorem asserts that for a characteristic function, the remainder term is better behaved than it is in such expansions generally.

EXAMPLE 7.12

(a)Prove that the characteristic function of the sum of two independent random variables, with the same characteristic function $C_X(t)$, is the product of the individual characteristic functions.

SOLUTION:

The characteristic functions $C_X(t)$ and $C_Y(t)$ of the two independent random variables

X and Y are defined as: $C_X(t) = E(e^{itx}) \, and \, C_Y(t) = E(e^{ity})$ respectively.

Page | 340

Let

Z=X+Y. Then $C_Z(t) = C_{X+Y}(t) = E(e^{itz}) = E(E^{it(X+Y)}) =$
$E(e^{itX}e^{itY}) = E(e^{itX})E(e^{itY})$

$=C_X(t)C_Y(t)$, since X and Y are independent.

EXAMPLE 7.13

Consider the distribution defined by the density e^{-x} for x>0. Obtain the characteristic function of this distribution.

SOLUTION:

The characteristic function is given as:

$$C_X(t) = E(e^{itX})$$
$$= \int_0^\infty e^{itX} dF(x) = \int_0^\infty Costx(e^{-x})dx$$
$$+ i \int_0^\infty Sintx(e^{-x}) dx$$

$$= \frac{1}{1+t^2} + i\frac{t}{1+t^2} = \frac{1}{(1-it)}$$

7.4.1.1 NOTE.
It is necessary to mention(without proofs) some other facts and statements of characteristic functions, which are not hard to understand.

7.4.2 The Inversion Formula

If x-h and x+h are any two points of continuity of F(x), the increment over the interval between them is given by the formula:

$$F(x + h) - F(x - h) = \lim_{T \to \infty} \frac{1}{\pi} \int_{-T}^T \frac{Sinht}{t} e^{-itx} C_X dt$$

7.4.3 The Uniqueness Theorem
.
To each characteristic function there corresponds a unique distribution having that characteristic function.

7.4.4 NOTE

The inverse formula, given in the general case in Theorem 7.4.2 above, provides not only the uniqueness claimed in theorem 7.4.3, but also a means of obtaining the cumulative distribution function(cdf.) of a distribution when that distribution is derived in terms of its characteristic function. In the case of a continuous distribution, the density

function is

$$f(x) = \lim_{h \to 0} \frac{F(x+h) - F(x-h)}{2h} =$$
$$\lim_{h \to 0} \lim_{T \to \infty} \frac{1}{\pi} \int_{-T}^{T} \frac{Sinht}{2ht} e^{-itx} C_X(t) dt.$$

If limits can be moved in and out at will, it would follow that

$$f(x) = \lim_{T \to \infty} \frac{1}{2\pi} \int_{-T}^{T} e^{-itx} C_X(t) dt.$$

This can be established and expresses the density function f(x) as the same kind of transform of $C_X(t)$ as $C_x(t)$ is of the density (the only essential difference is in the sign of the exponent). The functions f(x) and $C_X(t)$ are said to constitute a Fourier transform pair in which the first one is the (direct) Fourier transform of the second one , and the second one is the inverse Fourier transform of the first one (which one is direct and which one is inverse is arbitrary). The formula for f(x) and $C_X(t)$ in terms of each other are analogous to the formulae for the Fourier series of a periodic function and for the coefficients in the series.

EXAMPLE 7.14

Let X be a random variable with characteristic function given as :
$$C_X(t) = e^{-|t|}.$$
Then the density function
$$f(x) = \lim_{T \to \infty} \frac{1}{2\pi} \int_{-T}^{T} e^{-|t|-itx} \, dt = \lim_{T \to \infty} \frac{1}{T} \int_{0}^{T} e^{-t} Cos txdt = \frac{1/\pi}{1+x^2}.$$

Similarly, the characteristic function of the distribution defined by this density is

$$C_X(t) = \frac{1}{\pi} \int_{-\infty}^{\infty} \frac{\cos tx}{1+x^2} dx + \frac{i}{\pi} \int_{-\infty}^{\infty} \frac{\sin tx}{1+x^2} dx = e^{-|t|}.$$

(The evaluation of these integrals is not elementary, but their values can be found in tables of definite integrals). This characteristic function is not differentiable at t=0, corresponding to the fact that not even the first moment of the distribution exists.

EXAMPLE 7.15

A random variable X has characteristic function given by
$$C_X(t) = \begin{cases} 1 - |t|, & for \ |t| \le 1 \\ 0, & for \ |t| > 1 \end{cases}$$
Find the density function of X

SOLUTION:

It is obvious that the function $C_X(t)$ is absolutely integrable over the interval
$(-\infty < t < \infty)$. The density function
$$f(x) = \frac{1}{2\pi} \int_{-\infty}^{\infty} e^{-itx} C_X(t) dt = \frac{1}{2\pi} \int_{-1}^{0} (1+t) e^{-itx} dt +$$

$\frac{1}{2\pi} \int_{0}^{1} 1 - t e^{-itx} dt$. Now:

$$\int_{-1}^{0} (1+t) e^{-itx} dt = \frac{-1}{ix} - \frac{1}{(ix)^2} (e^{-ix} - 1), \text{ also } \int_{0}^{1} (1-t) e^{-itx} dt =$$
$\frac{1}{ix} + e^{-itx} - 1.$

We then have

$$f(x) = \frac{1}{2\pi x^2} (2 - e^{ix} - e^{-ix}) = \frac{1}{\pi x^2} \left(1 - \frac{e^{ix} + e^{-ix}}{2}\right) = \frac{1 - \cos x}{\pi x^2}.$$

7.5. Multivariate Generating Function

Multivariate moment generating and characteristic functions can be defined in a manner similar to those of the univariate cases. We first consider the Bivariate moment generating function of the distribution of (X,Y), defined as $M_{XY}(s,t) = E(e^{sx+ty})$.

The derivatives of this function calculated at (0,0) produce the various moments of the distribution – marginal and mixed.

$$\frac{\partial^{m+n}}{\partial s^m \partial t^n} C_{XY}(s,t)\bigg]_{s=t=0} = E(X^m Y^n).$$

Whenever the implied differentiation under the averaging operation is legitimate. Again these moments are coefficients in the expansion of the moment generating function. The moment generating function for the random vector $(X_1, X_2, \ldots \ldots X_n)$ is defined to be

$$M_{X_1, X_2, \ldots \ldots X_n}(t_1, t_2, \ldots \ldots t_n) = E[\exp(t_1 X_1 + t_2 X_2 + \cdots \ldots t_n X_n)]$$

and the characteristic function, similarly is

$$C_{X_1, X_2, \ldots \ldots X_n}(t_1, t_2, \ldots \ldots, t_n) = E[\exp(it_1 X_1 + it_2 X_2 + \cdots . + it_n X_n)].$$

The marginal distributions for subsets of the n components have moment generating or characteristic functions obtained by setting equal to 0(zero) those t's that correspond to the unused variables. For example, the marginal distribution of (X_2, X_4) has the moment generating function

$$M_{X_2, X_4}(t_2, t_4) = E[\exp(t_2 X_2 + t_4 X_4)] = M_{X_2 X_4}(0, t_2, 0, t_4, 0, \ldots \ldots 0),$$
and the

characteristic function

$$C_{X_2 X_4}(t_2, t_4) = E[exp(it_2 X_2 + it_4 X_4)] =$$
$$C_{X_2 X_4}(0, t_2, 0, t_4, 0 \ldots \ldots, 0).$$

7.6. Negative Binomial Distribution

Let us consider an experiment in which the properties are the same as those listed for a Binomial experiment, except that instead of performing a fixed or given number of trials, one performs independent Bernoulli trials repeatedly, until a given number of successes are observed and then stops. That is, instead of finding the probability of r successes in k trials where k is fixed, we are now interested in the probability that the r-th success occurs on the k-th trial. In this case, the total number of trials required is random, equal to the number of failures encountered before the given number of success plus that number of successes. Experiments of this kind are called negative Binomial experiments. The probability distribution of the negative Binomial variable X is called the negative Binomial distribution (also referred to as the Pascal distribution).

Before elaborating on the negative Binomial distribution, it will be necessary first to revisit the simple Geometric distribution described earlier in section 4.7.4 of

chapter 4. In this simple Geometric setting, we are interested in the number of failures encountered prior to the first success, and call this random variable X. The probability that k failures occur in a row followed by just one success is the probability that the random variable X takes on the value k: $P(X = k) = q \times q \times q \dots \dots \dots \times q \times p = q^k p$ for k=1,2,…….. The distribution defined by these probabilities is said to be geometric, since the probabilities are terms in a Geometric series. The factorial moment generating function is

$$F_X(t) = E(t^X) = \sum_{k=0}^{\infty} t^k P(X = k) = \sum_{k=0}^{\infty} t^k q^k p = \frac{p}{1 - tq}$$

$$= \frac{1}{1 - (t - 1)\frac{q}{p}}$$

$$= \sum_{k=0}^{\infty} \left(\frac{q}{p}\right)^k (t - 1)^k.$$

The Factorial moments, $F_X^k(1)$ are the coefficients of $\frac{(t-1)^k}{k!}$ in the Taylor's series expansion about t=1.

$E[(X)_k] = E[X(X - 1)(X - 2) \dots \dots (X - k + 1)] = k! \left(\frac{q}{p}\right)^k$, which gives the results obtained earlier in sections 4.7.4.1 and 4.7.4.2 of Chapter 4, that for a Geometric distribution, the mean

$E(X) = \frac{q}{p}$ and the variance $V(X) = \frac{q}{p^2}$.

We now proceed to the Negative Binomial distribution. In this setting, we are interested in observing r successes. Denote by S_r the number of failures encountered prior to the r-th success. Let
X_1=the number of failures encountered prior to the first success.
X_2=the number of failures encountered after the first success but prior to the second success.

X_3=the number of failures encountered after the second success but prior to the third success.
X_4=the number of failures encountered after the third success but prior to the fourth success.

$\vdots$

$\vdots$

$\vdots$

X_r=the number of failures encountered after the (r-1)st success but prior to the r-th success.
Then we denote the total number of failures as
$S_r=X_1+X_2+X_3+X_4+\ldots\ldots\ldots+X_r.$
This is a sum of r independent random variables, each having the Geometric distribution. The Factorial moment generating function is therefore the r-th power of the Factorial moment generating function of X (the Geometric distribution). Thus

$$F_{S_r}(t) = F_{\sum X_i}(t) = \left[\frac{p}{1-tq}\right]^r = p^r \sum_{k=0}^{\infty} \binom{-r}{k}(-qt)^k, \quad \text{where the}$$

negative Binomial

coefficient is defined to be

$$\binom{-r}{k} = \frac{(-r)(-r-1)(-r-2)\ldots\ldots(-r-k+1)}{k!} = (-1)^k \binom{r+k-1}{k}, \quad \text{giving}$$

$$F_{S_r}(t) = p^r \sum_{k=0}^{\infty} (-1)^k \binom{r+k-1}{k}(-q)^k t^k$$

$$= \sum_{k=0}^{\infty} \binom{r+k-1}{k} p^r q^k t^k = \sum_{k=0}^{\infty} t^k P(S_r = k).$$

The coefficient of t^k in the Factorial moment generating function is the probability of k

failures prior to the r-th success: $P(S_r = k) = \binom{r+k-1}{k} p^r q^k.$
The mean and variance of S_r can be obtained either from the mean and variance of

the X's that make it up, that is,
$$E(S_r) = E(X_1 + X_2 + \cdots \ldots + X_r) = rE(X)$$

$$= \frac{rq}{p},$$

$$Var(S_r) = \sum_{i=1}^{r} Var(X_i) = \frac{rq}{p^2},$$

or from the coefficients in the expansion of the Factorial moment generating function:

$$F_{S_r}(t) = \left[\frac{p}{1-tq}\right]^r \ about \ t = 1.$$ It should be clear that if r=1, S_r has the

Geometric distribution.

EXAMPLE 7.16

Find the probability that a person tossing 3(three) coins will get either all heads or all tails for the second time on the fifth toss.

SOLUTION:

Using the Negative Binomial distribution with the number of repetitions equal to 5(five) and successes equal to 2(two), and p=1/4, we have r=2, k=3. Hence

$$P(S_r = 3) = \binom{4}{3}\left(\frac{1}{4}\right)^2\left(\frac{3}{4}\right)^3 = \frac{4!}{1!3!} \times \frac{3^3}{4^5} = \frac{27}{256} \ .$$

7.7. The Multinomial Distribution

The Binomial distribution arose in connection with a sequence of independent Bernoulli trials, each of which has two possible outcomes, as the number of times one particular outcome occurs in a given number of trials. The Binomial experiment becomes a multinomial experiment if we let each trial have more than two possible outcomes.
Consider now a sequence of independent trials of an experiment that has k outcomes, E_1, E_2, ... E_k with corresponding probabilities p_1, p_2 ..., p_k , then the multinomial distribution will give the probability that E_1 occurs x_1 times, E_2 occurs x_2 times, ..., E_k occurs x_k times in n independent trials, where $x_1 + x_2 + ...+ x_k = n$. We shall denote this joint probability distribution by
$f(x_1, x_2, ... x_k; p_1, p_2, ... p_k; n)$. Clearly $p_1 + p_2 + p_3 + ... + p_k = 1$, so that if
k − 1 of them are specified, the remaining one is automatically determined.

For a particular sequence of n trials, of which x_1 result in E_1, x_2 in E_2, ..., and x_k in E_k, the probability is just the product of corresponding p's:

$$P_1^{x_1} \bullet P_2^{x_2} \bullet P_3^{x_3} \cdots P_k^{x_k}$$

But such a sequence of results can come in many patterns – the number of which is the number of ways of arranging n objects, x_1 of one kind, x_2 of another kind ... and x_k of the k–th kind, namely n! divided by a factorial for each group of like objects. The total probability for all sequences with the given frequencies is then

$$P\left(X_1 = x_1, X_2 = x_2, \cdots \cdots and\ X_k = x_k\right)$$

$$= \frac{n!}{x_1!\, x_2!\cdots\cdots x_k!}\ p_1^{x_1}\, p_2^{x_2} \cdots p_k^{x_k}\ ,\ \text{provided of course } x_1 + x_2 +$$

$\ldots + x_n = n.$

This is the joint probability function of the distribution.

The moment generating function of the distribution of $(X_1, \ldots , X_k)$ is

$$M_{\sum X_i}\left(t_1, \cdots\cdots t_k\right) = E\left[\exp\left(t_1\, x_1 + t_2\, x_2 + \cdots t_k\, x_k\right)\right]$$

$$= \left(p_1\, e^{t_1} + p_2\, e^{t_2} + \cdots p_k\, e^{t_k}\right)^n.$$

The marginal distribution moment generating function of $(X_1, X_2, \ldots X_1)$ is obtained by setting $t_k = 0$:

$$M_{\sum X_i}\left(t_1, t_2, \cdots t_{k-1} 0\right) = \left(p_1\, e^{t_1} + p_2 e^{t_2} + . + . p_{k-1} e^{t_{k-1}} + p_k\right)^n$$

This distribution has marginal distributions of the same type.

EXAMPLE   7.17

A pair of dice is tossed 6 times. What is the probability of obtaining a total of 7 or 11 twice, a matching pair once, and any other combination 3 times?

SOLUTION:

We have the following possible events:
E₁: A total of 7 or 11 occurs,
E₂: A matching pair occurs,
E₃: Neither a pair nor a total of 7 or 11 occurs.

The corresponding probabilities for a given trial are $P_1 = \dfrac{2}{9}, P_2 = \dfrac{1}{6}$,

and $P_3 = \dfrac{11}{18}$.

These values remain constant for all 6 trials. Using the multinomial distribution with $x_1 = 2$, $x_2 = 1$, and $x_3 = 3$, we obtain the required probability:

$$P\left(X_1 = 2,\ X_2 = 1\ and\ X_3 = 3\right) = \frac{6!}{2!\ 1!\ 3!}\left(\frac{2}{9}\right)^2\left(\frac{1}{6}\right)^1\left(\frac{11}{18}\right)^3 = 0.1127$$

7.7.1   NOTE:

In recording the value of a continuous random variable X, it is necessary to round off", a process that in effect divides the value space of X into mutually disjoint intervals $I_1, I_2..., I_k$. These constitute a partition of the value space, and as a succession of values of X are obtained, in repeated trials of the underlying experiment, one notes only the interval in which each value falls. The number of observations in a given interval I_j, called the frequency f_j, of that interval, is a Binomial variable if the observations are independent. The joint distribution of the frequencies $(f_1, f_2, ... f_k)$ is multinomial, with

$$p_j = P\left(X\ falls\ in\ I_j\right) = \int_{f_j} f(x)dx,$$

where $f(x)$ is the density of the distribution of X.

EXERCISE 7

7.1   (a) A continuous random variable X has density of the form

$$f(x) = \begin{cases} a\, e^{-ax} & ,\ for\ 0 \le x < \infty \\ 0 & ,\ for\ 0\ x < 0 \end{cases}$$

where a is a constant.

Find (a) the distribution function F(x) of X. (b) the moment generating function (mgf) of X and deduce the mean and variance of X.

7.2   Let X and Y be two independent Binomial distributed random variables with distributions X ~ Bin (10, P) and Y ~ Bin (5, q) respectively, where
p + q = 1. Obtain the characteristic function of the random variable Z = X – Y + 5, and hence deduce the distribution of Z.

7.2.1   The discrete random variable X has moment generating function given by

$$M_X(t) = E\left(e^{tx}\right) = \left(\frac{1}{3}\, e^{-5t} + \frac{2}{3}\, e^t\right)^8.$$

Let $Y = (X + 40)/6$.
(a) Obtain the moment generating function $M_Y(t)$ of Y.
(b)   Obtain the second central moment for Y (i.e. obtain Var (y)).

7.3   Let X and Y be two independent random variables such that Y is uniformly distributed in the interval (0, 1), and X has frequency density function given by

$$f_x(x) = \begin{cases} x\, e^{-x}, & x > 0 \\ 0, & otherwise \end{cases}$$

Define a new random variable Z by the product of X and Y, i.e. Z = XY, and
(a)   Obtain the frequency density function of Z.
(b)   Obtain the moment generating function of Z.

7.4　　A fair die is tossed repeatedly until a tail shows up the first time. Let X be the total number of trials required (including the toss which resulted in a tail) for such an experiment. Assume that P(Head) = P(tail) $= \dfrac{1}{2}$ and that the tosses are independent of each other.

(a)　　　　Obtain the factorial moment generating function for X.
(b)　　　　Obtain the variance of X.

7.5　　Obtain a formula for the moment generating function of
(i) Y = a X + b (ii) Y=a+bX (where a and b are constants) in terms of $M_X(t)$.

7.6　　Given the moment generation function, $M_X(t) = \left(1 - t^2\right)^{-1}$, compute the variance of X.

7.7　　(a) Compute the factorial moment generating function of the discrete distribution defined by the probability function of

$$f(k) = \frac{2}{2^{k+1}}, \, for \, k=0, \, 1, \cdots$$

(b) Determine the mean and variance of the distribution.

7.8　　Derive the moment generating function of the sum $Y = X_1 + X_2 + \ldots X_n$, where the X's are independent random variables each with density e^{-x}, for x>0.
x > 0. By recognising the result as the moment generating function of a previously encountered distribution deduce the distribution of Y.

7.9　　Let X be a Gamma distribution with parameters α and r. Obtain the moment generating function of X.

7.10　　Suppose that X has pdf given by f(x) = 2x, $0 \le x \le 1$

(a)　Determine the mgf of X
(b)　Using the mgf, evaluate E(X) and V(X)

7.11　　Suppose that X has the following pdf:
f(x) = $\lambda \, e^{-\lambda(x-a)}$, for x $\ge$ a (This is known as the two – parameter exponential distribution).
(a)　　　　Find the mgf of X
(b)　　　　Using the mgf, find E(X) and V(X)

7.12   Suppose that the continuous random variable X has pdf.

$f(x) = \dfrac{1}{2}e^{-|x|}$, $-\infty < x < \infty$ Obtain the mgf of X.

7.13   Suppose that the mgf of a random variable X is of the form

$M_X(t) = (0.4e^t + 0.6)^8$

Obtain the mgf of the random variable y = 3 X + 2, hence evaluate E(X).

7.14   If the random variable X has an (mgf.) given by $M_X(t) = \dfrac{3}{(3 - t)}$

, obtain the variance of X.

7.15   Find the mgf of a random variable which is uniformly distributed over the interval

(− 1, 2).

7.16   Determine the characteristic function of the distribution defined by the density

f(x) = $\lambda\, e^{-\lambda x}$, for x > 0, in terms of the given constant λ.

7.17   Determine the characteristic function of the distribution defined by the density function.

$f(x) = K\, \exp\left(-\dfrac{1}{2}x^2\right)$

7.18   Obtain the mean and variance of the negative Binomial distribution from the coefficients in the expansion of the factorial mgf

about t = 1 [Hint: use the form $\left(1 - [t-1]\dfrac{q}{P}\right)^{-r}$, and write the first

three terms in a negative Binomial expansion].

7.19   The probability that a person will install a black telephone in a residence is estimated to be 0.3. Find the probability that the tenth phone installed in a new subdivision is the fifth black phone.

7.20   A Scientist inoculates several mice, one at a time, with a disease germ until he finds 2 that have contracted the disease. If the possibility of contracting the disease is $\dfrac{1}{6}$, what is the probability that 8 mice are required?

7.21   Find the probability that a person flipping a coin gets the third head on the seventh flip.

7.22   The probability that a student driver passes the written test for his private driving license is 0.7. Find the probability that a person passes the test (a) on the third try (b) before the fourth try.

7.23   Find the probability of obtaining 2 ones, 1 two, 1 three, 2 fours, 3 fives, and 1 six in 10 rolls of a balanced die.

7.24   According to the theory of genetics a certain cross of guinea pigs will result in red, black, and white offspring in the ratio 8:4:4. Find the probability that among 8 such offspring 5 will be red, 2 black and 1 white.

7.25   Four independent observations are to be made on a random variable with density f(x) = 1 − | x | , − 1 < x < 1.

Suppose that the interval from − 1 to 1 is divided into four class intervals of equal length. What is the probability that one observation will fall in the leftmost class interval, one in the next, two in the next, and none in the right most class interval?

7.26   A continuous random variable X has density of the form

$$f(x) = \begin{cases} a\,e^{-ax} & \text{for} \quad 0 \le x < \infty \\ 0 & \text{otherwise} \end{cases}$$

Where a is a constant. Find

(a)   The cumulative distribution function, $F(x)$ of X
(b)   The moment generating function (mgf) of X and deduce
(c) $E(X)$ and $V(X)$

7.27. Let X and Y be two independent Binomial distributed random variables with parameters $X \sim \text{Bin}(x, p) = \text{Bin}(10, p)$ and $Y \sim \text{Bin}(n, p) = \text{Bin}(5, q)$, respectively, where $p + q = 1$. Obtain the characteristic function of $Z = X - Y + 5$ and hence the distribution of Z

CHAPTER 8
STOCHASTIC PROCESSES

In this chapter we shall briefly and concisely describe the concepts of a stochastic process. The exposition will serve as an elementary introduction to stochastic processes. Interested readers in this area may refer to (Feller, 1957, 1970, Karlin and Taylor, 1975, Kannan, 1979 etc.) Consider performing an experiment, and suppose this experiment can be described to take place in stages under presumably "identical' conditions, but with resulting outcomes at each stage different. Let the outcome at any particular stage be allowed to depend on the outcomes of the previous stages, so that the probability for each possible outcome at a particular stage is known when the outcomes of all previous stages are known. Thus, probability enters, but not in the same sense that each result of a random experiment determines only a single number. Instead, the random experiment determines the behaviour of some system for an entire sequence or interval of time. That is, the result of the random experiment is a sequence or series of values, a function, not just a single number.

The probability structure of the process is completely determined provided that the joint distribution or density function of each such set of random variables is determined. Basically, analysis of a stochastic process involves determining these joint distributions and using them to predict the behaviour of the process in the future given certain behaviour in the past.

8.1. Classification of Stochastic Processes

A Stochastic process involves the behaviour of a system over time. In defining such a process, it is necessary to specify the time set T involved. Moreover, it is assumed that at each point t of the time set T one can observe a measurement of X_t (the collection of numerical values that the random variable X_t can take on), termed the state space of the process which is also important.

Thus, the main elements distinguishing stochastic processes are in the nature of the state space, the index parameter T, and the dependence relations among the random variables X_t (i.e. the joint distribution function of every finite family $X_{t_1}, X_{t_2}, \cdots X_{t_n}$).

8.1.1 Index Parameter T

The index parameter T might be a sequence of times; in this case one speaks of a discrete parameter process, or it might be some time interval; in this case one speaks of a continuous parameter process. If T = {0, 1, 2, 3, ...} or T = {1, 2, 3, ...}, (sequence of consecutive integers), then one shall always say that X_t is a discrete time stochastic process.

Often when T is a discrete time stochastic one shall write X_n instead of X_t. If T = {t: $0 \leq t \leq \infty$} or T = {0, ∞}, then X_t is called a continuous time process.

8.1.2 State Space S

Each X_t, for t $\in$ T, is a function of time. The values of X_t are termed states, and the space in which the possible values of each X_t lie, is termed the state space S of the process. If S = {0, 1, 2, ...}, that is, if X_t represents a count, or a whole number of objects, one speaks of a discrete state process.

Discrete state spaces are frequently termed random chains. If S = the real line ($-\infty$, ∞), that is X_t represents some measurement (temperature, voltage, etc), we speak of a continuous state process.

8.2. Examples of Stochastic Processes

8.2.1. Markov Dependent Sequence: (Discrete Case)

Consider a random sequence {J_1, J_2, ...} of mutually independent random variables with individual probabilities $P_i = P_r[J_i = x_i]$ (i = 1,2,3, ...) respectively.

The random sequence is said to be a Markov dependent sequence or a Markov process if the conditional probability

$$P_r\left[J_n = x_n \big/ J_1 = x_1 , J_2 = x_2 , \cdots\cdots, J_{n-1} = x_{n-1}\right] \text{ depends only}$$

on the values of x_n and x_{n-1}.

That is,

$$P_r\left[J_n = x_n / J_1 = x_1, J_2 = x_2, \cdots\cdots, J_{n-1} = x_{n-1}\right]$$
$$= P_r\left[J_n = x_n /, J_{n-1} = x_{n-1}\right].$$

This means that, if the present stage (n-1) of the system is known, the behaviour of the future stage (n), does not depend on the past stages (1, 2, 3, … n − 2). To state it more vaguely, we say that in a Markov dependent sequence, knowledge of the present makes the future independent of the past.

For such a sequence, the joint probability function can be expressed in terms of conditional probabilities as

$$P_r\left[J_1 = x_1, J_2 = x_2, \cdots J_n = x_n\right]$$

$$= P\left[J_1 = x_1\right] P_r\left[J_2 = x_2 / J_1 = x_1\right] P_r\left[J_3 = x_3 / J_2 = x_2\right]\cdots$$

$$\cdots P_r\left[J_n = x_n / J_{n-1} = x_{n-1}\right]$$

The conditional probabilities

$P_r\left[J_n = x_n / J_{n-1} = x_{n-1}\right]$ are termed transition probabilities, and depend on the value of n.

There are cases, which will be discussed in a later section for which the conditional probability $P_r\left[J_n = x_n / J_{n-1} = x_{n-1}\right]$ does not involve the value of n. When this happens, we say that the sequence has stationary transition probabilities. In this the transition probabilities can be written as $P_{ij} = P_r\left[J_n = j / J_{n-1} = i\right]$, and the stationarity is expressed by the fact that P_{ij} does not depend on n. Thus P_{ij} represents the probability of a one-step transition from state i to state j at any given stage of the sequence. For such a sequence, the joint probability function can be written in terms of the initial probability density function $P_r[J_1 = x_1]$, and the transition function P_{ij} , that is,

$$P_r[J_1 = x_1, J_2 = x_2, ..., J_n = x_n] = P_r[J_1 = x_1] P_{x_1 x_2} P_{x_2 x_3} \cdots P_{x_{n-1} x_n}$$

All joint probability density functions are determined once the initial probability density $P_r[J_1 = x_1]$ and the transition function P_{ij}, are known.

EXAMPLE 8.1

Let $\{X_1, X_2, \ldots\}$ be an independent random sequence where each $P(X_i)$ takes only the values 1 and 0 with probabilities p and q and $p + q = 1$.

The joint density of $\{X_1, X_2, \ldots X_n\}$ can be obtained as follows:

$$P\left(X_1 = i_1, X_2 = i_2, \cdots X_n = i_n\right)$$

$$= P\left(i_1, i_2, \ldots, i_n\right) = \underbrace{P.P.P\ldots}_{\text{number of } 1's} \; \underbrace{q.q.q\ldots}_{\text{number of } 0's}$$

$$= p^{\sum_{k=1}^{n} i_k \bullet} \; q^{n - \sum_{k}^{n} i_k}$$

$$= p^t \, q^{n-t}, \text{ where } t = \sum_{k=1}^{n} i_k$$

EXAMPLE 8.2

If in Example 8.1 above, we let $S_k = \sum_{k=1}^{k} X_i \; \text{for } k = 0, 1, 2, \cdots$, then

the partial sums can be written as:

$S_1 = X_1$, $S_2 = X_1 + X_2 = S_1 + X_2$, $S_3 = X_1 + X_2 + X_3 = S_2 + X_3$
and $S_k = S_{k-1} + X_k$. Thus, the sequence $\{S_1, S_2 \ldots S_k\}$ is a Markov Dependent Sequence. The transition probabilities are given as

$$P(S_n = i_n/S_{n-1} = i_{n-1}, S_{n-2} = i_{n-2} \ldots \ldots, S_1 = i_1)$$

$$= P(S_n = i_n/S_{n-1} = i_{n-1}) = P(S_{n-1} + X_n = i_n/S_{n-1} = i_{n-1})$$

$$= P(i_{n-1} + X_n = i_n/S_{n-1} = i_{n-1}) = P(X_n = \alpha), \text{ where } \alpha = i_n - i_{n-1}$$

It follows that

$$p(i_n/i_{n-1}) = \begin{cases} p^\alpha q^{1-\alpha}, & \text{for } \alpha = 0 \text{ or } 1 \\ 0, & \text{otherwise} \end{cases}$$

8.2.2. Random Walk

Consider a gambler starting out with an initial fortune X_0. At the end of each game he
either wins one unit of currency (+ 1) with probability p, or losses one unit of currency (− 1) with probability q, where $0 \leq p \leq 1$, $0 \leq q \leq 1$ and p+q=1.

Each game is played independently of the other games. Interpret the J − sequence $\{J_1, J_2, J_3 \ldots\}$ as a sequence of wins and losses, and denote by J_n, $n \geq 1$ the gambler's winning in the n − th game. Then J_1, J_2 J_3, ... are independent and identically distributed random variables with the common distribution.

$$P(J_n = 1) = p, \quad P(J_n = -1) = q, \quad p + q = 1.$$

Suppose we now define

$$X_n = X_0 + j_1 + j_2 + \cdots \ldots j_n$$

Then $\{X_n, n \geq 0\}$ is a discrete − time, discrete − space stochastic process denoting the gambler's cumulative fortune after n trials of the game, $n \geq 0$.
If $X_0 = 0$, then $X_n = (J_1 + J_2 + J_2 + J_3 + \ldots + J_n)$ denotes the gamblers cumulative winnings after n trials of the game, $n \geq 0$.

8.2.3. Gambler's Ruin

Let X_0 be a fixed positive integer and J_n, $n \geq 1$ be the independent and identically distributed jump variables in a random walk
$\{X_n, n \geq 0)$ such that

$X_n = X_0 + J_1 + J_2 + \ldots \ldots + J_n.$
The random walk $\{X_n, n \geq 0\}$ is called a simple random walk provided that

$$J_n = \begin{cases} +1 & with \ probability \ p \ (win) \\ -1 & with \ probability \ q \ (lose) \\ 0 & with \ probability \ r \ (draw) \end{cases}$$

where $(p + q + r) = 1$, $0 \le p \le 1$, $0 \le q \le 1$, $0 \le r \le 1$. In the special case that $r = 0$, that is no draw occurs, and $p = \dfrac{1}{2} = q$

We call the random walk a symmetric random walk.

8.2.3.1. The Probability of Ruin

Consider the random walk discussed in section 8.2.2 above, where the gambler starts off with an initial fortune $X_0 = a$, say. At the end of each game he either wins one unit of currency (+ 1) with probability p or losses one unit of currency (– 1) with probability q, where $0 \le p \le 1$, $0 \le q \le 1$, and $p + q = 1$.

Assume now that he plays against an opponent with initial capital b, so that $(a + b)$ represents their total joint resources which remains fixed. The game ends if he or his opponent is ruined [i.e. if either his capital or his opponent's capital drops to 0 (zero), so that the final cumulative fortune of one of them is $a + b$].

Let the sequence $(J_1, J_2, J_3 \ldots\ldots)$ represent the gambler's capital after the 1st, and 2nd, ... etc trials and $X_n = X_0 + J_1 + J_2 + J_3 \ldots$ represent his total capital at the n–trial. We wish to compute the probability S_a that the gambler will be ruined if he starts with an initial capital a.

For a player with initial capital a, he either wins the first trial with probability p, so that with capital $(a + 1)$ he is subsequently ruined, or he losses the first trial with probability q, so that with capital $(a – 1)$ he is subsequently ruined. This line of reasoning would lead to the difference equation.

$$S_a = q S_{a-1} + p S_{a+1}, \qquad 0 < a < a + b$$

Since $(p + q = 1)$, the above equation can be re-written with $(p + q) S_a = S_a$, as

$(p + q)S_a = qS_{a-1} + pS_{a+1}$ or $q(S_a - S_{a-1}) = p(S_{a+1} - S_a)$, giving

$$S_{a+1} - S_a = \left(\frac{q}{p}\right)(S_a - S_{a-1}), \qquad 0 < a < a+b$$

But $\left(S_a - S_{a-1}\right) = \left(\dfrac{q}{p}\right)\left(S_{a-1} - S_{a-2}\right)$ by definition.

Also $\left(S_{a-1} - S_{a-2}\right) = \left(\dfrac{q}{p}\right)\left(S_{a-2} - S_{a-3}\right)$

Hence repeated applications of the formula would yield.

$$S_{a+1} - S_a = \left(\dfrac{q}{p}\right)^a \left(S_1 - S_0\right),\ 0<a<a+b$$

Since ruin is certain starting with zero capital, this implies $S_0 = 1$ and

$$S_{a+1} - S_a = \left(\dfrac{q}{p}\right)^a (S_1 - 1)\ldots\ldots\ldots\ldots\ldots\ldots\ldots\ldots.8.2.3.1.1$$

Since $S_r - S_0 = \{(S_r - S_{r-1}) + (S_{r-1} - S_{r-2}) + \cdots \ldots\ldots +(S_1 - S0$

it follows from equation 8.2.3.1.1 that

$$S_r - S_0 = \left\{\left(\left(\tfrac{q}{p}\right)^{r-1} + \left(\tfrac{q}{p}\right)^{r-2} + \cdots + 1(S_1 - 1)\right)\right\}$$

or

$$S_r - 1 = \dfrac{1-\left(\frac{q}{p}\right)^r}{1-\left(\frac{q}{p}\right)}(S_1 - 1),\ 0 < r < a + b \ldots\ldots\ldots.8.2.3.1.2$$

(Summing the geometric series above valid for $p \neq q$). To find S_{r-1}, set $r = a + b$ in 8.2.3.1.2 clearly, $S_{a+b} = 0$ (i.e. ruin is impossible if one starts with all the money). Therefore

$$S_{a+b} - 1 = -1 = \dfrac{1-\left(\frac{q}{p}\right)^{a+b}}{1-\left(\frac{q}{p}\right)}(S_1 - 1),\ \text{ or}$$

$$S_1 - 1 = -\dfrac{1-\left(\frac{q}{p}\right)}{1-\left(\frac{q}{p}\right)^{a+b}}\ ,\ \text{when substituted into\ equation 8.2.3.1.2 gives}$$

$$S$$

$$S_r - 1 = -\frac{1-\left(\frac{q}{p}\right)^r}{1-\left(\frac{q}{p}\right)^{a+b}} \qquad \dots\dots\dots\dots\dots\dots\dots\dots\dots\dots 8.2.3.1.3$$

If we replace r by a in   8.2.3.1.3   we have,

$$S_a = 1 - \frac{1-\left(\frac{q}{p}\right)^a}{1-\left(\frac{q}{p}\right)^{a+b}}, \qquad p \neq q.$$

This gives the probability of a player who starts with a units of currency being ruined provided $p \neq q$.

Suppose now $p = q$, then the Geometric series shown earlier, would give

$$S_a = 1 - \frac{a}{a+b} = \frac{b}{a+b}, \qquad if \ p = q.$$

8.2.3.2.   Theorem: (Probability of Gambler's Ruin)

Let a and b be the initial fortunes of players A and B respectively and $\{X_n, n \geq 0\}$ be the random walk corresponding to A's cumulative fortune. Let $r < 1$. If $p = q$, then

(i) $\qquad\qquad P[X_r = a + b] = P \text{ (B is ruined)} = S_a = \frac{a}{a+b}$

$$P[X_r = 0] = P[A \text{ is ruined}] = 1 - \frac{a}{a+b} = \frac{b}{a+b}$$

If $p \neq q$ then:

(ii) $\qquad\qquad P[X_r = a + b] = P[B \text{ is ruined}] = \frac{1-\left(\frac{q}{p}\right)^a}{1-\left(\frac{q}{p}\right)^{a+b}}$

(iii) $$P[X_r = 0] = P[\text{A is ruined}] = 1 - \frac{1-\left(\frac{q}{p}\right)^a}{1-\left(\frac{q}{p}\right)^{a+b}} =$$

$$\frac{\left(\frac{q}{p}\right)^a - \left(\frac{q}{p}\right)^{a+b}}{1-\left(\frac{q}{p}\right)^{a+b}}$$

This process with finite (a + b) differs from the random walk we considered in sections 8.2.2 and 8.2.3 above, in that the states 0(zero) and (a + b) are absorbing states. If the system ever enters either of these states, it never can leave.

Although we have discussed this process by using gambling game terminology, similar situations arise in many physical and real-world processes that concern random walks or random motions between absorbing barriers.

8.2.3.3. Proposition

Let $X = \{X_n : n \geq 0\}$ be a (simple) random walk and k > 0 a fixed integer. Suppose we define $Y = \{Y_n = X_{n+k} - X_k, n \geq 0\}$, then Y is also a simple random walk. This proposition says that a Random Walk starts from scratch at any given time k.

Proof:

At about start of the process, time k = 0. Therefore by definition,
$Y = \{Y_n = X_{n+0} - X_n, n \geq 0\}$
That is,
$Y = Y_0 = 0$ and

$Y_n = X_{n+k} - X_k = 0 + J_1^* + J_2^* + \cdots + J_n^*$, where $J_r^* = J_{r+k}$. Since the random variables J_n are independent and identically distributed, the J_k^* are also independent and identically distributed. Therefore, Y_n is the sum of n independent identically distributed random variables for every n. Hence $\{y_n, n \geq 0\}$ is a (simple) random walk.

8.2.4. Stationary Processes

A Stochastic process X_t for t in T [here T could be either discrete or continuous] is said to be strictly stationary if the joint distribution function of the families of random variables

$$\left(X_{t_1+h},\ X_{t_2+h},\ \cdots\cdots,\ X_{t_n+h}\right) \text{ and } \left(X_{t_1},\ X_{t_2},\ \cdots X_{t_n}\right)$$

are the same for all $h > 0$ and arbitrary selections $t_1, t_2, \ldots t_n$ from T. This means that the joint probabilities depend only on the intervals between the times at which the process is to be observed, and not on when the sequence of observations starts.

A Stochastic process X_t, for $t \in T$ is said to be covariance stationary if it possesses finite second moments and if

$$Cov\left[X_t,\ X_{t+h}\right] = E\left[X_t\ X_{t+h}\right] - E\left(X_t\right)E\left(X_{t+h}\right)$$

depends only on h for all $t \in T$. A stationary process that has finite second moments is covariance stationary. There are covariance stationary processes that are not stationary.

Stationary processes are appropriate for describing many phenomena that occur in communication theory, Astronomy, Biology, and sometimes Economics.

8.2.4.1. Stationary Independent Increments

Random changes of the form $X_{t+h} - X_t$, for fixed $h > 0$, are termed increments of the process. If each set of increments, corresponding to a non-overlapping collection of time intervals (t_r, t_{r+h}), is mutually independent, then X_t is said to be a process with independent increments. If in addition, the increments $X_{t+h} - X_t$ have a distribution that depends only on h, and not on t, the process X_t is said to have stationary independent increments.

When discussing a process with stationary independent increments, we usually make the additional assumption that $X_0 = 0$, that is, the process initially starts in state 0.

Let $\{X_n, n > 0\}$ be a sequence of independent, identically distributed random variables that each takes only the values ± 1. A discrete parameter process $\{S_n\}$ can be formed from any sequence $\{X_n\}$ of independent, identically distributed random variables

$$S_n = X_1 + X_2 + \ldots + X_n \text{ for } n = 1, 2, 3, \ldots$$

If this is done, then clearly the increments of this process are

$$S_{n+h} - S_n = X_{n+1} + X_{n+2} \ldots + X_{n+h} \text{ for integers } h.$$

Furthermore the $\{X_r\}$ are mutually independent and, hence, the increments

$$\left\{ S_{n+h_1} - S_h, \; S_{n+h_1+h_2} - S_{n+h_1} \cdots \right\}$$

are independent since the sums involve overlapping sets of X_r's. Furthermore, since the $\{X_r\}$ are identically distributed, the distribution function of each increment depends only on the number h of terms in the sum, and not on n. That is

$$S_{n+h} - S_n = X_{n+1} + X_{n+2} + \ldots \ldots X_{n+h}$$

has the same distribution function as

$$S_{r+h} - S_r = X_{r+1} + X_{r+2} + \ldots + X_{r+h}$$

for each r, since the distribution of the sum depends on the fact that there are h terms in each sum, and not on the fact that they are the particular terms given. Hence the process $\{S_n\}$ is a discrete parameter process with stationary independent increments.

An example of a continuous parameter process with stationary, independent increments is the Poisson Process which will not be discussed in this chapter. Continuous state-space examples are important in the theory of noise, prediction, Brownian movement, and many other areas of application.

Neither the Poisson Process nor the Brownian motion process is stationary. In fact, no non constant process with stationary independent increments is stationary. However, if $\{X_t, t \in [0, \infty)\}$ is Brownian motion or a Poisson Process, then $Y_t = X_{t+h} - X_t$ is a stationary process for any fixed $h \geq 0$.

8.3. Markov Chains: (Discrete Parameter Case)

8.3.1. Introduction.

In this section, discussion will be focussed on certain systems (physical, Biological and Social etc) that evolve with time. Such systems may evolve in a discrete space S, where the state space S consists of a set of integers {0, 1, 2, ...} (countable), with respect to discrete time T = {0, 1, 2, ...}: the set of time points indexing a sequence of random variables {Xₙ} = {X₀, X₁, X₂, ..., Xₙ}. The state space S is the set of all possible values of Xₙ. The value of Xₙ is referred to as the outcome of the n – th step (or trial). The statement "Xₙ = i" can be read "the system is in state i after n steps." The state i may refer to the Gambler's capital after n trials of a gambling game, or i may refer to the number of individuals in a waiting line after the n – th customer completes service etc. Thus, i is in fact a count.

Let us now recall the concept of the Markov dependent random sequence: if the present state of the system is given, then the past and the future states are conditionally independent of each other. Such a process is termed a Markov chain. The study of Markov chains is the study of probabilities associated with possible transitions between the states of an experimental system.

8.3.2. Definition: (Markov Chain Discrete Parameter Case)

A sequence of trials of an experiment is a Markov chain if the outcome of the n-th trial depends only on the outcome of the (n-1)st trial and not on the outcomes of earlier trials.

8.3.3 One-Step Transition Probability

A discrete-time Stochastic process {Xₙ, n ≥ 0} is called a Markov Chain (MC) if for any sequence {r₀,r₁,r₂,............rₖ}

$$P\left(X_0 = r_0, X_1, r_1, X_2 = r_2, ..., X_{k-1} = r_{k-1}, X_k = r_k\right)$$
$$= P(X_0 = r_0, X_1 = r_1, X_2 = r_2, ..., X_{k-1} = r_{k-1})P(X_k = r_k)$$
$$= P(X_0 = r_0)P(X_1 = r_1/X_0 = r_0)P(X_2 = r_2/X_1 = r_1)....P(X_k$$
$$= r_k/X_{k-1} = r_{k-1})$$

for each k and for some (r₀, r₁, r₂, ... rₖ) of states.

If this relation holds for each k, repeated use of the identity

$$P(X_k = r_k, X_{k-1} = r_{k-1})$$
$$= P(X_{k-1} = r_{k-1})P(X_k = r_k/X_{k-1} = r_{k-1})$$

can be used to express the joint probability

$$P(X_0 = r_0, X_1 = r_1, \ldots \ldots, X_k = r_k)$$

as a product of simple conditional probabilities:- i.e.

$$P(X_0 = r_0, X_1 = r_1, X_2 = r_2, \ldots \ldots \ldots \ldots, X_{k-1} = r_{k-1}, X_k = r_k)$$
$$= P(X_0 = r_0, X_1 = r_1, X_2 = r_2, \ldots \ldots \ldots, X_{k-1} = r_{k-1})P(X_k = r_k)$$

$$=$$
$$P(X_0 =$$
r0PX1=r1X0=r0PX2=r2X1=r1···P(Xk=rkXk−1=rk−1)

The simple conditional probabilities $P(X_k = r_k/X_{k-1} = r_{k-1})$ are referred to as one-step transition probabilities, and the probability $P(X_0 = r_0)$ is termed the initial probability of the process. The one-step transition probabilities are seen to depend on the initial and final stages and also on the time of transition n.

8.3.4 Stationary Transition Probabilities

For the most part, however, we study only the Markov process in which the one-step transition probabilities do not depend on the time n (or the step n which is being considered). In this case we have

$$P(X_k = r/X_{k-1} = s) = P(X_m = r/X_{m-1} = s)$$

for each k and m (provided that X_{k-1} and X_{m-1} each has non-zero probability which equals s). This process is said to have stationary transition probabilities. In this section, only discrete Markov processes having stationary probabilities will be considered. For such processes, the one-step transition probability can be denoted by

$$P_{ij} = P(X = j/X_{k-1} = i),$$

where i does not depend on the value of k. The initial probability can also be written as

$$P_i = P(X_0 = i) \text{ for i, j} = 0, 1, 2, \ldots \ldots$$

Thus, the joint probability density of the process can be written as

$$P(X_0 = j_0, X_1 = j_1, X_2 = j_2 \ldots \ldots X_{k-1} = j_{k-1}, X_k = j_k)$$

$$= P(X_0 = j_0) P_{j_0 j_1} P_{j_1 j_2} P_{j_2 j_3} \ldots \ldots \ldots \ldots \ldots P_{j_{k-1} j_k}.$$

Clearly, we have $\sum_i P_{ij} = 1$ for each i, since by starting in state i after k−1 steps, the process will have to be in some state j after the next step.

8.3.5 Stochastic (or Probability) Matrix

The transition probabilities can be treated as the elements of a matrix. The matrix (P_{ij})=M whose entries are the one-step transition probabilities.

$$M = \left(P_{ij} \right) = \begin{pmatrix} P_{00} & P_{01} & \cdots \cdots \\ P_{10} & P_{11} & \cdots \cdots \\ \vdots & \vdots & \cdots \cdots \\ \vdots & \vdots & \cdots \cdots \end{pmatrix}$$

is termed the transition matrix of the process, and the condition $\sum_j P_{ij} = 1$ simply asserts that the sum of the probabilities in each row of M equals 1. This type of matrix with non-negative probabilities in each row, and with the row-sums equalling 1 (one) is termed a Stochastic (or Probability) matrix. The vector

$$\underline{P}^0 = (P_1, P_2, P_3, \ldots \ldots \ldots \ldots \ldots P_{k-1}, P_k)$$

is termed the initial probability vector.

EXAMPLE 8.3

The following are examples of Stochastic (or probability) matrices.

(i) $\begin{pmatrix} 1 & 0 \\ \frac{3}{4} & \frac{1}{4} \end{pmatrix}$ (ii) $\begin{pmatrix} \frac{1}{2} & \frac{1}{2} \\ \frac{1}{4} & \frac{3}{4} \end{pmatrix}$ (iii) $\begin{pmatrix} \frac{1}{4} & \frac{1}{4} & \frac{1}{2} \\ \frac{1}{3} & \frac{1}{3} & \frac{1}{3} \\ 0 & 1 & 0 \end{pmatrix}$ (iv) $\begin{pmatrix} 0 & 0 & 1 \\ 0 & 1 & 0 \\ \frac{1}{3} & \frac{1}{4} & \frac{5}{12} \end{pmatrix}$

ASSIGNMENT 8.1

Use the matrices of Example 8.3 above to show that,
(i) if $\underline{P}$ is an $(n \times n)$ Probability matrix, then
   $P^2, P^3, \ldots \ldots \ldots, P^n$ are probability matrices
(ii) If P, Q are probability matrices, then (PQ) is a probability matrix,
for all combinations of P and Q.

EXAMPLE 8.4

Consider the one-dimensional random walk of section 8.2.2 above. This
is a Markov chain whose state space $S = (0, 1, 2, \ldots)$ is a finite or infinite
subset of the integers, in which the process, if it is in state x, can in a
single transition either stay in x with probability equal to r or move to
state $x - 1$ with probability equal to q, or move to state
$x + 1$ with probability equal to p.
i.e. $P(x \to x) = r$, $P(x \to x - 1) = q$ and $P(x \to x + 1) = p$.
The transition matrix of this random walk with state space
$S = (0, 1, 2, \ldots.)$ has the form

$$
M(i, j) =
\begin{bmatrix}
 & 0 & 1 & 2 & 3 & 4 & \cdots & \cdots \\
0 & r & p & 0 & 0 & 0 & \cdots & \cdots \\
1 & q & r & p & 0 & 0 & \cdots & \cdots \\
2 & 0 & q & r & p & 0 & \cdots & \cdots \\
3 & 0 & 0 & q & r & p & \cdots & \cdots \\
4 & 0 & 0 & 0 & q & r & \cdots & \cdots \\
\vdots & \vdots & \vdots & \vdots & \vdots & \vdots & \vdots & \vdots \\
\vdots & \vdots & \vdots & \vdots & \vdots & \vdots & \vdots & \vdots
\end{bmatrix}
$$

where $0 \le p \le 1$, $0 \le q \le 1$, $0 \le r \le 1$,
and $p + q + r = 1$. $(i \in S)$.
Specifically, if the cumulative fortune of player A at the nth trial equals i,

(i.e. $X_n = i, i \ge 1$), then

$$P(X_n = i + 1/X_{n-1} = i) = p \,,$$

$$P(X_n = i - 1/X_{n-1} = i) = q \,, \quad \text{and}$$

$$P(X_n = i/X_{n-1} = i) = r$$

EXAMPLE 8.5

Consider again the Random Walk of section above. Two players A and B play a game that consists of a sequence of independent trials for stakes of 1 unit of currency per game. Let the probability of A winning any particular game be denoted by p, and the probability of B winning be denoted by q. If draw games are allowed (in which no money changes hands), with probability, r, then p + q + r =1.
Suppose that player A starts with a units of currency, B starts with b units of currency, and thus a total of $(a + b)$ units of currency is available for play. We are interested in the behaviour of A's cumulative fortunes.

Let X_n = the amount of money that A has after n games ($X_0 = a$). Then, if $X_n = x$, X_{n+1} must equal x with probability r or equal $x - 1$ with probability q or equal $x + 1$ with probability p, regardless of the values of X_i for $i < n$. Thus $\{X_n\}$ forms a Markov chain with one-step transition probabilities given by

$$P_{x,\,x} = P_r\left[X_{n+1} = x/X_n = x\right] = r$$
$$P_{x,\,x-1} = P_r\left[X_{n+1} = x-1/X_n = x\right] = q$$
$$P_{x,\,x+1} = P_r\left[X_{n+1} = x+1/X_n = x\right] = p$$

for $0 \leq x \leq a + b$, $P_{00} = P_{a+b,\,a+b} = 1$ and all other $P_{ij} = 0$.

Obviously, the game termites if A's cumulative fortune ever reaches 0 (zero) or
$(a + b)$.

The initial probabilities are given by
$$p_m = P[X_0 = m] = 1$$
$$p_l = P[X_0 = l] = 0, \; for \; l \neq m$$

The one-step transition matrix of this random walk with state space $S = \{0, 1, 2, \ldots a + b - 1, \; a + b\}$ has the form

$p(i,j)=$		0	1	2	3	...	a+b−1	a+b
	0	1	0	0	0	...	0	0
		q	r	p	0	...	0	0
	1		q	r	p	...	0	0
	2	⋮	⋮	⋮				
a+b−1		0	0	0	...	...	r	p
a+b		0	0	0			0	1

The initial probability vector is $p^0 = (0,0 \ldots \ldots \ldots \ldots 0,1,0, \ldots)$
where the
1 occupies the m–th position (counting the initial term as the zero-th term).

In this example, the transition matrix contains absorbing states 0 and $a + b$ with transition probability, $P_{00} = P_{a+b,\ a+b} = 1$. This means that one can make a transition into these states, but not out of them.
This type of Markov chain (which has an absorbing state(s)) is said to be non-irreducible.

In Example 8.4 above, however, the transition matrix does not have absorbing states, and for such a chain transitions from any state to any other state are possible, although they may not be possible in one step. This type of Markov chain is referred to as being irreducible.

8.3.6 Absorbing State

A state X_i of a Markov chain is absorbing if, once the system reaches the state X_i on some trial, the system remains in the state X_i on all future trials.

8.3.7 Absorbing Markov Chain

A Markov chain is absorbing if it has one or more absorbing states and if it is possible to reach an absorbing state from every non-absorbing state.

If the state X_i is absorbing, the probability of transition from X_i to X_i is 1. In other words, the state X_i is absorbing if and only if $P_{ii}=1$. The non-absorbing states of an absorbing Markov chain are called transient states. The probability that the system is in a transient state decreases as the number of trials increases.

EXAMPLE 8.6

The matrix P given below has 3(three) states X_1, X_2, and X_3.

$$P = \begin{array}{c} \\ X_1 \\ X_2 \\ X_3 \end{array} \begin{pmatrix} X_1 & X_2 & X_3 \\ 0 & \dfrac{1}{2} & \dfrac{1}{2} \\ \dfrac{1}{2} & 0 & \dfrac{1}{2} \\ 0 & 0 & 1 \end{pmatrix}$$

The states X_1 and X_2 are transient states and X_3 is an absorbing state. We note that it is possible to reach X_3 from either X_1 or X_2, and therefore the Markov chain is absorbing.

EXAMPLE 8.7

The matrix Q given below is the transition matrix of a Markov chain with 3(three) states X_1, X_2, X_3.

$$Q = \begin{array}{c} \\ X_1 \\ X_2 \\ X_3 \end{array} \begin{pmatrix} X_1 & X_2 & X_3 \\ \dfrac{1}{2} & 0 & \dfrac{1}{2} \\ 0 & 1 & 0 \\ \dfrac{1}{3} & \dfrac{1}{3} & \dfrac{1}{3} \end{pmatrix}$$

The second state (X_2) is absorbing. It can be reached from X_3 on one trial and from X_1 after two trials(first go from X_1 to X_3 with probability $\dfrac{1}{2}$ and from X_3 to X_2 with probability $\dfrac{1}{3}$).

ASSIGNMENT. 8.2

Which of the following matrices are the transition matrices of absorbing Markov chains? Determine the number of absorbing states in each example.

$$\text{(i)}\begin{pmatrix} 1 & 0 & 0 \\ 0 & 0 & 1 \\ 0 & 1 & 0 \end{pmatrix} \quad \text{(ii)}\begin{pmatrix} 0 & 1 & 0 \\ 1 & 0 & 0 \\ 0 & 0 & 1 \end{pmatrix} \quad \text{(iii)}\begin{pmatrix} \frac{1}{3} & \frac{1}{3} & \frac{1}{3} \\ 0 & 0 & 1 \\ \frac{1}{3} & \frac{1}{3} & \frac{1}{3} \end{pmatrix}$$

$$\text{(iv)}\begin{pmatrix} \frac{1}{2} & \frac{1}{2} & 0 & 0 \\ 1 & 0 & 0 & 0 \\ 0 & 1 & 0 & 0 \\ 0 & 0 & \frac{1}{2} & \frac{1}{2} \end{pmatrix} \quad \text{(v)}\begin{pmatrix} \frac{1}{2} & \frac{1}{2} & 0 & 0 \\ \frac{1}{2} & \frac{1}{2} & 0 & 0 \\ 0 & 0 & 1 & 0 \\ 0 & 0 & 0 & 1 \end{pmatrix}$$

8.3.8 n – step Transition Probabilities

The n-step transition probabilities $P_{ij}^{(n)}$ is defined by,

$$P_{ij}^{(n)} = P(X_{k+n} = j/X_k = i) \quad \text{for } i,j = 0, 1, 2 \ldots, \text{ and } n \ge 0. \text{ The}$$

probability $P_{ij}^{(n)}$

is the probability of the system ending up in state j after n transitions, stating from state i. That is, $P_{ij}^{(n)}$ is the probability that the process goes from state i to state j in n steps. When n equals one, we see that

$$P_{ij}^{(1)} = P_{ij} = P(X_{k+1} = j/X_k = i)$$

is the transition probability of the system ending up in state j after a single transition, starting from state i.

The one step transition probability matrix, $\mathbf{P} = (P_{ij})$, has been defined before. $P_{ij}^{(n)}$ may be regarded as the entries in the matrix $\mathbf{P}^n$, the n–th power of $\mathbf{P}$.

Because of the Markovian property, $P_{ij}^{(n)} = P(X_{k+n} = j/X_k = i)$ can be expressed in terms of the matrix (P_{ij}) as follows: $P_{ij}^{(n)} = \sum_{r=0}^{\infty} P_{ir}^k P_{rj}^{n-k}$

for any fixed non-negative integers k and n. That is, if the one-step transition probability matrix of a Markov chain is P = (P$_{ij}$), then $P_{ij}^n = \sum_{r=0}^{\infty} P_{ir}^k P_{rj}^{n-k}$,

where $P_{ij}^0 = \begin{cases} 1, & if \quad i = j \\ 0, & if \; i \; \neq j. \end{cases}$

From the theory of matrices, we recognize the above relation as just the formula for the multiplication of matrices.

The argument can now be carried out in the case $n = 2$. The event of going from state i to state j in two steps can be realized first through some intermediary state k (in several mutually exclusive possible ways: k = 0, 1, 2, ...) in the first step and then going from state k to state j in the second step.

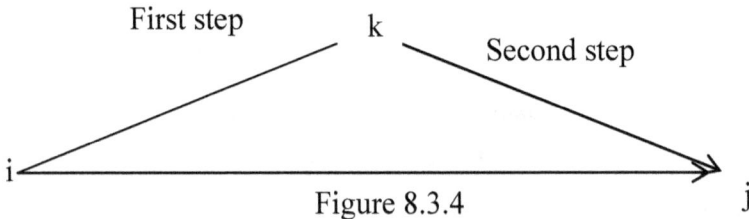

Figure 8.3.4

(Diagram above showing one of several possible paths from state i to state j in two steps). Thus for $n = 2$, the Markovian relation gives the probability matrix as,

$$P_{ij}^{(2)} = \sum_{r=0}^{\infty} P_{ir}^k P_{rj}^{2-k}$$

If $\mathbf{P}^0$ is defined as the identity matrix 'I', then for all non-negative integers m, n, one can deduce from the matrix relation.

$\mathbf{P}^{(n+m)} = \mathbf{P}^n \, \mathbf{P}^m$ an important relation about the elements of $\mathbf{P}^n$:

$$P_{ij}^{(n+m)} = \sum_{k=0}^{\infty} P_{ik}^n P_{kj}^m.$$

The above equations are termed the Chapman – Kolmogorov equations, and form one of the basic relations in Markov process theory.

The event of going from state i to state j in $(n + m)$ steps can be realized in several mutually exclusive ways of going to some intermediate step k $(k = 0, 1, 2, \ldots)$ in n steps and then going from state k to state j in the remaining m steps. Because of the Markovian property the probability of going from state i to state k in n steps is P_{ik}^n. Similarly, because of the Markovian property, the probability of going from state k to state j in m steps is P_{kj}^m. Hence the probability of going from state i to state j in $(n + m)$ steps by following the paths i, k, j is $P_{ik}^n P_{kj}^m$. The probability of going from i to j in several mutually exclusive ways that are possible, is then

$$\sum_{k=0}^{\infty} P_{ik}^n P_{kj}^m = P_{ij}^{n+m}$$

EXAMPLE 8.8

Suppose, the doubly stochastic matrix (that is, matrix with the sum of probabilities in each row, equal to 1; and the sum of probabilities in each column equal to 1) is $\mathbf{P}$, given below with one-step transition probabilities P_{ij}, and state space $S = \{0, 1\}$. The matrix of two-step transition probabilities, $\mathbf{P}^2$, is formed by multiplying the matrix of one-step transition probabilities $\mathbf{P}$ by itself twice. Similarly, the matrix of three-step transition probabilities, $\mathbf{P}^3$, is formed by multiplying the matrix of one-step transition probabilities by itself three times. In general, the matrix of n–step transition probabilities is formed by multiplying the matrix of one-step transition probabilities by itself n-times.

Thus, if

$$
\begin{array}{c}
\text{Next} \\
\rightarrow\text{State}
\end{array}
$$

$$
\begin{array}{c}
\text{Present} \\
\downarrow\text{State}
\end{array}
\qquad
P = \begin{pmatrix} & 0 & 1 \\ 0 & \frac{1}{4} & \frac{3}{4} \\ 1 & \frac{3}{4} & \frac{1}{4} \end{pmatrix}.
$$

Direct multiplication gives

$$
P^2 = \begin{pmatrix} & 0 & 1 \\ 0 & \frac{10}{16} & \frac{6}{16} \\ 1 & \frac{6}{16} & \frac{10}{16} \end{pmatrix}, \text{ and } \quad P^3 = \begin{pmatrix} & 0 & 1 \\ 0 & \frac{28}{64} & \frac{36}{64} \\ 1 & \frac{36}{64} & \frac{28}{64} \end{pmatrix}
$$

$$P^4 = \begin{pmatrix} & 0 & 1 \\ 0 & \dfrac{136}{256} & \dfrac{120}{256} \\ 1 & \dfrac{120}{256} & \dfrac{136}{256} \end{pmatrix}, \text{ and } P^5 = \begin{pmatrix} & 0 & 1 \\ 0 & \dfrac{496}{1024} & \dfrac{528}{1024} \\ 1 & \dfrac{528}{1024} & \dfrac{496}{1024} \end{pmatrix}$$

In the matrix $\mathbf{P}^2$, for example, the two-step transition probability

$P^2_{01} = P(X_{k+2} = 1/X_k = 0)$ gives the probability that the system is now is state 1 given

that two-steps earlier it was in state 0. Similarly,

$P^5_{10} = P(X_{k+5} = 0/X_k = 1)$ gives the conditional probability that the system is now in

state 0 given that $5 -$ steps earlier, it was in state 1, and so on for other entries of the matrices $\mathbf{P}^2$, $\mathbf{P}^3$, $\mathbf{P}^4$, and $\mathbf{P}^5$ respectively.

The matrix $\mathbf{P}^4$ can also be formed by multiplying $\mathbf{P}^2$ by itself twice: That is

$$P^4 = P^2 P^2 \begin{pmatrix} & 0 & 1 \\ 0 & \dfrac{10}{16} & \dfrac{6}{16} \\ 1 & \dfrac{6}{16} & \dfrac{10}{16} \end{pmatrix} \begin{pmatrix} & 0 & 1 \\ 0 & \dfrac{10}{16} & \dfrac{6}{16} \\ 1 & \dfrac{6}{16} & \dfrac{10}{16} \end{pmatrix} = \begin{pmatrix} & 0 & 1 \\ 0 & \dfrac{136}{256} & \dfrac{120}{256} \\ 1 & \dfrac{120}{256} & \dfrac{136}{256} \end{pmatrix}$$

Also,

$$P^5 = P^3 P^2 = P^2 P^3 = \begin{pmatrix} & 0 & 1 \\ 0 & \dfrac{496}{1024} & \dfrac{528}{1024} \\ 1 & \dfrac{528}{1024} & \dfrac{496}{1024} \end{pmatrix}$$

Also,

$$P^9 = P^5 P^4 = P^4 P^5 = \begin{pmatrix} & 0 & 1 \\ 0 & ? & ? \\ 1 & ? & ? \end{pmatrix}$$

(This is left as an exercise to the student).

On a certain day, a person is either healthy or ill. If the person is healthy today, the probability that he will be healthy tomorrow is estimated to be 98 per cent. If the person is ill today, the probability that he will be healthy tomorrow is 30 per cent. Describe the sequence of states of health as a Markov chain. What is the transition matrix? If the person is ill today, what are the probabilities that he will recover tomorrow, two days from, and three days from now?(This is left as an exercise to the student).

Assume now, that the matrices given above are students who attempt the advanced level Mathematics examination n-times before succeeding. Define X to be a random variable with values 0, 1 denoting fail or pass respectively.

That is,

$$X = \begin{cases} 0, \text{ if the student fails the examination on any attempt} \\ 1, \text{ if the student passes the examination} \end{cases}$$

Thus, the state space of the process, S = {0, 1}. Suppose, the student fails the examination on the first attempt. Then, for example, the probability that the student passed on the 5th trial (the 4th additional trial to the first) is easily obtained from the matrix P^4. The entry

$$P_{01}^4 = \frac{120}{256} = \qquad$$ the probability that the student passed the examination on the fifth attempt (the 4th additional attempt to the first), given that the first attempt was a fail.

Similarly,

$$P_{01}^5 = \frac{528}{1024} = \qquad$$ the probability that student passed the examination the 6th attempt (the fifth additional attempt to the first), given that the first attempt was a fail.

Several other situations can be interpreted similarly as done above, using transition probabilities obtained from matrices of higher order than P^1.

ASSIGNMENTS 8.3

(1)   Calculate the two-step transition matrices for the following transition matrices with state spaces $S_1=(0,1)$ and $S_2=(0,1,2)$ respectively.

(a)
$$\begin{pmatrix} & 0 & 1 \\ 0 & 1 & 0 \\ 1 & \frac{1}{2} & \frac{1}{2} \end{pmatrix}$$

(b)
$$\begin{pmatrix} & 0 & 1 \\ 0 & \frac{1}{2} & \frac{1}{2} \\ 1 & \frac{1}{2} & \frac{1}{2} \end{pmatrix}$$

(c)
$$\begin{pmatrix} & 0 & 1 \\ 0 & \frac{2}{3} & \frac{1}{3} \\ 1 & \frac{1}{3} & \frac{2}{3} \end{pmatrix}$$

(d)
$$\begin{pmatrix} & 0 & 1 & 2 \\ 0 & \frac{1}{3} & \frac{1}{3} & \frac{1}{3} \\ 1 & \frac{1}{3} & \frac{1}{3} & \frac{1}{3} \\ 2 & \frac{1}{3} & \frac{1}{3} & \frac{1}{3} \end{pmatrix}$$

(e)
$$\begin{pmatrix} & 0 & 1 & 2 \\ 0 & \frac{1}{2} & \frac{1}{3} & \frac{1}{6} \\ 1 & \frac{1}{6} & \frac{1}{2} & \frac{1}{3} \\ 2 & \frac{1}{3} & \frac{1}{6} & \frac{1}{2} \end{pmatrix}$$

(f)
$$\begin{pmatrix} & 0 & 1 & 2 \\ 0 & \frac{2}{3} & \frac{1}{6} & \frac{1}{6} \\ 1 & \frac{1}{6} & \frac{2}{3} & \frac{1}{6} \\ 2 & \frac{1}{6} & \frac{1}{6} & \frac{2}{3} \end{pmatrix}$$

(2)   Interpret each entry of the two-step transition matrices calculated above.

8.3.9 The n-th Step Transition Probability

Suppose, the probability of the process initially being in some state j is P_j, that is, if X_0 is this initial state, then $P[X_0 = k] = P_k$. It is necessary to find an expression for the probability of the process being in state j at time n. Since the process starts in any state k (with probability $P[X_0 = k] = P_k$), there are several mutually exclusive ways that can lead to state j, regardless of the initial state. Adding all these ways, then the probability of the process being in state j at time n is

$$P_j^n = \sum_{k=0}^{\infty} P_k P_{kj}^n = P(X_n = j)$$

when $n = 1$

$$P_j^1 = P(X_1 = j) = \sum_k P_k P_{kj} = \sum_k P(X_0 = k)P(X_1 = j/X_0 = k)$$

Thus, if the system starts in some state j at the start (time 0), P_j^1 is the unconditional probability of the process being in state j after one-step (at time 1). Writing it in matrix form,

$P^{(1)} = P^{(0)}P$, where $P^{(0)}$ is the initial probability vector, and P is the one step transition probability matrix.

When $n = 2$

$$P_j^2 = P(X_2 = j) = \sum_k P(X_0 = k)P(X_2 = j/X_0 = k) = \sum_k P_k P_{kj}^2$$

and in matrix form

$$P^{(2)} = P^{(0)} \cdot P^2$$

Similarly, when $n = 3$

$$P_j^3 = P(X_3 = j) = \sum_k P(X_0 = k)P(X_3 = j/X_0 = k) = \sum_k P_k P_{kj}^3 \, ,$$

and in matrix form

$$P^{(3)} = P^{(0)} P^3$$

Proceeding in this manner, it follows by induction that

$$P^{(n)} = P^{(0)}P^n$$

Thus, the matrix of state probabilities after n steps is given as a product of the initial probability matrix and the n-th power of the one-step transition probability matrix.

NOTE:

There is a clear distinction between the n-step transitions probabilities discussed in section 8.3.4 above, that is

$$P_{ij}^{(n)} = \sum_{k=0}^{\infty} P_{ik}^r P_{kj}^{(n-r)}$$

(which is the conditional probability of being in state j at present, given that n steps earlier the system was in state i), and the n-th step transition probabilities discussed presently, that is

$$P_j^n = \sum_{k=0} P_k P_{kj}^n$$

(which is the unconditional probability of being in state j after n steps, regardless of the initial state).

In some applications of Markov chains, we may have information concerning the initial state of the system. For example, we may know that the system is initially in state X, or we may know the probabilities that the system begins in the n possible states: $X_1, X_2, X_3, \ldots, X_n$. A natural problem is then to determine the probability that the system is in state X_j after n steps. To solve this problem, we first introduce the following definitions.

8.3.9.1 Initial Probability Distribution (Vector).

The initial probability distribution of a Markov chain is the n-component row vector $\underline{P}^{(0)} = \left(P_1^{(0)}, P_2^{(0)}, P_3^{(0)}, \ldots \ldots \ldots \ldots, P_n^{(0)} \right)$,

where $P_i^{(0)}$ is the probability that the system begins in state X_i for $I = 1, 2, 3, \ldots, n$.

8.3.9.1.1 NOTE:

If the experimental system is initially in state X_1, the initial probability distribution is $P^{(0)} = (1, 0, 0, 0 \ldots \ldots \ldots, 0)$. Similarly, if the experimental system is initially in state X_2, the initial probability distribution is
$P^{(0)} = (0, 1, 0, 0 \ldots \ldots \ldots, 0)$, and so on. If the experimental system is equally likely to be in any of the n states initially, then the initial probability distribution is

$P^{(0)} = \left(\frac{1}{n}, \frac{1}{n}, \frac{1}{n}, \ldots \ldots \ldots \ldots, \frac{1}{n} \right)$. It is important to note that the initial probability

distribution is a probability vector.

EXAMPLE 8.9

Consider again, the problem discussed in section 8.3.4.1 above, where the student attempts the Mathematics examination n-times. Suppose that as a final test before the examination the student is encouraged by his teacher to take (under strict examination conditions) 20 past papers in Mathematics. If the student passes in 5 test papers and fails the remaining 15, and in the absence of any other test scores, it may be appropriate to assess his chances in the examination as:

Probability of passing $= P(X = 1) = \frac{1}{4}$, and the probability of failing$=P(X = 0) = \frac{3}{4}$.

If his initial probability vector is now taken to be $P^{(0)} = \left(\frac{3}{4} \quad \frac{1}{4}\right)$, then the probability

that the student; for instance passes the examination the 5–th attempt, regardless of the initial state, is given by

$$P^{(4)} = P^{(0)}.P^4 = \left(\frac{3}{4} \quad \frac{1}{4}\right)\begin{pmatrix} \frac{136}{256} & \frac{120}{256} \\ \frac{120}{256} & \frac{136}{256} \end{pmatrix} = \left(\frac{528}{1024} \quad \frac{496}{1024}\right)$$

Also, the probability that he passes the examination the 6–th attempt, regardless of the initial state, is given by

$$P^{(5)} = P^{(0)}P^5 = \left(\frac{3}{4} \quad \frac{1}{4}\right)\begin{pmatrix} \frac{496}{1024} & \frac{528}{1024} \\ \frac{528}{1024} & \frac{496}{1024} \end{pmatrix} = \left(\frac{2016}{4096} \quad \frac{2080}{4096}\right)$$

EXAMPLE 8.10

Imagine 5 boys (numbered 1-5) passing a ball from one to another. Let each boy, having received the ball, pass it on with probabilities as follows:

Boy 1 always passes to 5
Boy 2 passes to boy 3 with probability ½ and to boy 4 with probability ½,
Boy 3 passes to boy 1 with probability 1/3, to boy 2 with probability 1/3, and to boy 5 with probability 1/3.
Boy 4 always passes to boy 1.
Boy 5 passes to any of the others, with probability ¼ each.
We now have a system whereby a ball moves among 5 states (boys), the one-step transition matrix being:

$$M = (P_{ij}) = \begin{array}{c} \\ S_1 \\ S_2 \\ S_3 \\ S_4 \\ S_5 \end{array} \begin{pmatrix} S_1 & S_2 & S_3 & S_4 & S_5 \\ 0 & 0 & 0 & 0 & 1 \\ 0 & 0 & \dfrac{1}{2} & \dfrac{1}{2} & 0 \\ \dfrac{1}{3} & \dfrac{1}{3} & 0 & 0 & \dfrac{1}{3} \\ 1 & 0 & 0 & 0 & 0 \\ \dfrac{1}{4} & \dfrac{1}{4} & \dfrac{1}{4} & \dfrac{1}{4} & 0 \end{pmatrix}$$

To start the process off, let the referee give the ball to one of the boys. He will choose a boy according to personal preference (or possibly at random). Suppose that he will give the ball to boy 1 with probability ½ and to boy 3 with probability ½. Then the initial probability vector is:

$$P^{01} = \left(\tfrac{1}{2}, 0, \tfrac{1}{2}, 0, 0\right)$$

If selected at random,

$$P^{02} = \left(\tfrac{1}{5}, \tfrac{1}{5}, \tfrac{1}{5}, \tfrac{1}{5}, \tfrac{1}{5}\right)$$

Once P^{01} and P^{02} are given, the Markov Chain is completely defined. It is then possible to assign probabilities to all possible paths of the ball for any sequence of n throws.

ASSIGNMENTS 8.4

Calculate the two-step probability distributions for the following transition matrices and initial probability distributions

(a) $P^{(0)} = \left(\tfrac{1}{2} \quad 0 \quad \tfrac{1}{2}\right)$, $P = \begin{pmatrix} \tfrac{1}{2} & \tfrac{1}{2} & 0 \\ 0 & \tfrac{1}{2} & \tfrac{1}{2} \\ \tfrac{1}{2} & \tfrac{1}{2} & 0 \end{pmatrix}$

(b) $P^{(0)} = (1,0,0)$, $P = \begin{pmatrix} \frac{1}{3} & \frac{1}{3} & \frac{1}{3} \\ 0 & 1 & 0 \\ \frac{1}{3} & \frac{1}{3} & \frac{1}{3} \end{pmatrix}$

(c) $P^{(0)} = \left(\frac{1}{4} \quad \frac{1}{4} \quad \frac{1}{4} \quad \frac{1}{4} \right)$, $P = \begin{pmatrix} \frac{1}{2} & \frac{1}{4} & \frac{1}{4} & 0 \\ 0 & \frac{1}{2} & \frac{1}{4} & \frac{1}{4} \\ 0 & 0 & 1 & 0 \\ \frac{1}{4} & \frac{1}{4} & \frac{1}{4} & \frac{1}{4} \end{pmatrix}$

(a) Prove by Mathematical induction that the n-th power of the transition

matrix $\quad P = \begin{pmatrix} \frac{1}{2} & \frac{1}{4} & \frac{1}{4} \\ 0 & 1 & 0 \\ 0 & 0 & 1 \end{pmatrix}$ is $P^n = \begin{pmatrix} \frac{1}{2^n} & \frac{2^n-1}{2^{n+1}} & \frac{2^n-1}{2^{n+1}} \\ 0 & 1 & 0 \\ 0 & 0 & 1 \end{pmatrix}$

What is (i) the limit: $\lim_{n \to \infty} P^n$? (ii) What is the probability of going from the first state to the third state in exactly n steps if n is very large?

(e) Prove by mathematical induction, that the n-th power of the transition

matrix $\quad P = \begin{pmatrix} \frac{1}{2} & 0 & \frac{1}{2} \\ 0 & \frac{1}{2} & \frac{1}{2} \\ 0 & 0 & 1 \end{pmatrix}$ is $P^n = \begin{pmatrix} \frac{1}{2^n} & 0 & 1 - \frac{1}{2^n} \\ 0 & \frac{1}{2^n} & 1 - \frac{1}{2^n} \\ 0 & 0 & 1 \end{pmatrix}$

What is the limit : $\lim_{n \to \infty} P^n$?

(f) For the transition matrix

$P = \begin{pmatrix} \frac{3}{4} & \frac{1}{4} \\ \frac{1}{4} & \frac{3}{4} \end{pmatrix}$, prove that $P^n = \begin{pmatrix} \frac{1}{2} + \frac{1}{2^{n+1}} & \frac{1}{2} - \frac{1}{2^{n+1}} \\ \frac{1}{2} - \frac{1}{2^{n+1}} & \frac{1}{2} + \frac{1}{2^{n+1}} \end{pmatrix}$

and deduce that

$$\lim_{n \to \infty} P^n = \begin{pmatrix} \dfrac{1}{2} & \dfrac{1}{2} \\ \dfrac{1}{2} & \dfrac{1}{2} \end{pmatrix}$$

(g) For the transition matrix $P = \begin{pmatrix} \dfrac{4}{7} & \dfrac{3}{7} \\ \dfrac{3}{7} & \dfrac{4}{7} \end{pmatrix}$, determine the exact

formula for

P^n. Use this formula to evaluate $\lim_{n \to \infty} P^n$.

The determination of the joint probability distributions of the process for all times is usually a formidable task. It is frequently of interest to find the asymptotic behaviour of P_{ij}^n *as* $n \to \infty$. One might expect that the influence of the initial state recedes in time and that consequently, as $n \to \infty$, P_{ij}^n approaches a limit which is independent of the initial state i.

In order to analyse precisely the asymptotic behaviour of the process one needs to introduce some principles of classifying states of a Markov chain.

8.3.10   Classification of States of Discrete Markov Chain

Suppose that the process starts in some state i   If j is some other state, the state j is said to be accessible from state i if for some integer $n \geq 0$, $P_{ij}^n > 0$. That is, state j is accessible from state i if there is positive probability that in a finite number of transitions state j can be reached from state i. Similarly, state i is accessible from j if for some integer

$m \geq 0$, there is positive probability $P_{ji}^m > 0$, that in a finite number of transitions state i can be reached from state j. Two states i and j, each accessible to each other are said to communicate and we write $i \leftrightarrow j$. If two states i and j do not communicate, then either $P_{ij}^n = 0 \ for \ n \geq 0 \ or \ P_{ji}^m = 0 \ for \ m \geq 0$ or both relations are true.

The concept of communication is an equivalence relation. With the following properties:

(i)   Reflexive Property: State i communicates with itself (i.e. i →

i) with a positive probability $P_{ii}^n > 0$, for some integer n > 0.

(ii)   Symmetric Property: If state i communicates with state j
(i.e.

i → j) with positive probability $P_{ij}^n > 0$, for some integer n > 0,

then state j communicates with state i (i.e. j → i) with positive
probability $P_{ji}^m > 0$ for some integer m > 0.

(iii)   Transitivity: If state i communicates with state j (i.e. i → j)
with positive probability $P_{ij}^n > 0$, for n > 0, and state j in turn
communicates with state k (i.e. j → k) with positive probability
$P_{jk}^m > 0$ for m > 0, then state i communicates with state k (i.e.

i → k) with positive probability $P_{ik}^{(n+m)} > 0$.
Consequently, by the Chapman – Kolmogorov equation which

was discussed earlier,

$$P_{ik}^{(n+m)} = \sum_{j=0}^{\infty} P_{ij}^n P_{jk}^m \geq P_{ij}^n P_{jk}^m > 0.$$

8.3.11   Decomposition of State Space

(a) The totality of states can be partitioned into equivalence classes. The
states in an equivalence class are those which communicate with each
other. It may be possible, starting in one class, to enter some other class
with positive probability; if so, however, it is clearly not possible to
return to the initial class, or else the two classes would together form a
single class.

(b) If C is a set of states so that no state outside C can be reached from
any state in C, the set C is said to be closed. C has the property that
once the process enters it, it cannot leave. If each pair of states in C
communicate, C is termed a closed communicating class.

(c) If a closed set contains only one state, that state is termed an
absorbing state. It is necessary, however, to consider the possibility that
an absorbing state might actually be a class of several states with the
property that transitions are possible between states within the class, but
not from states inside the class to states outside. Such a state is termed a
closed communicating class.

(d)   A discrete Markov chain is said to be irreducible if the chain contains no closed sets with the exception of the set of all states (the equivalence class). In other words, a process is irreducible if all states communicate with each other.

EXAMPLE 8.11

Determine the classes of the various states $S = \{1,2,3,4\}$ for a Markov chain with transition probability matrices M_1 and M_2 given below:

$$M_1 = \begin{array}{c} \\ 1 \\ 2 \\ 3 \\ 4 \end{array} \left(\begin{array}{cccc} 1 & 2 & 3 & 4 \\ 0 & 0 & 1 & 0 \\ 1 & 0 & 0 & 0 \\ \dfrac{1}{2} & \dfrac{1}{2} & 0 & 0 \\ \dfrac{1}{3} & \dfrac{1}{3} & \dfrac{1}{3} & 0 \end{array} \right)$$

SOLUTION (For M_1)

The state space decomposes into classes C_1 and C_2 as follows:

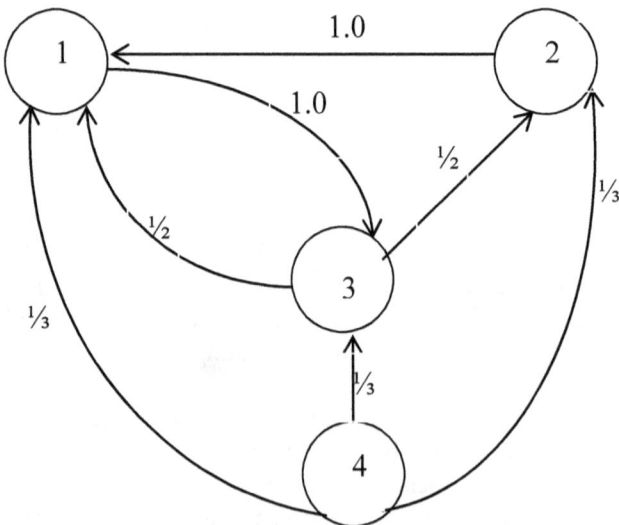

$C_1 = \{1, 2, 3\}$ and $C_2 = \{4\}$

C_1 is a closed communicating class, and C_2 is not closed.

SOLUTION (For M_2).

$$M_2 = \begin{array}{c}\\1\\2\\3\\4\end{array}\begin{pmatrix}1 & 2 & 3 & 4\\0 & 1 & 0 & 0\\0 & 0 & 0 & 1\\0 & 1 & 0 & 0\\\dfrac{1}{3} & 0 & \dfrac{2}{3} & 0\end{pmatrix}$$

The Markov chain given above, with state space $S=\{1,2,3,4\}$ contains no closed set with the exception of all states

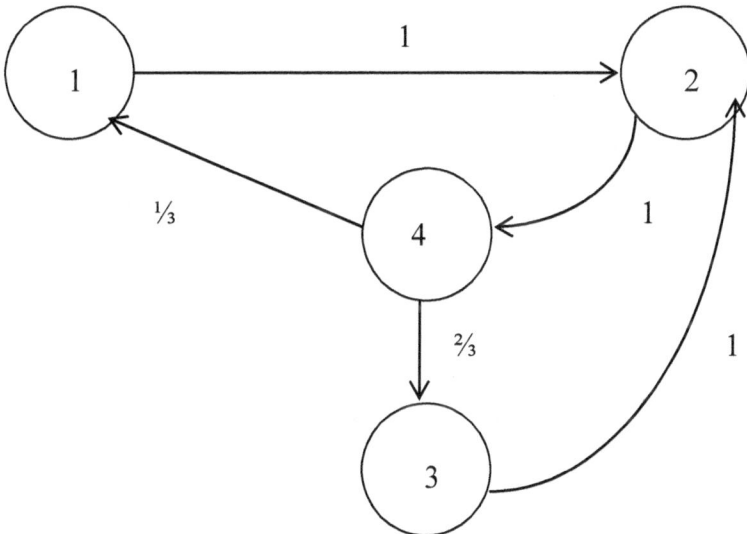

The set $C = \{1, 2, 3, 4\}$, is irreducible since every state can be reached from every other state.

EXAMPLE 8.12

Random walk with absorbing barriers at 0 and b

$$
\begin{pmatrix}
 & 0 & 1 & 2 & 3 & \cdots & b-2 & b-1 & b \\
0 & 1 & 0 & 0 & 3 & \cdots & 0 & 0 & 0 \\
1 & q & r & p & 0 & \cdots & o & 0 & 0 \\
2 & 0 & q & r & p & \cdots & 0 & 0 & 0 \\
3 & 0 & 0 & q & r & \cdots & 0 & 0 & 0 \\
\vdots & \vdots & \vdots & \vdots & \vdots & \vdots & \vdots & \vdots & \vdots \\
b-2 & 0 & 0 & 0 & 0 & \cdots & r & p & 0 \\
b-1 & 0 & 0 & 0 & 0 & \cdots & q & r & p \\
b & 0 & 0 & 0 & 0 & \cdots & 0 & 0 & 1
\end{pmatrix}
$$

The state space $S = \{0,1,2,3,4,\ldots\ldots,b-2,b-1,b\}$ decomposes into $C_1 = \{0\}$, $C_2 = \{1, 2, 3, \ldots b-1\}$ and $C_3 = \{b\}$. The states $\{0\}$ and $\{b\}$ are closed absorbing states.

EXAMPLE 8.13

Markov chain with state space $S = \{1,2,3,4,5\}$.

$$
\begin{pmatrix}
 & 1 & 2 & 3 & 4 & 5 \\
1 & \dfrac{1}{2} & 0 & 0 & \dfrac{1}{2} & 0 \\
2 & \dfrac{1}{2} & \dfrac{1}{2} & 0 & 0 & 0 \\
3 & \dfrac{1}{4} & \dfrac{1}{2} & 0 & 0 & \dfrac{1}{4} \\
4 & 0 & 0 & 0 & 1 & 0 \\
5 & 0 & 0 & \dfrac{1}{2} & \dfrac{1}{4} & \dfrac{1}{4}
\end{pmatrix}
$$

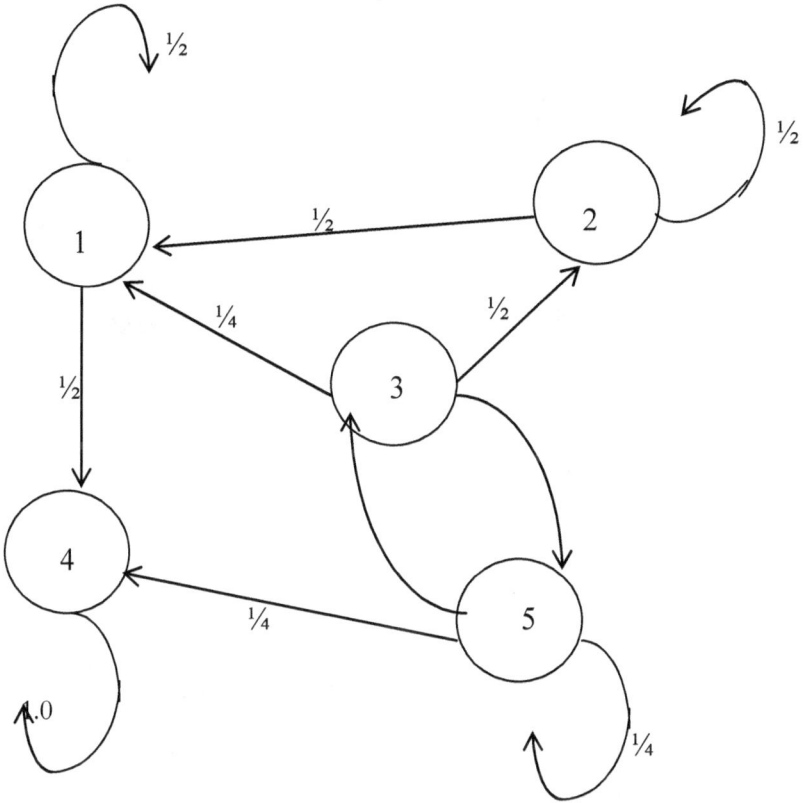

The state space decomposes into
$C_1 = \{1, 2, 3, 5\}$, and $C_2 = \{4\}$. The state $C_2 = \{4\}$ is a closed absorbing
state.

EXAMPLE 8.14

Markov chain with state space $S = \{1, 2, 3, 4, 5\}$

$$
\begin{pmatrix}
 & 1 & 2 & 3 & 4 & 5 \\
1 & 0 & \frac{1}{3} & 0 & \frac{1}{3} & \frac{1}{3} \\
2 & \frac{1}{3} & \frac{1}{3} & 0 & \frac{1}{3} & 0 \\
3 & 0 & 0 & \frac{2}{3} & 0 & \frac{1}{3} \\
4 & \frac{1}{4} & \frac{1}{4} & 0 & \frac{1}{4} & \frac{1}{4} \\
5 & 0 & 0 & \frac{1}{3} & 0 & \frac{2}{3}
\end{pmatrix}
$$

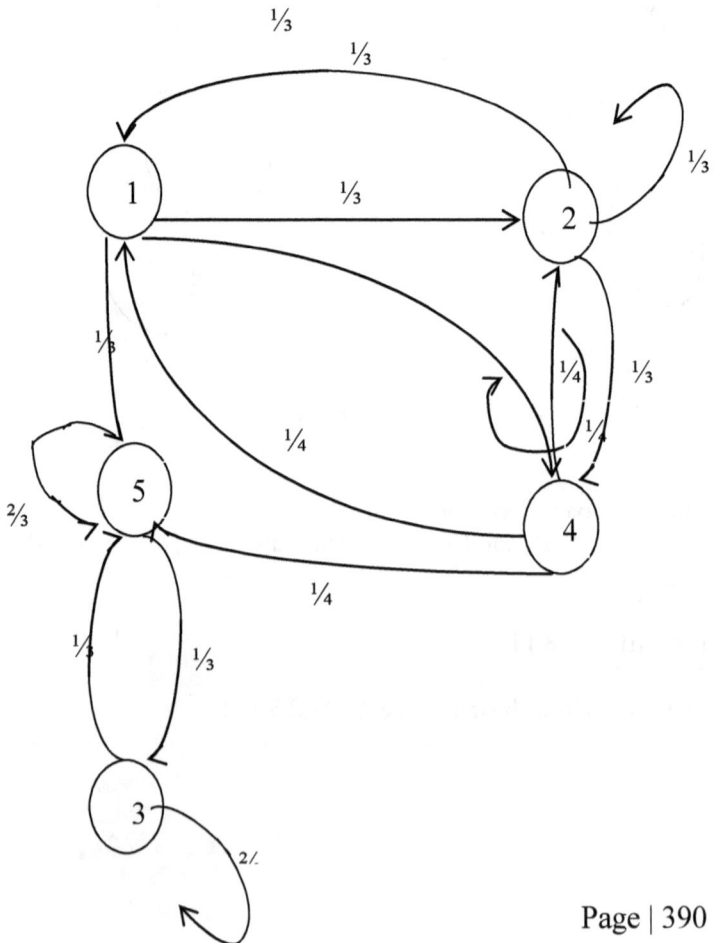

The state space decomposes into

$C_1 = \{3, 5\}$ and $C_2 = \{1, 2, 4\}$. C_1 is a closed communicating class.

ASSIGNMENT 8.5

Determine the classes of the various states for the Markov chains with transition probability matrices given below:

(i)

	1	2	3	4	5
1	0.2	0.3	0.5	0	0
2	0.7	0.3	0	0	0
3	0	1.0	0	0	0
4	0	0	0	0.4	0.6
5	0	0	0	1.0	0

(ii)

	1	2	3	4
1	0	0	1.0	0
2	1.0	0	0	0
3	0.3	0.7	0	0
4	0.6	0.2	0.2	0

(iii)

	1	2	3	4
1	0	0	1	0
2	1	0	0	0
3	$\frac{1}{2}$	$\frac{1}{2}$	0	0
4	$\frac{1}{3}$	$\frac{1}{3}$	$\frac{1}{3}$	0

(iv)

	1	2	3	4
1	0	1	0	0
2	0	0	0	1
3	0	1	0	0
4	$\frac{1}{3}$	0	$\frac{2}{3}$	0

(v)

	0	1	2	3	4
0	0	$\frac{1}{3}$	0	$\frac{1}{3}$	$\frac{1}{3}$
1	$\frac{1}{3}$	$\frac{1}{3}$	0	$\frac{1}{3}$	0
2	0	0	$\frac{2}{3}$	0	$\frac{1}{3}$
3	$\frac{1}{4}$	$\frac{1}{4}$	0	$\frac{1}{4}$	$\frac{1}{4}$
4	0	0	$\frac{1}{3}$	0	$\frac{2}{3}$

8.3.12 Periodicity of a Markov Chain

Let $\{X_n, n \geq 0\}$ be a Markov chain, with state space $S = \{0, 1, 2, \ldots n\}$. A state $i \in S$ is said to be a periodic state if its period, $d(i)$, is defined by

d(i) = {GCD of all integers n ≥ 1 for which $P_{ii}^n > 0$} is greater than 1. [GCD = greatest common divisor]. If d(i)=1, then state I is said to be aperiodic.
[A Markov chain in which each state has period 1(one) is called aperiodic.]

$$[\text{If } P_{ii}^n = 0 \text{ for all } n \geq 1 \text{ define } d(i) = 0].$$

We state without proof some basic properties of the period of a state.

8.3.12.1 Theorem

Let $\{X_n, n \geq 0\}$ be a Markov chain with state space S and i ∈ S be a periodic state. Then for some state j ∈ S.
(i)   If i ↔ j then $d(i) = d(j)$. That is, if state i communicates with state j, then the period of i equals the period of j.

This assertion shows that the period is a constant in each class of communicating states.
(ii)   If state i has period d(i) then there exists an integer N depending on i such that for all integers

$$n \geq N \quad P_{ii}^{nd(i)} > 0$$

This asserts that a return to state i can occur at all sufficiently large multiples of the period d(i).
(iii)   If $P_{ji}^m > 0$, then $P_{ji}^{m+nd(i)} > 0$ for all m, n (positive integers) sufficiently large.
(iv)   An irreducible Markov chain is aperiodic if $P_{i,i} > 0$ for some i ∈ S.

EXAMPLE 8.15

Consider the Markov chains with transition probability matrices given

below.

$$
\begin{pmatrix}
 & 1 & 2 & 3 & 4 \\
1 & 0 & 0 & 1 & 0 \\
2 & 1 & 0 & 0 & 0 \\
3 & \dfrac{1}{2} & \dfrac{1}{2} & 0 & 0 \\
4 & \dfrac{1}{3} & \dfrac{1}{3} & \dfrac{1}{3} & 0
\end{pmatrix}
$$

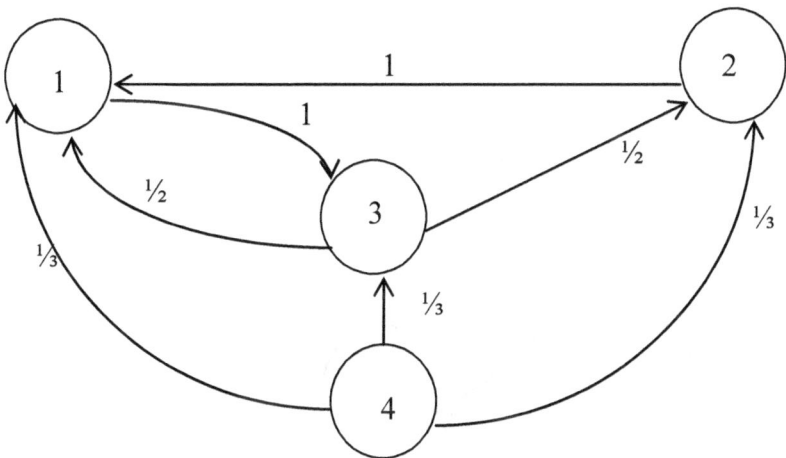

From the transitions shown in the graph above, $1 \to 3 \to 1$ of state 1 back to itself and $2 \to 1 \to 3 \to 2$ of state 2 back to itself, the period of state 1 is 2 and the period of state 2 is 3. Since state 1 communicates with state 3, i.e. $1 \leftrightarrow 3$, by Theorem 8.3.12.1 (i) above, the period of state 3 is also 2.

State 4 has period 0 since $P_{44}^{n} = 0 \; for \; all \; n \geq 1$.

(b)

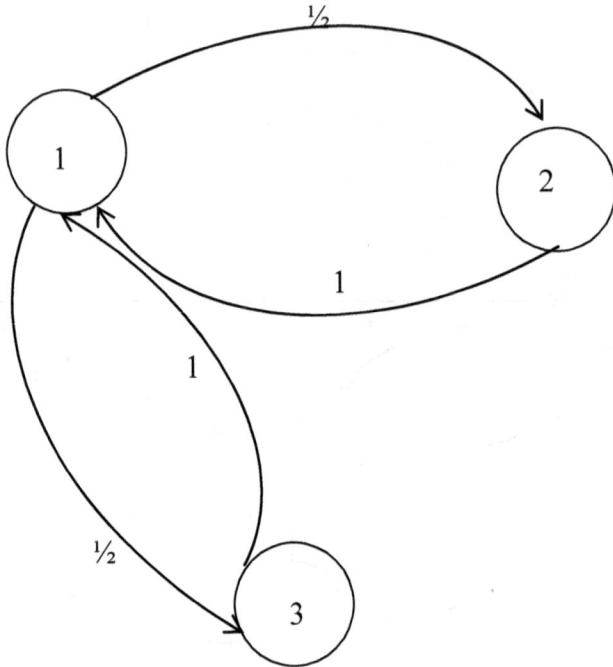

From the transitions shown in the graph above,

$1 \to 2 \to 1$ (from state 1 back to itself)

$1 \to 3 \to 1$ (from state 1 back to itself)

$2 \to 1 \to 2$ (from state 1 back to itself)

$3 \to 1 \to 3$ (from state 1 back to itself)

the period of state 1 is 2, the period of state 2 is 2 and the period of state 3 is also 2. As the Markov chain is irreducible, it follows that the MC is periodic with period 2.

EXAMPLE 8.16

In the finite Markov chain $\{X_n, n \geq 0\}$ with state space $S = \{0, 1, 2, ..$

n–1, n\},

$$\begin{pmatrix} & 0 & 1 & 2 & \cdots & n-1 & n \\ 0 & 0 & 1 & 0 & \cdots & 0 & 0 \\ 1 & 0 & 0 & 1 & \cdots & 0 & 0 \\ \vdots & \vdots & \vdots & \vdots & \vdots & \vdots & \vdots \\ \vdots & \vdots & \vdots & \vdots & \vdots & \vdots & \vdots \\ n-1 & 0 & 0 & \cdots & \cdots & \cdots & 1 \\ n & 1 & 0 & 0 & \cdots & 0 & 0 \end{pmatrix}$$

It follows from the transitions

$$0 \to 1 \to 2 \ldots \ldots \to n{-}1 \to n \to 0$$

with probability $P_{ii}^n = 1, n \geq 1$, that state 0 has period n. Since the chain is irreducible, it follows that each state has period n.

8.3.13 Recurrence and Transience

Consider an arbitrary, but fixed, state i. We define for each integer $n \geq 1$,

$$f_{ii}^n = P(X_n = i, X_v \neq i, v = 1,2, \ldots \ldots = i\, n - 1/X_0 = i).$$

In other words, f_{ii}^n is the probability that, starting from state i, the first return to state i occurs at the n-th transition. Clearly $f_{ii}^1 = P_{ii}$ and f_{ii}^n may be calculated recursively according to

$$P_{ii}^n = \sum_{k=0}^n f_{ii}^k P_{ii}^{n-k}, \ n \geq 1 \ \ldots\ldots\ldots\ldots\ldots\ldots\ldots\ldots 8.3.13.1,$$

Where we define $f_{ii}^0 = 0,\ for\ all\ i.$ Consider all the possible realizations of the process for which $X_0 = i, X_n = i$ and the first return to state I occurs at the k-th transition. Call this event E_k. The event E_k (k=1,2,…………,n) are clearly mutually exclusive. The probability of the event that the first return is at the k-th transition is by definition f_{ii}^k. In the remaining n-k transitions, we are dealing only with those realizations for which $X_n = i$. Using the Markov property, we have,

$$P(E_k) = P(first\ return\ is\ at\ the\ k-th\ transition/X_k = i) *$$
$$P(X_n = i/X_k = i)$$
$$= f_{ii}^k P_{ii}^{n-k}, = 1 \leq k \leq n.\ Recall\ that\ P_{ii}^0 = 1.\ Hence,$$
$$P(X_n = i/X_0 = i) = \sum_{k=1}^n P(E_k) = \sum_{k=1}^n f_{ii}^k P_{ii}^{n-k} = \sum_{k=0}^n f_{ii}^k P_{ii}^{n-k},$$

Since by definition $f_{ii}^0 = 0.$

3.13.1 Definition (Recurrent and Transient States)

The generating function
$P_{ij}(s)$ *of the sequence* $\{P_{ij}^n\}$ *is*

$$P_{ij}(s) = \sum_{n=0}^{\infty} P_{ij}^n S^n \text{ for } |S| < 1 \dots \dots \dots 8.3.13.1.1$$

In a similar manner we define the generating function of the sequence $\{f_{ij}^n\}$ *when* $i \neq j$.

$$F_{ij}(s) = \sum_{n=0}^{\infty} f_{ij}^n S^n \text{ for } |S| < 1 \dots \dots \dots \dots \dots .8.3.13.1.2$$

If
$A(s) = \sum_{k=0}^{\infty} a_k S^k, \text{ and } B(s) = \sum_{n=0}^{\infty} b_l S^l \dots \dots \dots 8.3.13.1.3$
Then $A(s)B(s) = C(s) = \sum_{r=0}^{\infty} C_r S^r \text{ for } |S| < 1 \dots .8.3.13.1.4$
where
$C_r = a_0 b_r + a_1 b_{r-1} + \dots \dots a_r b_0 \dots \dots \dots \dots \dots 8.3.13.1.5$
If we identify the the
$a_k^{\prime s}$ *with the* $f_{ii}^{k,s}$ *and the* $b_l^{\prime s}$ *with the* $P_{ii}^{n \prime s}$, then comparing
equations 8.3.13.1 and 8.3.13.1.5 we obtain
$F_{ii}(s)P_{ii}(s) = P_{ii}(s) - 1 \text{ for } |S| < 1 \dots \dots \dots \dots .8.3.13.1.6$
or $P_{ii}(s) = \frac{1}{1 - F_{ii}(s)} \text{ for } |S| < 1 \dots \dots \dots \dots \dots 8.3.13.1.7$
Subtracting the constant 1(one) in 8.3.13.1.6 is necessary since (1) is not valid for n=0. By argument analogous to that which led to 8.3.13.1 we obtain

$$P_{ij}^n = \sum_{k=0}^{n} f_{ik}^k P_{kj}^{n-k}, \quad i \neq j, \ n \geq 0 \dots \dots \dots \dots \dots 8.3.13.1.8$$

Where f_{ij}^k is the probability that the first passage from state i to state j occurs at the k-th transition. Again we define $f_{ij}^0 = 0$ for all i and j. It follows from 8.3.13.1.8 ,if we refer to 8.3.13.1.4 that
$P_{ij}(s) = F_{ij}(s)P_{jj}(s) \text{ for } |S| < 1 \dots \dots \dots \dots \dots .8.3.13.1.9$
We say that a state i is recurrent if and only if $\sum_{n=1}^{\infty} f_{ii}^n = 1$. This says that a state i is recurrent if and only if, starting from state i, the

probability of returning to state i after some finite length of time is 1(one). A non-recurrent state is said to be transient.

8.3.13.2 Definition

A state $i \in S$ is called recurrent if $\sum_{n=0}^{\infty} f_{ii}^n = 1$ that is, state i is recurrent if with probability 1(one) the Markov chain having started from i, will eventually return to i. A state i is transient if $\sum_{n=0}^{\infty} f_{ii}^n < 1$. Recurrent states are also called Persistent states, and transient states are also called nonrecurrent states.

We state without proof the following theorems.

8.3.13.3 Theorem

Let i be recurrent state, and $\mu(i)$ =the expected number of returns. Then if g(i,i)=1, and $\mu(i, i) = \infty$, the particle starting at a recurrent state i returns to i infinitely often. If $\sum f_{ui}^n > 0$ $then$ $\mu(u, i) = \infty$, that is if a particle starting from a state u hit the recurrent could state i at some time, it visits i infinitely often.

8.3.13.4 Definition.

A recurrent state i is positive recurrent if $\mu(i) < \infty$ and i is null recurrent if $\mu(i) = \infty$

8.3.13.5 Theorem

(i) Let $i, j \in S$ and $i \leftrightarrow j$. If i is positive recurrent, then j is also positive recurrent.
(ii) Every irreducible finite Markov chain is a positive recurrent chain.
(iii)A positive recurrent aperiodic state is called ergodic.

8.3.13.5.1 NOTE: A Markov chain with an absorbing state (or closed communicating class) is said to be non-irreducible. In a finite, non-irreducible Markov chain, any state that does not belong to a closed communicating class is termed a transient state. Clearly an irreducible finite Markov chain can have no transient states since the class of all states forms a closed communicating class.

Since a process can move from a transient state into a closed communicating class, but not vice versa, it seems reasonable to say that, if a finite Markov $\{X_n, n \geq 0\}$ starts in a transient state (X_0 transient), then the chain is certain (with probability 1) eventually to enter some closed communicating class.

EXAMPLE 8.17

Let $\{X_n, n \geq 0\}$ be a finite Markov chain with state space
$S = \{0, 1, 2, \ldots 6\}$ and one step transition matrix given below.
Determine the recurrent and transient states.

$$
P = \begin{array}{c}
\textit{current state} \downarrow \\
0 \\
1 \\
2 \\
3 \\
4 \\
5 \\
6
\end{array}
\left(
\begin{array}{ccccccc}
0 & 1 & 2 & 3 & 4 & 5 & 6 \\
0.4 & 0 & 0.2 & 0.2 & 0.2 & 0 & 0 \\
0 & 0 & 1.0 & 0 & 0 & 0 & 0 \\
0 & 0 & 0 & 1.0 & 0 & 0 & 0 \\
0 & 1.0 & 0 & 0 & 0 & 0 & 0 \\
0 & 0 & 0 & 0 & 0.7 & 0 & 0.3 \\
0 & 0 & 0 & 0 & 0.5 & 0.5 & 0 \\
0 & 0 & 0 & 0 & 0 & 0.4 & 0.6
\end{array}
\right)
$$

From the transitions shown in the graph below, it is clear that the state 0
leads to every other state i in S, whereas no state $x \neq 0$ leads to 0.
Also, $C_1 = \{1,2,3\}$ and $C_2 = \{4,5,6\}$ are irreducible closed sets of states.
Hence 0 is transient and the rest of the states are recurrent.

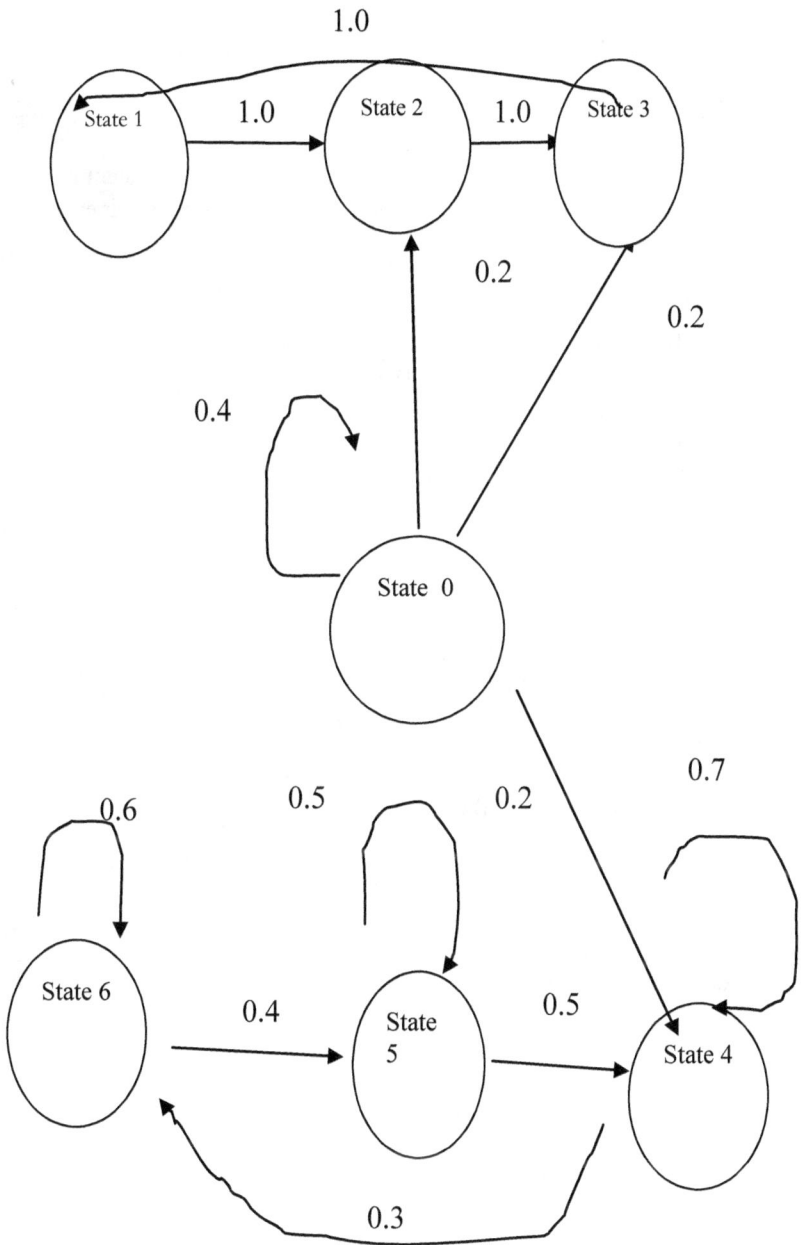

ASSIGNMENTS 8.6

(i)   Determine the classes and classify the states into transient and recurrent states for the Markov chain with transition matrices given below:

(ii)

(a)
$$
\begin{array}{c|ccccc}
 & 0 & 1 & 2 & 3 & 4 \\
\hline
0 & 0.0 & 1.0 & 0.0 & 0.0 & 0.0 \\
1 & 1.0 & 0.0 & 0.0 & 0.0 & 0.0 \\
2 & 0.1 & 0.1 & 0.3 & 0.5 & 0.0 \\
3 & 0.3 & 0.7 & 0.0 & 0.0 & 0.0 \\
4 & 0.3 & 0.4 & 0.2 & 0.0 & 0.1
\end{array}
$$

(b)
$$
\begin{array}{c|cccc}
 & 1 & 2 & 3 & 4 \\
\hline
1 & 0.2 & 0.0 & 0.0 & 0.8 \\
2 & 1.0 & 0.0 & 0.0 & 0.0 \\
3 & 0.0 & 1.0 & 0.0 & 0.0 \\
4 & 0.0 & 0.0 & 1.0 & 0.0
\end{array}
$$

(c)
$$
\begin{array}{c|cccccc}
 & 1 & 2 & 3 & 4 & 5 & 6 \\
\hline
1 & 0.1 & 0.9 & 0.0 & 0.0 & 0.0 & 0.0 \\
2 & 1.0 & 0.0 & 0.0 & 0.0 & 0.0 & 0.0 \\
3 & 1.0 & 0.0 & 0.0 & 0.0 & 0.0 & 0.0 \\
4 & 0.0 & 0.2 & 0.8 & 0.0 & 0.0 & 0.0 \\
5 & 0.3 & 0.0 & 0.0 & 0.0 & 0.3 & 0.4 \\
6 & 0.0 & 0.0 & 0.0 & 0.7 & 0.3 & 0.0
\end{array}
$$

(d)
$$
\begin{array}{c|ccccc}
 & 1 & 2 & 3 & 4 & 5 \\
\hline
1 & 0.4 & 0.5 & 0.1 & 0.0 & 0.0 \\
2 & 0.0 & 0.3 & 0.2 & 0.5 & 0.0 \\
3 & 0.0 & 1.0 & 0.0 & 0.0 & 0.0 \\
4 & 1.0 & 0.0 & 0.0 & 0.0 & 0.0 \\
5 & 0.3 & 0.7 & 0.3 & 0.0 & 0.0
\end{array}
$$

8.3.14 Stationary Distribution

Consider a Markov particle beginning its motion from an arbitrary state i. In studying the subsequent motion, we would like to ask what could be said about the motion after the elapse of a large number of steps. In physical terms, this is a question about the asymptotic stability of the motion, that is, we want to look at the limiting steady –state distribution, irrespective of the initial position of the particle.

Let $p_n(i) = P(X_n = i), i \in S, n \geq 0.$ Then
$p_n(i) \geq 0$ and $\sum_{i \in S} p_n(i) = 1, n \geq 0.$ Also
$p_{n+m}(i) = \sum_{j \in S} p_n(j) p_m(j, i), j \in S, n, m, \geq 0 \dots \dots .8.3.14.1$

If the absolute distributions $p_n(i)$ are independent of n, say
$p(i) = P(X_n = i), i \in S, \ n \geq 0,$ then the probability distribution
$(p(j), j \in S)$ is called the steady distribution of the Markov chain
$(X_n, \ n \geq 0).$ From 8.3.14.1
$p(i) = \sum_{j \in S} p(j) p^n(j, i), \ i \in S, \ n \geq 0 \dots \dots \dots \dots \dots 8.3.14.2$

Naturally, the existence of a steady state distribution is related to a nontrivial and non-negative solution of
$\pi(i) = \sum_{j \in S} \pi(j) p(j, i), \ i \in S \dots \dots \dots \dots \dots \dots \dots ..8.3.14.3$

8.3.14.1 Definition.

If $\pi(i) \geq 0$ for all $i \in S,$ and $\sum_{i \in S} \pi(i) = 1$ and satisfies
$\pi(i) = \sum_{j \in S} \pi(j) p(j, i), \ \{\pi(i), i \in S\}$ is then called a stationary distribution.

EXAMPLE 8.17

Let $X = (X_n, n \geq 0)$ be a Markov Chain with state space $S = (1,2,3,4,5)$ and transition matrix M given below.

$$M = \begin{pmatrix} \text{current state} \downarrow & 1 & 2 & 3 & 4 & 5 \\ 1 & 0.0 & 0.5 & 0.5 & 0.0 & 0.0 \\ 2 & 0.0 & 0.0 & 0.0 & 0.2 & 0.8 \\ 3 & 0.0 & 0.0 & 0.0 & 0.4 & 0.6 \\ 4 & 1.0 & 0.0 & 0.0 & 0.0 & 0.0 \\ 5 & 1.0 & 0.0 & 0.0 & 0.0 & 0.0 \end{pmatrix}$$

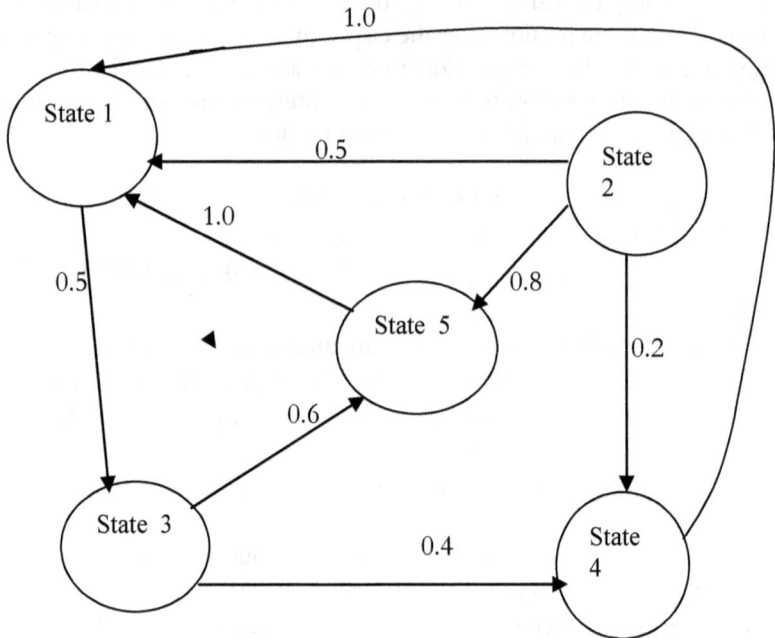

It follows from the transitions shown in graph above:

$1 \to 2 \to 4 \to 1, \quad 1 \to 3 \to 4 \to 1, \quad and \quad 1 \to 3 \to 5 \to 1$

That the Markov chain X is an irreducible chain. Since S is finite, X is a recurrent chain. From the transitions shown from state 1 back to itself, it is clear that the period of state 1 is 3. As it is irreducible, it follows from Theorem 8.3.8.1 above, that X is periodic with period 3.

To find the stationary distribution $\pi(i)$ of X, we use the defining relations of

$\pi(i): \pi(j) = \sum_i \pi(i)p(i,j)$. Then from matrix M

$\pi(1) = \pi(1)p(1,1) + \pi(2)p(2,1) + \pi(3)p(3,1) + \pi(4)p(4,1) + \pi(5)p(5,1)$

$\qquad = \pi(4) + \pi(5)$

$\pi(2) = 0.5\pi(1), \quad \pi(3) = 0.5\pi(1), \quad \pi(4) = 0.2\pi(2) + 0.4\pi(3),$

$\qquad\qquad \pi(5) = 0.8\pi(2) + 0.6\pi(3)$

Solving this set of equations along with $\pi(1) + \pi(2) + \pi(3) + \pi(4) + \pi(5) = 1$ since π is a probability distribution, we obtain

$\pi = \left(\frac{1}{3},\frac{1}{6},\frac{1}{6},\frac{1}{10},\frac{7}{30}\right) = \{\pi(i),\ i \in S\}$ as the stationary probability distribution.

EXAMPLE 8.18

Find the stationary distribution concentrated on each of the irreducible closed sets of the Markov chain with state space $S=(0,1,2,3,4,5,6)$ and transition matrix is M given below:

$$M = \begin{pmatrix} currentstate \downarrow & 0 & 1 & 2 & 3 & 4 & 5 & 6 \\ 0 & 0.1 & 0.1 & 0.2 & 0.2 & 0.4 & 0.0 & 0.0 \\ 1 & 0.0 & 0.0 & 0.5 & 0.5 & 0.0 & 0.0 & 0.0 \\ 2 & 0.0 & 0.0 & 0.0 & 1.0 & 0.0 & 0.0 & 0.0 \\ 3 & 0.0 & 1.0 & 0.0 & 0.0 & 0.0 & 0.0 & 0.0 \\ 4 & 0.0 & 0.0 & 0.0 & 0.0 & 0.5 & 0.0 & 0.5 \\ 5 & 0.0 & 0.0 & 0.0 & 0.0 & 0.5 & 0.0 & 0.5 \\ 6 & 0.0 & 0.0 & 0.0 & 0.0 & 0.0 & 0.5 & 0.5 \end{pmatrix}$$

From the transitions shown in graph below, it is seen that the two irreducible closed sets are $C_1=(1,2,3)$ and $C_2=(4,5,6)$.

On C_1, the distribution $\{\pi(i)\}$ satisfies $\pi(1) = \pi(3)$, $\pi(2) = 0.5\pi(1)$, $\pi(3) = 0.5[\pi(1) + \pi(2)]$ from which the stationary distribution $\pi(1)$ corresponding to C_1 is given by

$\pi(1) = \left(0,\frac{2}{5},\frac{1}{5},\frac{2}{5},0,0,0\right)$. Similarly, the stationary distribution corresponding to class C_2 is given by $\pi(2) = (0,0,0,0,\frac{1}{3},\frac{1}{3},\frac{1}{3})$

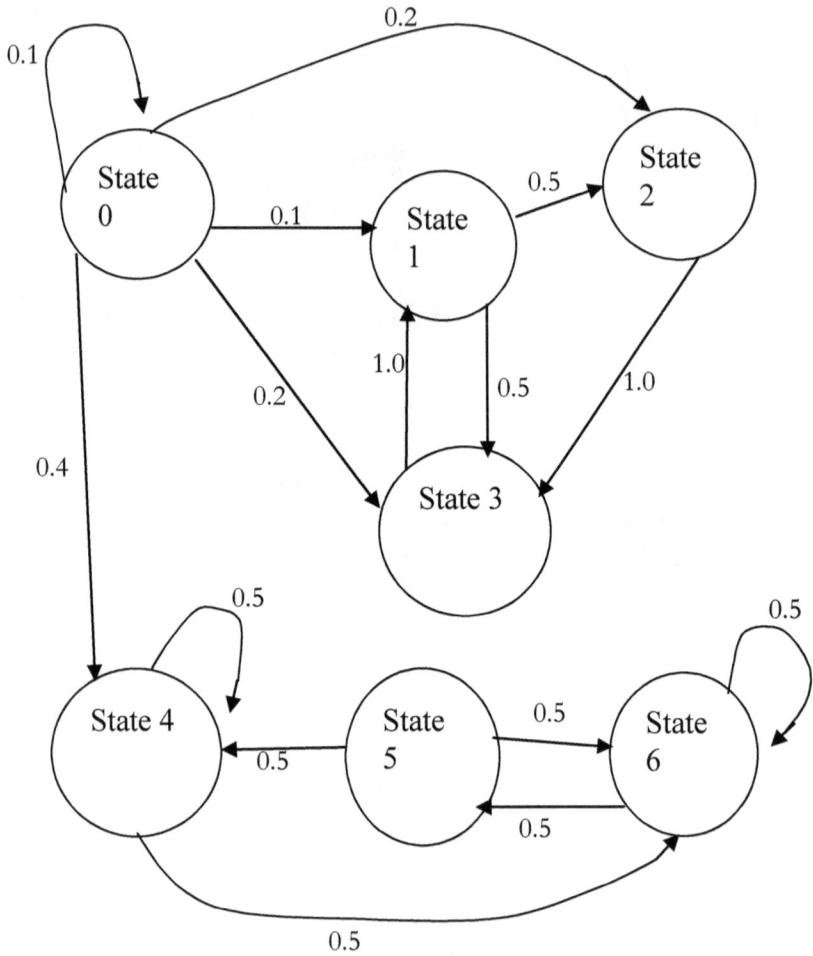

Find the stationary distribution of each of the Markov chain whose transition matrices are given by:

(i) $M_1 =$

current state ↓	1	2	3	4	5
1	1.0	0.0	0.0	0.0	0.0
2	0.3	0.4	0.3	0.0	0.0
3	0.0	0.3	0.7	0.0	0.0
4	0.0	0.0	0.2	0.0	0.8
5	0.0	0.0	0.0	0.0	1.0

(ii) $M_2 =$

current state ↓	1	2	3	4	5
1	0.1	0.1	0.1	0.3	0.4
2	0.0	0.7	0.3	0.0	0.0
3	0.0	0.0	0.1	0.2	0.7
4	0.3	0.1	0.1	0.4	0.1
5	0.0	0.0	0.3	0.3	0.4

8.4. Stochastic Processes (Continuous Time Parameter Case

In the previous sections, we discussed the discrete-valued stochastic processes $\{X_n; n = 0, 1, 2, \ldots \ldots\}$ developing in discrete time. In the following sections we turn our attention to discrete processes $X(t): t \geq \geq 0\}$ developing in continuous time. Specifically, we shall be looking at situations where the random variable X(t) takes non-negative integer values only: for example X(t) may be the number of individuals alive at time t in an evolving continuity, or the number of individuals in a community who by time t have caught a disease etc.
Some Stochastic Processes are non-decreasing, for instance: (i) counting arrivals at a particular location where individuals arrive according to a Poisson Process. The Poisson Process is the simplest of the various integer-valued continuous time process. There are many applications where it can be used as a model: In counting radioactive emissions from an isotope, recording arrivals at a bank or supermarket or at a doctor's surgery etc. In the Poisson Process we may be concerned with either the arrival times or the corresponding realization, which shows the total number of arrivals so far. Usually it is the second process of the two that is considered, the integer-valued process $\{X(t); t \geq 0\}$ which is a non-decreasing process developing in continuous time.(ii)another example is the Simple Birth Process: individuals born into a community

themselves independently generate further offsprings according to a
Poisson Process, or a combination of both. The more individuals there
are alive at any time, the more frequent the incidence of births is likely to
be.

Some Stochastic Processes are non-increasing, for instance Death
Processes- in this a population which may be initially quite large
eventually vanishes once everybody in it has died. There are no births or
arrivals from outside to balance the deaths, so this is a non-increasing
process.

Many Stochastic Processes are subject to decrease as well as increase.
For instance, if departures occur as well as arrivals, then the number of
individuals present at a location can fall as well as rise. Examples of this
type include the number of aircrafts on the ground at the airport, or the
number of customers present in a supermarket, or the number of
patients in a surgery waiting to see the doctor. Another example of this
type is a Birth and Death Process, members of a community generate
offspring as in a Birth Process, but an added feature is that individuals
do not live forever—they die, at some random age, and so the
population size goes down as well as up.

8.4.1. Independent and Stationary Increments

A Stochastic Process$\{X(t): t \geq 0\}$, is said to have independent
increments if for $n \geq 1$ and time points $0 \leq t_0 < t_1 < \cdots \ldots . . t_n$ the
increments

$$X(t_0) - X(0), X(t_1) - X(t_0), \ldots \ldots . . X(t_n) - X(t_{n-1}) \text{ are}$$
stochastically independent

random variables. A process $X(t)$ is said to possess stationary
increments if the distributions of the increments X(t) - X(r), r<t, depend
only on the length $(t - r)$ of the time interval [r,t] over which we have
taken the individual increment.

8.4.2 The Simple Poisson Process

A counting process$\{X(t): t \geq 0\}$, denoting the number of events that have occurred by time t, is said to be a Poisson Process with rate $\lambda > 0$ if:

(i) X(0)=0, that is the system starts from rest at time t=0, (ii) X(t) is a process with independent stationary increments, and (iii) the number of events in any interval of length t is Poisson distributed with rate λt, that is, for all $r, t \geq 0$,

$$P\{X(t + r) - X(r) = x\} = e^{-\lambda t} \cdot \frac{(\lambda t)^x}{x!}, \qquad x = 0,1,2, \dots \dots \dots \dots \dots$$

This definition of the Poisson Process is equivalent to the following postulates:

8.4.2.1 Postulates of the Poisson Process

(i) The probability that (exactly) one event occurs in any small time interval $[t, t + \delta t]$ is
equal to $\lambda \delta t + 0(\delta t)$

$$ie. P[X(t + \delta t) - X(t) = 1/X(t) = x] = \lambda \delta t + 0(\delta t) \ as \ \delta t$$
$$\to 0, for \ x = 0, 1, 2, \dots \dots \dots \dots .8.4.2.1.1$$

Equation 8.4.2.1.1 above, could also be written as:

$$P[X(t + \delta t) = x + 1/X(t) = x] = \lambda \delta t + 0(\delta t) .$$

$$as \ \delta t \to 0, for \ x = 0, 1, 2, \dots \dots \dots \quad 8.4.2.1.2$$

$$P[X(t + \delta t) = x + 1/X(t) = x] = \lambda \delta t + 0(\delta t) .$$

$$as \ \delta t \to 0, for \ x = 0, 1, 2, \dots \dots \dots \quad 8.42.1.2$$

The precise interpretation of equations 8.4.2.1.1 and 8.4.2.1.2 is the relationship

$$\lim_{\delta t \to 0+} \frac{P\{X(t + h) - X(t) = 1/X(t) = x\}}{\delta t} = \lambda$$

The $0(\delta t)$ *symbol* means that if we divide this term by δt then its value tends to zero. Notice that the right hand side is independent of x. That is, a function f(x) is said to be

of order $0(\delta t)$ *if* $\lim_{\delta t \to 0} \frac{f(\delta t)}{\delta t} = 0.$

(ii)   As a consequence of (i), the probability that no event occurs in any small time

interval $[t, t + \delta t]$ *is equal to* $1 - \lambda \delta t + 0(\delta t)$ *as* $\delta t \to 0.$
  ie. $P[X(t + \delta t) - X(t) = 0 / X(t) = x] = 1 - \lambda \delta t + 0(\delta t)$ *as* $\delta t \to 0$

(iii)The probability that two or more events occur in any small time interval
$[t, t + \delta t]$ is equal to
  $0(\delta t)$ *ie.* $P[X(t + \delta t) - X(t) = 2 / X(t) = x] = 0(\delta t)$

(iv) $X(0) = 0$, the process starts from rest at time t=0 with zero count.

(v) The occurrence of events after any time t is independent of the occurrence of events before time t.
The probability $P[X(t) = x]$ is denoted by

  $p_x(t)$; *or more precisely* $P[X(t) = x / X(0) = 0] = p_x(t)$

8.4.2.2   The Differential–difference equations for the Poisson Process

Suppose we start observing a Poisson process at time 0(zero) and continue for a fixed time t. Then, the number of events that occur in the interval [0, t] is the random

Variable X(t). Denote the probability function by $p_x(t)$, *so* $p_x(t) = P[X(t) = x]$

The probability that no event occurs in [0,t] is:

$p_0(t) = P[X(t) = 0]$= probability that no event occurs in [0,t].
Consider the interval [0,t] and a further interval $[t, t + \delta t]$; the probability that no events have occurred by the end of the second interval is denoted by $p_0(t + \delta t)$. Since by one of the Poisson postulates, the occurrence of events after any time interval t is independent of what happened before time t, so

$p_0(t + \delta t) = P[no\ events\ occurred\ by\ time\ t + \delta t]$
$= P\{[no\ events\ in\ [0, t]\} \times P\{no\ events\ in\ [t, t + \delta t]\}$
$= p_0(t)$
$\times [1 - prob(1\ event\ in\ (t, t + \delta t) - p(two\ or\ more\ events\ in$
$(t, t + \delta t)]$
$= p_0(t)[1 - (\lambda\delta t + 0(\delta t) - 0(\delta t)] = p_0(t)[1 - \lambda\delta t + 0(\delta t)]$
$\therefore p_0(t + \delta t)) = p_0(t)\{1 - \lambda\delta t + 0(\delta t),$ hence

$$\frac{p_0(t+\delta t) - p_0(t)}{\delta t} = -\lambda p_0(t) + \frac{0(\delta t)}{\delta t} \quad and$$

$$\lim_{\delta t \to 0} \frac{p_0(t+\delta t) - p_0(t)}{\delta t} = p_0^{'}(t) = \frac{dp_0(t)}{dt}.$$

Since the

$$limit: \lim_{\delta t \to 0} \frac{0(\delta t)}{\delta} \to 0, \quad we\ have$$

$$\frac{dp_0(t)}{dt} = -\lambda p_0(t) \dots \dots \dots \dots \dots \dots \dots \dots \dots \dots \dots \dots \dots .8.4.2.2.1$$

Integrating both sides of 8.4.2.2.1 and using the initial condition that at time zero (t=0) the process starts from rest (x=0), and the probability $p_0(0) = (no\ occurrence\ in\ the\ interval\ [x = 0, t = 0]) = 1,$ we obtain

$ln\ p_0(t) = -\lambda t\ or\ p_0(t) = e^{-\lambda t} \dots \dots \dots 8.4.2.2.2.$

For x=1,2,........., the differential equation is derived similarly.

$p_x(t + \delta t) = P[\ x\ events\ by\ time\ t + \delta t]$
$= P[x\ in\ (0, t)and\ none\ in\ (t, t + \delta t)]$
$\cup P[(x - 1)\ events\ in\ (0, t)\ and\ one\ in\ (t, t + \delta t)]$
$\cup P[fewer\ than\ (x$
$- 1)\ events\ in\ (0, t)\ and\ more\ than\ one\ event\ in$

$$(t, t + \delta t)]$$

Thus in the union of three mutually exclusive events, we can add the separate probabilities:

$$p_x(t + \delta t) = P[x \text{ events in } (0, t) \cap none \text{ in } (t, t + \delta t)]$$
$$+P[(x - 1) in (0, t) \cap (1 \text{ in } (t, t + \delta t)]$$
$$+P[fewer \text{ than } (x - 1) \text{ events in } (0, t)$$
$$\cap (more \text{ than one event } in(t, t + \delta t)]$$

This gives:

$$p_x(t + \delta t) = p_x(t) \times [1 - \lambda\delta t + 0(\delta t)] + p_{x-1}(t) \times [\lambda\delta t + 0(\delta t)] + 0(\delta t)$$
$$= p_x(t) + \lambda p_{x-1}(t) - p_x(t)\delta t + 0(\delta t)$$

Rearranging gives

$$\frac{p_x(t+\delta t)-p_x(t)}{\delta t} = \lambda p_{x-1}(t) - \lambda p_x(t) + \frac{0(\delta t)}{\delta t}$$

Taking limits on both sides as
$\delta t \to 0 \text{ leads to the differential equation}$ gives,

$$\frac{dp_x(t)}{dt} = p_x'(x) = \lambda p_{x-1}(t) - \lambda p_x(t), \quad x=1,2,\text{..........................}8.4.2.2.3$$

Equations 8.4.2.2.1 and 8.4.2.2.3 are called the differential-difference equations.
[differential because the equation involves derivatives; difference because it expresses the probability
$p_x(t) \text{ in terms of the probability } p_{x-1}(t).$

For x=1, equation 8.4.2.2.3 gives,

$$\frac{dp_1(t)}{dt} = \lambda p_0(t) - \lambda p_1(t) \qquad \text{......}8.4.2.2.4$$

Substituting $p_0(t)$ from 8.4.2.2.2 into 8.4.2.2.4 gives,

$$\frac{dp_1(t)}{dt} + \lambda p_1(t) = \lambda e^{-\lambda t} \quad \text{.... ... ...} \quad 8.4.2.2.5$$

Using the integrating factor $e^{\lambda t}$, to integrate 8.4.2.2.5, and using the conditions that when t=0, no events have occurred, so that that the probability
$p_1(0) = P[X(0) = 1] = 0$ gives,

$$p_1(t) = \lambda t e^{-\lambda t} \qquad \text{.........} \qquad 8.4.2.2.6$$

Continuing in this way by using x=2,3,.... In 8.4.2.2.3, we can obtain $p_2(t)$, $p_3(t)$ etc.
as $p_2(t) = (\lambda t)^2 e^{-\lambda t}$, $p_3(t) = (\lambda t)^3 e^{-\lambda t}$, and so on.
Recursively, we could obtain

$$p_x(t) = P[X(t) = x] = \begin{cases} \dfrac{e^{-\lambda t}(\lambda t)^x}{x!}, \end{cases}$$
$$for \; x = 1,2,3, \text{... ... ...} \; 8.4.2.2.7$$

To analyse the differential-difference equations 8.4.2.2.1 and 8.4.2.2.3 above, we use the probability generating, say $F(s,t)$, of the random variable X(t).

$$F(S,t) = E\left(S^{X(t)}\right) = \sum_{x=0}^{\infty} P(X(t) = x)S^x = \sum_{x=0}^{\infty} p_x(t) S^x,$$
$$\text{..............} 8.4.2.2.8$$

which is a function of S and t.

8.4.2.3 The Probability generating Function of the Differential-Difference equations

We start with the differential-difference equations of the Poisson Process:

$$\frac{d}{dt} P_0(t) = -\lambda P_0(t), \quad for \; x = 0 \text{... ... ...} 8.4.2.3.1$$
$$\frac{d}{dt} P_x(t) = -\lambda P_x(t) + \lambda P_{x-1}(t) \text{... ...} for \; x = 1,2, \text{...} 8.4.2.3.2$$

Multiplying 8.4.2.3.1 and 8.4.2.3.2 by S^x and noting that $S^0 = 1$ gives,

$$\frac{d}{dt}p_0(t) = -\lambda p_0(t) \dots \dots \dots \dots \dots \dots \dots 8.4.2.3.3$$

$$\frac{d}{d}p_x(t)S^x = -\lambda p_x(t)S^x + \lambda p_{x-1}(t)S^x, \qquad for\ x$$
$$= 1,2, \dots \dots \dots .8.4.2.3.4$$

Adding equations 8.4.2.3.3 and 8.4.2.3.4 gives,

$$\frac{d}{dt}p_0(t) + \sum_{x=1}^{\infty}\frac{d}{dt}p_x(t)S^x$$

$$= -\lambda p_0(t) - \lambda\sum_{x=1}^{\infty}p_x(t)S^x$$

$$+ \lambda\sum_{x=1}^{\infty}p_{x-1}(t)S^x \dots \dots \dots .8.4.2.3.5$$

i.e.

$$\sum_{x=0}^{\infty}\frac{d}{dt}p_x(t)S^x \ on\ LHS = \qquad -\lambda\sum_{x=0}^{\infty}p_x(t)S^x +$$
$$\lambda S \sum_{x=1}^{\infty}p_{x-1}(t)S^{x-1} \ on\ RHS \dots \dots .8.4.2.3.6$$

Let the probability Generating Function of the random variable X(t) be denoted by,

$$F(S,t) = \sum_0^{\infty}p_x(t)S^x \qquad for\ x=1,2,\dots\dots\dots\dots\dots\dots\dots.4.2.3.7$$

Making the change of variable $r = x-1$ in the second term of the RHS of equation 8.4.2.3.6 and using 8.4.2.3.7, means that

$$\lambda S \sum_{r=0}^{\infty}p_r(t)S^r = \lambda SF(S,t)$$

Equation 8.4.2.3.6 can now be rewritten as:

$$\frac{\partial}{\partial t}F(S,t) = -\lambda(1-S)\pi \ \dots \dots \dots \dots \dots \dots \dots \dots \dots 8.4.2.3.8$$

Equation 8.4.2.3.8 is the partial differential equation (p.d.e) for $\pi(S, t)$, the probability generating function of the total number of events that occur by time t in a Poisson Process.
It will be assumed at this point that the student is knowledgeable about partial differential equations (pde).
Regarding S to be constant and applying the rules of ordinary differential equations
(the Lagrangian form), one obtains after a few manipulations the expression:

$Log\ F(S, t) = -\lambda(1 - S)t + f(s)$, where f(s) is an arbitrary function of S......8.4.2.3.9

Taking exponentials of both sides of 8.4.2.3.9 will give the general solution of [the(p.d.e)] equation 8.4.2.3.8 as,

$$F(S, t) = f(S)e^{-\lambda t(1-S)} \quad\quad\quad\quad\quad 8.4.2.3.10$$

Since the process starts from rest at time t=0, the initial condition is X(0)=0 with probability 1(one), and $F(S, 0) = 1$. Putting t=0 in equation 8.4.2.3.10, gives f(S)=1, hence the particular solution of 8.4.2.3.8 is,

$$F(S, t) = e^{-\lambda t(1-S)} \quad\quad\quad\quad\quad 8.4.2.3.11$$

Equation 8.4.2.3.11 is the probability generating function of a Poisson distribution with mean
$\lambda t,\ and\ so\ the\ random\ variable\ X(t) \sim Poisson(\lambda t)$.

EXAMPLE 8.18

The secretary of a big department receives telephone calls at a constant rate, λ, and in a manner that is assumed to be a Poisson process. Find the probability that (i) the second telephone call comes in the third minute (ii) the third call comes in the second minute.

SOLUTION:
(i)   The second call comes in the third minute means that :
[no call comes in the 1st minute and none in the 2nd and two or more in the 3rd] or [no call in the 1st and one in the 2nd and one or more calls in the third] or [one call in the 1st and none in the 2nd and one or more in the third].

$$=p_0(1).p_0(1)[\sum_{n=2}^{\infty}p_n(1) + p_0(1).p_1(1)[\sum_{n=1}^{\infty}p_n(1)] + \\ p_1(1).p_0(1)[\sum_{n=1}^{\infty}p_n(1)]$$

$$= [p_0(1)]^2[1 - p_0(1) - p_1(1)] + [p_0(1).p_1(1)][1 - p_0(1)]$$

$$= e^{-2\lambda}[1 - e^{-\lambda} - \lambda e^{-\lambda}] + e^{-\lambda}.\lambda e^{-\lambda}[1 - e^{-\lambda}] = e^{-2\lambda}(1 + \lambda) - \\ e^{-3\lambda}(1 + 2\lambda)$$

(ii)   The third call comes in during the second minute means that:
[no calls in the first minute and three or more calls in the second minute]
or [one call in the first minute and two or more calls in the second minute] or
[two calls in the first minute and one or more calls in the second minute]

$$=p_0(1)[\sum_{n=3}^{\infty}p_n(1)] + p_1(1)[\sum_{n=2}^{\infty}p_n(1)] \ p_2(1)[\sum_{n=1}^{\infty}p_n(1)]$$

$$=p_0(1)[1 - p_0(1) - p_1(1) - p_2(1)] + p_1(1)[1 - p_0(1) - p_1(1)] + \\ p_2(1)[1 - p_0(1)]$$

$$=e^{-\lambda}\left[1 - e^{-\lambda} - \lambda e^{-\lambda} - \frac{\lambda^2}{2}e^{-\lambda}\right] + e^{-\lambda}[1 - e^{-\lambda} - e^{-\lambda}] + \\ \frac{\lambda^2}{2}e^{-\lambda}[1 - e^{-\lambda}]$$

$$=e^{-\lambda}[\left(1 + \lambda + \frac{\lambda^2}{2}\right) - e^{-\lambda}(1 + 2\lambda + 2\lambda^2)]$$

8.4.2.4 Waiting Time (T) Between Successive Poisson Events

The waiting between successive events, T_1, exceeds t only if no events occur in the time interval (0,t). That is the probability of T_1 exceeding t:
$P(T_1>t)=P[$no events occur in (0,t)$]=P[$the Poisson random variable $X(t)=0] = e^{-\lambda t}$,

since X(t) has distribution: $X(t)\sim Poisson(\lambda t)$.

Denote the cumulative distribution of T_1 by F(t). Then by definition of the cumulative distribution function,

$$F(t)=P[TT_1 \leq t] = 1 - P[T_1 > t] = 1 - e^{-\lambda t}, \qquad for \ t \geq 0.$$

The probability density function of T_1 is therefore $f(t) = \frac{d}{dt} F(t) = \lambda e^{-\lambda t}, \ for \ t \geq 0.$

Therefore T_1 has an exponential distribution with parameter λ, in short $T \sim \exp(\lambda)$.

Also the distribution of the time T_s, of the other inter- events for s=2,3,..................

is exponential with parameter λ, in short $T_s \sim \exp(\lambda)$, for $s = 1,2,3, ...$

Since the waiting time between successive events has an exponential distribution with parameter λ, it follows that the average (mean) time between successive events is the reciprocal of λ, ie. $average(or \ mean \ time) = \frac{1}{\lambda}.$

8.4.3 Multivariate Poisson Process

A multivariate Poisson Process is a process in which each of the sequence of events is just one of several types of events. For instance if the sequence comprises events of type 1, type 2, and so on; events of type 1 occur as a Poisson process in time at rate λ_1, events of type 2 occur as a Poisson process in time at rate λ_2, ...etc. events of type k occur as a Poisson process in time at rate λ_k. It is assumed that the processes occur independently of each other, and if the events are observed along the same time axis, then the overall sequence of events occurs as a Poisson process at a rate

$$\lambda = \lambda_1 + \lambda_2 + \lambda_3 + \cdots \dots \dots \dots + \lambda_k = \sum_{i=1}^{k} \lambda_i.$$

In any such process, the probability that an event is of type i is given by

$$p_i = \frac{\lambda_i}{\lambda}, \ where \ \sum_{i=1}^{k} p_i = 1.$$

As an example of a multivariate Poisson process, a traffic police may observe an event to be the passage of a vehicle through the point where he stands along the road. The vehicles may be classified as private cars, commercial cars (taxis), vans, heavy trucks, mini buses and coaches. So there are several types of events, each event in the process being just one type. Events of type i occur as a Poisson process in time at rate $\lambda_i = \lambda p_i$.

Consider the occurrence of events of type I in the small interval $[t, t + \delta t]$, then the probability:

P{one event of type I occurs in $[t, t + \delta t]$}
=P{ one event occurs in [t, t+δt] and is of type i}+P{more than one event
in $[t, t + \delta t]$ and one of them is of type i}.
=P{one event in $t, t + \delta t]$} × P{an event is of type i} +0(δt)
=$\lambda \delta t + 0(\delta t) \times p_i + 0(\delta t) = \lambda p_i \delta t + 0(\delta t)$.
So P{one event occurs of type i in $[t, t + \delta t] = \lambda p_i \delta t + 0(\delta t)$.

EXAMPLE 8.19

Vehicles pass a traffic policeman standing at the side of a main street according to a Poisson process with an average rate of 50 vehicles per 30 minutes. Of these vehicles 50 per cent are private cars, 30 per cent re taxis, 10 per cent are buses and 10 per cent are trucks.
(i)   At what rate do taxis pass the policeman?
(ii)   What is the probability that in a 5 minute interval more than 3 private cars pass the policeman?
(iii)   What is the average waiting time between taxis?

SOLUTION:

Proportion of private cars(p_1)=0.5, proportion of taxis(p_2)=0.3, proportion of buses (p_3)=0.1, proportion of tucks p_4=0.1
Given that the constant rate of the overall Poisson process= $\lambda =$ $50\ per\ 30\ minutes,$ we have for the independent Poisson events:

$\lambda_1 = rate\ for\ private\ cars = \lambda p_1 = 50 \times 0.5 =$ $25\ per\ 30\ minutes,\ \lambda_2 = rate\ for\ taxis = \lambda p_2 = 50 \times 0.3 =$ $15\ per\ 30\ minutes,$ similarly

$\lambda_3 = rate\ for\ buses = 5\ per\ 30\ minutes,\ and\ \lambda_4 = rate\ for\ trucks = 5\ per\ 30\ minutes.$

Hence,

(i) Taxis pass the policeman at a rate $\lambda_2 = 15\ per\ 30\ minutes.$

(ii) The expected number of private cars passing the policeman in 5 minutes is

$\lambda_1 \times \frac{5}{30} \approx 4.$ The probability that more than 3 private cars pass the policeman in 5 minutes equals

$P(C > 3) = 1 - P(C \leq 3),$ where $C \sim Poisson(4).$ ie.
$1 - P(C \leq 3) = 1 - \{P(X = 0) + P(X = 1) + P(X = 2) + P(X = 3),$ where
$P(X = x) = e^{-\lambda} \frac{(\lambda)^x}{x!}, \quad for\ x = 0,1,2,3.$

Hence,
$$P(C > 3) = 1 - (C \leq 3)$$
$$\approx 1 - e^{-4}\left\{1 + 4 + \frac{4^2}{2!} + \frac{4^3}{3!} = 1 - \left(\frac{71}{3}\right)e^{-4}\right\}$$

(iii) The waiting time between successive taxis has an exponential distribution with parameter $\lambda_2 = 15\ per\ 30\ minutes.$ So the averge time between successive taxis is given by
$\mu = \frac{1}{\lambda_2} = \frac{1}{15}\ per\ 30\ minutes = 2\ minutes.$

EXAMPLE 8.20

Customers arrive at a restaurant according to a Poisson process at an average rate of 10 per minute. Fifty (50) per cent of them simply wish to buy fried chicken and take away (category 1), forty (40) per cent of them wish to order soft drinks and sandwiches and eat in (category 2) and the remaining ten (10) per cent wish to order a variety of food and eat in (category 3).
(i) What is the probability that less than five (5) customers arrive in a time interval of 30 seconds.

(ii)What is the probability that, in two minutes, twenty (20) customers of category 2 arrive?
(iii)What is the probability that, in two minutes, ten(10) customers of category 1, and at most three(3) of category 3 arrive?

SOLUTION:

(i)Since customers arrive at the restaurant at an average rate of 10 per minute, the expected number of customers (Y) to arrive in a time duration of t= 30 seconds is
$\lambda t = 10 \times 0.5 = 5$. Hence $P(Y \leq 5) = P(Y = 0) + P(1) + P(2) + P(3) + P(4) + P(5)$

$$= e^{-5}\left(1 + 5 + \frac{5^2}{2!} + \frac{5^3}{3!} + \frac{5^4}{4!} + \frac{5^5}{5!}\right) \approx 0.616$$

arrival $= \lambda_2 = \lambda p_2 = 10 \times 0.4 = 4\ per\ minute$. Therefore in a time duration of 2 minutes, the expected number of customers Y to arrive= 8 per 2 minutes. The probability

$$P(Y=20) = e^{-8} \times \frac{8^{20}}{20!}$$

(iii)Proportion of category 1 customers arriving at the restaurant $=$ p_1=0.5, giving a rate of
arrival=$\lambda_1 = \lambda p_1 = 10 \times 0.5 = 5\ per\ minute.\ Similarly,\ \lambda_3 = \lambda p_3 = 1\ per\ minute.$
Therefore in a time duration of 2 minutes, the expected number of customers Y_1 of category 1 and Y3 of category 3 to arrive are: 10 and 2 respectively. Hence,

For category 1: $P(Y_1=10)=e^{-10} \times \frac{10^{10}}{10!}$, $and\ for\ category\ 3\ P(Y_3 = $
$3=e-2 \times 233!$

The required probability is therefore equal to:$P(Y_1 = 10) \times P(Y_3 \leq 3)$

$$= \left(e^{-10} \times \frac{10^{10}}{10!} \right) \times \left\{ e^{-2} \left(1 + 2 + 2 + \frac{4}{3} \right) \right\}$$

$$= \left\{ e^{-10} \times \frac{10^{10}}{10!} \right\} \times \left\{ e^{-2} \left(\frac{19}{3} \right) \right\}$$

8.4.4 The Compound Poisson Process

One of the postulates of the Poisson process is that (exactly) one event occurs in a

given small time interval $[t, t + \delta t]$ *with probability* $\lambda \delta t + 0(\delta t)$.

The other postulate is that, the probability of two or more events occurring in any small interval of time $[t, t + \delta t]$ *is equal to* $0(\delta t)$.

This means that there are no multiple occurrences in a Poisson process. In contrast, multiple occurrences are allowed in a Compound Poisson process. In the latter case, events occur according to a Poisson process, and associated with each of these events is another event which has multiple occurrences. Each event of the Poisson is referred to as compound event. The number of occurrences associated with each event in the Poisson process is an observation on a discrete random variable Y. Let X(t) represent the number of compound events to occur in the time interval (0, t), and let Y_i be the number of occurrences associated with each compound event (i), then if R(t) represents the total number of occurrences associated with all the compound events in question in the time interval (0,t), we can write

$$R(t) = Y_1 + Y_2 + Y_3 + \ldots\ldots\ldots\ldots\ldots\ldots\ldots + Y_{X(t)}.$$

8.4.4.1 Deriving an expression for the MEAN of R(t)

The mean of R(t) is given by: $E[R(t)] = \sum_{r=0}^{\infty} rP[R(t) = r]$.
Since X(t) takes just one of the values x=0,1,2,3,........., by the Theorem of Total Probability,

$$P[R(t) = r] = \sum_{r=1}^{\infty} P[R(t) = r/X(t) = x]P[X(t) = x].$$

Thus we can write

$$E[R(t)] = \sum_{r=0}^{\infty} r \sum_{x=0}^{\infty} P[R(t) = r/X(t) = x]P[X(t) = x].$$

Interchanging the order of integration on the right hand side gives:

$$E[R(t)] = \sum_{x=0}^{\infty} P[X(t) = x] \sum_{r=0}^{\infty} r\, P[R(t) = r/X(t) = x]$$
........................8.4.4.1.1

The sum over r is simply the conditional expectation of R(t) given X(t)=x. We can therefore write 8.4.4.1.1 as,

$$E[R(t)] = \sum_{x=0}^{\infty} P[X(t) = x] E[R(t)/X(t) = x]$$
    ....8.4.4.1.2

When X(t)=x, R(t) =fixed=Y_1+Y_2+Y_3+..........+Y_x. Since the Y_i (i=1,2,3,.......,x) are identically distributed with mean μ, the expected value:

E(Y_1+Y_2+.........+Y_x)=$E[R(t)/X(t) = x] = x\mu$.

Substituting in 8.4.4.1.2 above gives,

$$E[R(t)] = \sum_{x=0}^{\infty} P[X(t) = x] \times (x\mu) = \mu \sum_{x=0}^{\infty} xP[X(t) = x] = \mu E[Xt].$$

Since X(t) is the number of events occurring in time t in a Poisson process with rate

λ, $X(t) \sim poisson(\lambda t)$ and so $E[(X(t)] = \lambda t$.

Thus, the expected number of events in an interval of length t is,

$$E[R(t)] = \mu\lambda t$$
    ......8.4.4.1.3

where λt Compound events are expected in the interval (0,t) and μ is the mean number of events associated with each compound event.

8.4.4.2. Deriving an expression for the VARIANCE of R(t).

The variance of R(t) is defined as $E\{R(t)^2\} - \{E(R(t)\}^2$

But $E\{S(t)^2\} = \sum_{x=0}^{\infty} P\{X(t) = x\}E\{R(t)^2/X(t) = x\}$. If X(t) =x, then

R(t)=Y_1+Y_2+Y_3+........................+Y_x, so that

$$E\{R(t)^2\}/X(t) = x\} = E\{(Y_1 + Y_2 + \cdots \ldots + Y_x)^2\}$$

$$= E\{Y_1^2 + (Y_1Y_2 + Y_1Y_3 + \cdots \ldots Y_1Y_x) + Y_2^2$$
$$+ (Y_2Y_1 + Y_2Y_3 + \cdots \ldots Y_2Y_x) +$$
$$+\vdots$$
$$\vdots$$
$$+Y_x^2 + (Y_xY_1 + Y_xY_2 + \cdots \ldots \ldots \ldots +Y_xY_{x-1})\}$$

In the above sum there are x terms of the $E(Y_i^2)$ and x(x-1) terms of the form $E(Y_iY_j)$ where $i \neq j$. Since Y_i and Y_j are independent with mean μ it follows that

$$E(Y_iY_j) = E(Y_i)E(Y_j) = \mu^2.\ \text{Also since}\ E(Y_i = \mu)\ and\ V(Y_i) = \sigma^2,$$

the definition of the variance

$$V(Y_i) = E(Y_i^2) - \{E(Y_i)\}^2,\ gives\ E\{Y_i\}^2 = \sigma^2 + \mu^2.$$
$$\therefore\ E\{R(t)^2/X(t) = x\} = xE(Y_i^2) + x(x-1)E(Y_iY_j) =$$
$$x(\sigma^2 + \mu^2) + x(x-1)\mu^2$$

$$= x\sigma^2 + x^2\mu^2 \quad \ldots\ldots\ldots\ldots\ldots\ldots\ldots\ldots\ldots\ldots\ldots\ldots\ldots.8.4.4.2.1$$

Using equation 8.4.4.2.1 above, we obtain

$$E\{R(t)^2\} = \sum_{x=0}^{\infty} P\{X(t) = x\}E\{R(t)^2/X(t) = x\}$$

$$= \sum_{x=0}^{\infty} P\{X(t) = x\}\{x\sigma^2 + x^2\mu^2\}$$

$$= \sigma^2 \sum_{x=0}^{\infty} xP\{X(t) = x\} + \mu^2 \sum_{x=0}^{\infty} x^2 P\{X(t) = x\}$$

$= \sigma^2 E[X(t)] + \mu^2\{V(X(t)) + [E(X(t))]^2\}$, Since X(t) has a Poisson distribution with mean (λt), we know that for a Poisson distribution the mean and variance are equal,

Hence E[X(t)]=V[X(t)]=λt, so

$$E\{R(t)^2\} = \sigma^2 \lambda t + \mu^2\{\lambda t + (\lambda t)^2\},$$
hence

$$V\{R(t)\} = E\{R(t)^2\} - \{E(R(t))\}^2$$
$$= \sigma^2 \lambda t + \mu^2\{\lambda t + (\lambda t)^2\} - (\mu \lambda t)^2$$

$=\sigma^2 \lambda t + \mu^2 \lambda t + \mu^2 (\lambda t)^2 - (\mu \lambda t)^2 = \lambda t(\sigma^2 + \mu^2)$............8.4.4.2.2

Hence we have derived that if R(t) is the number of events expected to occur in a time interval (0,t) in a compound Poisson process in which compound events occur at the rate of λ, *then*
    the expected value, $E\{R(t)\} = \mu \lambda t$;
and the variance $V\{R(t)\} = \lambda t(\sigma^2 + \mu^2)$,

where $\mu \ and \ \sigma^2$ are the mean and variance of of the number of events associated with each compound event.

EXAMPLE 8.20

Various commercial trucks, loaded with cement, arrive at a large building site according to a Poisson process at an average rate of one truck every ten (10) minutes from 9am to 12 noon. The number of trucks arriving at the site over an interval of observation of duration t hours, say, has a Poisson distribution with mean $\lambda = 6t$ per 't' hours.
Suppose that the number of bags of cement offloaded from each truck is an independent observation on a discrete random variable 'Y' with geometric distribution $G_1(p) = G_1(0.5)$. There will often be more than one bag of cement offloaded from each truck. Then over the interval of observation (9am-12 noon) =3 hours, the total number of bags of cement offloaded from all the trucks is given by

$$R(t) = Y_1 + Y_2 + Y_3 + \cdots \ldots \ldots \ldots + Y_{X(t)},$$

where Y_i denotes the number of bags of cement offloaded from each truck, and Y_i has a geometric distribution.

Denoted the mean and variance of Y, by μ and σ^2 respectively :
[ie. $E(Y_i = \mu,$ and $V(Y_i = \sigma^2)$].

If X(3) is the number of trucks that arrive in the three hour period, then the total number of bags of cement offloaded t the site over the 3-hour period is

$$R(3) = Y_1 + Y_2 + Y_3 + \cdots \ldots \ldots \ldots \ldots Y_{X(3)},$$ where X(3) has a Poisson distribution with parameter 18 (i.e.

$\lambda = 6$ trucks per hour, on average and $t == 3$ hours)

We wish to determine (i) the mean of Y_i [ie. $E(Y_i)$] and the variance of Y_i [ie $V(Y_i)$] of the discrete observation Y and (ii) the mean and variance of the compound Poisson process R(t) over the 3 hour time interval.

SOLUTION

(i)The number of bags of cement offloaded, at the building site, from each truck is an observation on a discrete geometric random variable $Y_i \sim G_1(p) = G_1(0.5)$. so

$$\mu = E(Y_i) = \frac{1}{p} = \frac{1}{0.5} = 2,$$

$$\text{and } \sigma^2 = V(Y_i) = \frac{1-p}{p^2} = \frac{1-0.5}{0.5^2} = 2$$

(ii)The trucks arrive at the building site according to a Poisson process with rate $\lambda = 6$ per hour, on average, in the time interval t=3hours (9am to 12 noon). Referring to equations 8.8.1.3 and 8.8.2.2 above, respectively, for a compound Poisson process;

$$E[R(t)] = \mu\lambda t, \text{ and } V[R(t)] = \lambda t(\sigma^2 + \mu^2).$$

Since $\lambda = 2, t = 3$, the mean and variance of R(2), the total number of bags of cement offloaded at the site over the three-hour period, are
E[R(3)]=2× 6 × 3 = 36, and $V[R(3)] = 6 \times 3 \times (2 + 4) = 108.$

EXAMPLE 8.20

Customers enter a big shopping centre according to a Poisson process at an average rate of ten (10) per minute. The customers buy Y items independently of each other. The discrete random variable Y has the probability distribution given in table below:

y_i	0	1	2	3	4
$P(Y = y_i)$	0.1	0.3	0.2	0.3	0.1

(i)Find the mean and variance of Y.
(ii)Use the results of (i) to find the mean and variance of the total number of items R(5) bought over a five (hour) period.

SOLUTION.
The expected value of Y is

$$\mu = E(Y) = \sum_{i=1}^{5} y_i P(Y = y_i) = 2.0, \quad and \; E(Y^2)$$

$$= 0.3 + 0.8 + 2.7 + 1.6 = 5.4$$
$$\therefore V(Y) = \sigma^2 = E(Y^2) - \{E(Y)\}^2 = 5.4 - 2.0 = 3.4$$

(ii)The customers enter the shopping centre at a rate of 10 per minute or 600 per hour. The interval of observation is t=5 (five) hours. Hence, If R(5)= the total number of items bought during the five hour period, the mean of

R(5), $E[R(5)]=\mu\lambda(t = 5) = 2.0 \times 600 \times 5 = 6000$, $and \; V[R(5)] = 22,200$

8.4.5 The Simple Birth Process

8.4.5.1 Postulates of the simple Birth Process

This is an example of a non-decreasing Stochastic Process in which there are no deaths. Individuals alive in the population independently generate further offsprings according to a Poisson process at rate κ. The population therefore increases with time. If the population has size x at time t, the waiting time to the additional birth(s) has an exponential distribution with parameter κx.

(i)   The probability that the population increases by an individual within the time interval $(t, t + \delta t)$ is equal to $\kappa x \delta t + 0(\delta t)$,

(ii)   The probability of more than one birth in the interval $(t, t + \delta t$ equals $0 \delta t$,

(iii)   For any individual alive at time t, the occurrence of births after any time t is independent of the occurrence of births before time t

By postulate (i) we can simply write,

$$P\{X(t + \delta t) = x + 1/X(t) = x\} = \kappa x \delta t + 0(\delta t).$$

This equation is analogous to equation 8.4.2.1.1 of the Poisson process described above.

If the population is to be sustained, then by time $t + \delta t$, individuals must be present to give birth to new individuals. This requires that the population size

$$[X(t + \delta t)] \neq 0. \text{ Let } [X(t + \delta t) = x], \; for \; x = 1,2,3, ...,$$

at least one individual is alive up to time $t + \delta t$. If there is exactly one individual at time
$t + \delta t$, we simply write $[X(t + \delta t) = 1]$.

The probability of this likely population size is,

$$p_1(t + \delta t) = P[X(t + \delta t) = 1] =$$
$$\sum_{x=0}^{\infty} \{X(t + \delta t) = 1\}/\{X(t = x\} \{P(X(t) = x\},$$

by applying the Law of Total Probability. The only possibility of having one individual alive by time $t + \delta t$, is that one individual is alive upto time t and this individual gives no birth in the interval $t + \delta t$. Hence, we write

$$p_1(t + \delta t) = p_1(t)[1 - \kappa \delta t + 0(\delta t)]$$
..........................8.4.5.1

Rearranging equation 8.4.5.1 and taking the limits as $\delta t \to 0$ on both sides gives the first differential equation

$$p_1'(t) = \frac{dp_1(t)}{dt} = -\kappa p_1(t) \qquad \text{for x=1,}\ldots\ldots\ldots\ldots\ldots 8.4.5.2$$

of the Simple Birth process.

If the population should grow in size, we must consider $[X(t + \delta t) = x]$, the population size at time $t + \delta t$ for other values of x=2,3,4,.....................

The only ways we can have a population with size x at time $t + \delta t$ is to have

(i) size equal to x up to time t and there are no births in the interval $(t, t + \delta t)$, or

(ii) size (x-1) up to time t and there is exactly one birth in the interval $(t, t + \delta t)$. hence for x=2,3................ we can write in terms of probabilities,

$$p_x(t + \delta t) = p_x(t)\{1 - \kappa x \delta t + 0(\delta t)\} + p_{x-1}(t)\{\kappa(x - 1)\delta t + 0\delta t + 0\delta t. \qquad \ldots\ldots\ldots\ldots 8.4.5.3$$

We see that by putting x=1 in equation 8.4.5.3, we obtain equation 8.4.5.1, so one of these equations is not considered separately as in the case of the Poisson process.

Again rearranging equation 8.4.5.3 and taking limits as $\delta t \to 0$ for x=2,3......... we have the the general differential-difference equation of the Simple Birth process:

$$\frac{d}{dt}p_x(t) = -\kappa x p_x(t) + \kappa(x - 1)p_{x-1}(t), \ for \ x = 1,2,3 \ldots..,$$
.... 8.4.5.4

The probability generating function
$\pi(s, t) of \ the \ random \ variable \ X(t)$ is defined

as: $F(s, t) = \sum_{x=0}^{\infty} p_x(t) S^x = p_0(t) + \sum_{x=1}^{\infty} p_x(t) S^x.$

The probability of no individuals being alive up to time t ,$(p_t(t))$, is equal to zero for all t.

Multiplying equation 8.4.5.4 above by S^x and sum over all x=1,2,.........gives,

$$\sum_{x=1}^{\infty} \frac{d}{dt} p_x(t) S^x = -\kappa \sum_{x=1}^{\infty} x p_x(t) S^x + \kappa \sum_{x=1}^{\infty} (x-1) p_{x-1}(t) S^x$$
....8.4.5.5

The left hand side of equation 8.4.5.4 gives

$$\sum_{x=1}^{\infty} \frac{d}{dt} p_x(t) S^x = \frac{\partial}{\partial t} \pi(s,t).$$

We could make the change of variable $r=x-1$ in the right hand side of equation 8.4.5.5 and obtain a new expression

$$-\kappa \sum_{x=1}^{\infty} x p_x(t) S^x + \kappa S \sum_{r=1}^{\infty} r p_r(t)$$

which could simply be written as,

$$-\kappa(1-S) \sum_{x=1}^{\infty} x p_x(t) S^x = -\kappa S(1-S) \frac{\partial \pi(S,t)}{\partial t}$$
........................8.4.5.6

Thus the partial differential-difference equation equation for the Simple Birth process

is obtained as:

$$\frac{\partial F(S,t)}{\partial t} = -\kappa S(1-S) \frac{\partial F(S,t)}{\partial t} \qquad .... 8.4.5.7$$

ASSIGNMENT 8.8 (FOR THE STUDENT).

The partial differential equation for the probability generating function $\pi(S,t)$ of the number of people alive at time t in a Simple Birth process, is given in equation 8.9.1.7 above. Find the general solution of this partial differential equation, and show that the particular solution corresponding to the case X(0)=1 is

$$F(S,t) = \frac{S e^{-\kappa t}}{1 - S(1 - e^{-\kappa t})}.$$ More generally, a simple birth process may be initiated not

by a single individual but several, so that $X(0)=x_0>1$. In this case each individual generates independent family trees, so the total number of individuals alive t time t is the sum of x_0 independent identically distributed geometric variables. Hence F(S, t), the probability generating function (p.g.f) of the total number of individuals alive at

time t, is the product of x_0 identical p.g.f's and is given by $F(S,t) =$
$$\left\{\frac{Se^{-\kappa t}}{1-S(1-e^{\kappa t})}\right\}^{x_0}$$

8.4.6 The Immigration – Birth Process

The immigration –birth process comprises two components by means of which the population size increases:
(i) The immigrants: arrivals occur in the population according to a Poisson process with rate λ. and (ii) individuals already in the population who independent of the arrivals, give births to new members according to the Simple Birth model with birth rate κ ; so that if $X(t)=x$ is the size of the population at time t, the instantaneous birth rate is κx. Assume the population increases by one individual by time $t + \delta t$:this increase is either through arrivals (immigrants) or through births. The probability of this likely population size $X(t + \delta t) = x + 1$ at $time$ $t + \delta t$, is given as,

$$P\{X(t + \delta t) = x + 1/X(t) = x\} = (\lambda + \kappa x)\delta t + 0(\delta t).$$

The population has size zero by time $t + \delta t$ means that the size is zero in the interval (0,t) and no further increase in the interval $(t, t + \delta t)$. Hence, in probabilistic terms:

$$p_0(t + \delta t) = p_0(t)[1 - \lambda\delta t + 0(\delta t)],$$

which after rearrangement gives

$$\frac{dp_0(t)}{dt} = -\lambda p_0(t). \dots\dots\dots\dots\dots\dots\dots\dots\dots\dots\dots.8.4.6.1$$

The population has size x at time $t + \delta t$ means that it either has size x by time t and no further increase in interval $[t, t + \delta t]$, or it has size x-1 by time t and an increase by one(1) in the interval $[t, t + \delta t]$. In probabilistic terms,

$$p_x(t + \delta t) = p_x(t)[1 - (\lambda + \kappa x)\delta t + 0(\delta t)]$$
$$+ p_{x-1}(t)\left[\left(\lambda + \kappa(x - 1)\right)\delta t + 0(\delta t)\right] + 0(\delta t)$$

for values of x=1,2,3............................

Which after rearranging gives,

$$\frac{dp_x(t)}{dt} = -(\lambda + \kappa x)p_x(t) + \left(\lambda + \kappa(x - 1)\right)p_{x-1}(t)............ 8.4.6.2$$

Multiplying both sides of 8.4.6.1 and 8.4.6.2 by S^x and adding gives:

On the left hand side

$$\sum_{x=0}^{\infty} \frac{dp_x(t)}{dt} S^x = \frac{dp_0(t)}{dt} + \sum_{x=1}^{\infty} p_x(t)S^x$$

$$= -\lambda p_0(t) - \sum_{x=1}^{\infty}(\lambda + \kappa x)p_x(t)S^x + \sum_{x=1}^{\infty}\left(\lambda + \kappa(x - 1)\right)p_{x-1}(t)S^x$$
$$= -\lambda \sum_{x=0}^{\infty} p_x(t)S^x - \kappa \sum_{x=1}^{\infty} xp_x(t)S^x + \lambda S \sum_{x=1}^{\infty} p_{x-1}(t)S^{x-1} +$$
$$\kappa S \sum_{x=1}^{\infty}(x - 1)p_{x-1}(t)S^{x-1}$$

$$......8.4.6.3$$

Substituting r=x-1 in 8.4.6.3 gives,

$$= -\lambda \sum_{x=0}^{\infty} p_x(t)S^x - \kappa \sum_{x=0}^{\infty} xp_x(t)S^x + \lambda S \sum_{r=0}^{\infty} p_r(t)S^r +$$
$$\kappa S \sum_{r=0}^{\infty} rp_r(t) S^r$$

Since by definition of the probability generating function (see equation 8.4.2.2.8.

$$F(S, t) = \sum_{x=0}^{\infty} p_x(t)S^x, \quad \text{then} \quad S\frac{\partial F}{\partial S} = \sum_{x=0}^{\infty} xp_x(t)S^x, \quad \text{so that}$$
equation 8.4.6.3 could now be written as,

$$\frac{\partial F(S, t)}{\partial t} = -\lambda F(S, t) - \kappa S\frac{\partial F(S, t)}{\partial S} + \lambda SF(S, t) + \kappa S\left(S\frac{\partial F(S, t)}{\partial S}\right)$$

$$= \lambda(1-S)F(S,t) - \kappa\, S(1-S)\frac{\partial F(S,t)}{\partial S}, \quad so \quad \kappa\, S(1-S)\frac{\partial F(S,t)}{\partial S} +$$

$$\frac{\partial F(S,t)}{\partial t} = -\lambda\,(1-S)F(S,t)$$

$$............8.4.6.4$$

We can find F(S, t) by using the Lagrangian method to solve the partial differential equation 8.4.6.4 above. The trio equations are given as follows:

$$p = \kappa S(1-S), \quad q = 1, \quad r = -\lambda(1-S)F(S,t). \text{ Hence,}$$

$$(i)\,\frac{dS}{\kappa S(1-S)} = \frac{dt}{1}, \quad (ii)\,\frac{dF(S,t)}{-\lambda(1-S)F(S,t)} = \frac{dS}{\kappa S(1-S)}, (iii)\,\frac{dt}{1} = \frac{dF(S,t)}{-\lambda(1-S)F(S,t)}$$

$$...\quad 8.4.6.5$$

Integrating 8.4.6.5 (i) gives $\frac{S}{1-S}e^{-\kappa t} = k_1(constant)$, similarly integrating
8.4.6.5(II)

gives $S^{\lambda}[F(S,t)]^{\kappa} = k_2(constant)$. A general solution can be obtained by regarding k_2 as function of k_1. [i.e. $S^{\lambda}[F(S,t)]^{\kappa} = f\left[\left(\frac{S}{1-S}\right)e^{-\kappa t}\right]$, Where f is an arbitrary function. The general solution is:

$$F(s,t) = S^{-\lambda/\kappa}f\left\{\left(\frac{S}{1-S}\right)e^{-\kappa t}\right\}.......................\,....8.4.6.6$$

When t=0, the population size $X(t=0) = x_0$, there are no arrivals at t=0, so the individuals in the population are through the Simple Birth process. Using the result of

the Simple Birth process Assignment 8.8 above where

$$F(S,t) = \left\{\frac{Se^{-\kappa t}}{1-s(1-e^{-\kappa t})}\right\}^{x_0}, \quad F(s,0) = s^{x_0}.$$

Putting t=0 in the general solution given in 8.4.6.6, we have

$$S^{x_0} = S^{\lambda/\kappa} f\left(\frac{S}{1-S}\right), so\ f\left(\frac{S}{1-S}\right) = S^{x_0+\lambda/\kappa}.$$ To find f(x), put
$$x = \frac{S}{1-S},\ then$$

$$S = \frac{x}{x+1}\ and\ obtain\ f(x) = \left(\frac{x}{1+x}\right)^{x_0+\lambda/\kappa}.$$

Substituting this into the general solution in equation 8.4.6.6 to find the particular solution for $X(0) = x_0$: we have

$$F(S,t) = S^{-\lambda/\kappa} f\left\{\left(\frac{S}{1-S}\right)e^{-\kappa t}\right\} =$$

$$S^{-\lambda/\kappa}\left\{\frac{\left(\frac{S}{1-S}\right)e^{-\kappa t}}{1+\left(\frac{S}{1-S}\right)e^{-\kappa t}}\right\}\left(\frac{Se^{-\kappa t}}{1-s+Se^{-\kappa t}}\right)^{x_0+\lambda/\kappa} \quad \dots\dots\dots\dots\dots \quad 8.4.6.7$$

[Hint for the student who is trying Assignment 8.8: The results of the Pure-Birth process is just a special case of the Immigration-Birth process with

$$X(0) = x_0 = 1, and\ \lambda = 0.]$$

8.4.7 The Pure Death Process

This is an example of non-increasing processes. Individuals continue to die in the population and there are no births or arrivals from outside to balance the deaths, as a result a population which may be quite large initially, eventually vanishes once everybody in it dies.
It is assumed that the probability that a specified individual alive at time t dies during the short time interval $(t, t + \delta t)$ is $\alpha\delta t + 0(\delta t)$. If the population is of the size x at time t, then the probability of a death occurring in this short time interval is $\alpha x\delta t + 0(\delta t)$. If a death occurs at time $t + \delta t$, then a population that has size x at time t [ie. $X(t) = x$] decreases by one individual and we simply write the probability as,

$$P\{X(t + \delta t) = x - 1/X(t) = x\} = \alpha x\delta t + 0(\delta t).$$

We now attempt to derive the differential-difference equations for the Pure Death process. Since the population changes size only through deaths, there are basically two ways to have x survivors at time $t + \delta t$:

(i)(x+1) individuals are alive at time t, and one of them dies in the interval $(t, t + \delta t)$, or

(ii)x individuals are alive at time t and all of the survive to time $t + \delta t$.

It follows therefore that for x=0,1,2,..........., the probability

$$p_x(t + \delta t) = p_{x+1}(t)[\alpha(x + 1)\delta t + 0(\delta t)] \\ + p_x(t)[(1 - \alpha x \delta t + 0(\delta t)] \qquad + 0(\delta t).$$

...............................8.4.7.1

Rearranging equation 8.4.7.1 and letting $\delta t \to 0$ gives,

$$\frac{dp_x(t)}{dt} = -\alpha x p_x(t) + \alpha(x + 1)p_{x+1}(t),$$
$$for\ x = 0,1,2 \dots \dots \dots \dots .8.4.7.2$$

Recalling the definition of the probability generating function $F(S, t)$, and multiplying both sides of 8.4.7.2 by S^x and summing over all values of x=0,1,2,............................yields,

$$\frac{\partial F(S,t)}{\partial t} = -\alpha \sum_{x=0}^{\infty} x p_x(t) S^x + \alpha \sum_{x=0}^{\infty} (x + 1)p_{x+1}(t) S^x.$$
.... 8.4.7.3

If we substitute $r = x + 1$ in equation 8.11.3 we have,

$$= -\alpha \sum_{x=0}^{\infty} x p_x(t) S^x + \alpha \sum_{x=0}^{\infty} r p_r(t) S^{r-1}$$

which simplifies to

$$\frac{\partial F(S,t)}{\partial t} = -\alpha S \frac{\partial F(S,t)}{\partial S} + \alpha \frac{\partial F(S,t)}{\partial S} = \alpha(1 - S) \frac{\partial F(S,t)}{\partial S}$$
.... 8.4.7.4

as the partial differential equation of the Pure Death process.

8.4.8 The General Birth and Death Process

A Markov Process in which both births and deaths may occur is known as a Birth and Death process.

We recall from the previous sections, other Markov processes such as the Poisson process and the Simple Birth process on the one hand , in which changes occur by an increase of one unit only (say, a birth), and the Pure Death process on the other hand, in which changes may occur by a decrease of one unit (say, a death). The Birth and Death process embraces both aspects mentioned above.

Suppose at time t the size of a certain population is denoted by x, and let the random variable $X(t)$ representing this size (i.e. $X(t) = x$). Then for a further change in time δt, the possible changes in the small time interval $(t, t + \delta t)$ are a birth which increases the population size by one [i.e. x to x+1] at time $(t + \delta t)$, and has probability:

$$p_{x,x+1}(t, t + \delta t) = \kappa_x \delta t + 0(\delta t) \dots \dots \dots \dots \dots \dots \dots \dots .8.4.8.1$$

and a death which decreases the population size by one [i.e. x to x-1] at time $(t + \delta t)$ and has probability:

$$p_{x,x-1}(t, t + \delta t) = \alpha_x \delta t + 0(\delta t) \dots \dots \dots \dots \dots \dots \dots \dots 8.4.8.2$$

κ_x and α_x are the overall birth and death rates respectively of the process, and depend on x, t and δt. Since the population size is a non-negative integer, the death rate $\alpha_0 = 0(zero)$, and equations 8.4.8.1 and 8.4.8.2 hold for x=0, 1, 2, …………..

The probability that two or more events occur in the interval $(t, t + \delta t)$ is $0(\delta t)$. If a birth and a death occur at the same time in the interval above, these are considered as two events and the probability is of the order zero.

The probability that no event occurs [i.e. neither a birth nor a death] in the interval

$(t, t + \delta t)$ is given as:

$$p_{x,x}(t, t + \delta t) =$$
$$1 - \kappa_x \delta t - \alpha_x \delta t + 0(\delta t) \dots \dots \dots \dots \dots \dots \dots \dots \dots .8.4.8.3$$

The birth and death processes we will focus attention on are said to be homogeneous, so that both the birth rate κx and the death rate αx are independent of time.

8.4.8.1 The Chapman-Kolmogorov Equations (General Birth and Death process)

Assume the population of size x at time t [i.e. $X(t) = x$], and at time t=0, X(0)=i. We wish to calculate the transition probability $p_{i,x}(t)$ of the population changing size from i to x in the time interval $(0, t + \delta t)$. At some intermediate time t, the process must be in some state (x-1, x, or x+1). The population changes from size i at time 0(zero) to size (x-1, x or x+1) at time t and then to size x at time $(t + \delta t)$. Because of the Markovian property, the probability of the change from (x-1, x, or x+1) does not depend on the initial sixe I, hence $0 < t < t + \delta t)$. Hence the Chapman-Kolmogorov equations can be written as,

$$p_{i,x}(t + \delta t) = \sum_r p_{i,r}(t)p_{r,x}(t, t + \delta t)) \dots \dots \dots \dots \dots .8.4.8.1.1$$

The transition probability $p_{r,x}(t, t + \delta t)$ is negligible for all values of r except for r=x-1, r=x+1, r=x.
From equations 8.4.8.1 to 8.4.8.3 above, we obtain the transition probabilities for the different values of r as,

$$p_{x-1,x}(t, t + \delta t) = \kappa_{x-1}\delta t + 0(\delta t) \dots \dots \dots \dots \dots \dots 8.4.8.1.2$$

corresponding to the population size r=(x-1) at time t and a birth occurring in the small time interval $(t, t + \delta t)$, and

$$p_{x+1,x}(t, t + \delta t) = \alpha_{x+1}(t, t + \delta t) + 0(\delta t) \dots \dots \dots \dots \dots .8.4.8.1.3$$

corresponding to the population size r=x+1 at time t and a death occurs in the small interval $(t, t + \delta t)$ and,

$$p_{x,x}(t, t + \delta t) = 1 - \kappa_x\delta t - \alpha_x\delta t + 0(\delta t) \dots \dots \dots \dots .8.4.8.1.4$$

corresponding to r=x at time t and no change occurring in the small time interval $(t, t + \delta t)$.
By substituting equations 8.4.8.1.2 to 8.4.8.1.4 into equation 8.4.8.1.1 we obtain

$$p_{i,x}(t + \delta t) = p_{i,x-1}(t)\kappa_{x-1}\delta + p_{i,x+1}(t)\alpha_{x+1}\delta t$$
$$+ p_{i,x}(t)[1 - \kappa_x \delta t - \alpha_x \delta t] +$$

$0(\delta t)$ 8.4.8.1.5

Rearranging 8.4.8.1.5, we obtain for values of x=1,2,3...........

$$\frac{p_{i,x}(t+\delta t)-p_{i,x}(t)}{\delta t} = \kappa_{x-1}p_{i,x-1}(t) + \alpha_{x+1}p_{i,x+1}(t) - (\kappa_x +$$
$$\alpha_x)p_{i,x}(t) + \frac{0(\delta t)}{\delta t} \ldots\ldots\ldots$$

....8.4.8.1.6

When the population size x=o, the death rate $\alpha_0 = 0$, and in the first term of equation 8.4.8.1.6, if we define $p_{i,-1}(t) = 0\ for\ x = 0$, and assuming that at time t=0 the population size X(0)=I, and taking limit as $\delta t \to 0$ we obtain,

$$\frac{dp_{i,x}(t)}{dt} = \kappa_{x-1}p_{i,x-1}(t) + \alpha_{x+1}p_{x+1}(t) - (\kappa_x + \alpha_x)p_x(t),$$
$$for\ x = 1,2, \ldots \ldots \ldots \ldots \ldots\ldots\ldots\ldots\ldots\ldots\ldots\ldots.8.4.8.1.7$$

Equations 8.4.8.1.7 are the differential-difference equations of the general Birth and Death Process, and they are called the Chapman-Kolmogorov Forward equations.
We will now obtain the differential-difference equations for the Simple Birth process by making certain substitutions into equations 8.4.8.1.7 above.

8.4.9 The Differential-difference Equations for the Simple Birth and Death process

The Kolmogorov forward equations for the Simple Birth process can be derived by simply substituting $\kappa_x = \kappa x, and\ \alpha_x = \alpha x,$ in equation 8.4.8.1.7 above, giving

$$\frac{dp_{i,x}(t)}{dt} = \kappa(x - 1)p_{i,x-1}(t) + \alpha(x + 1)p_{x+1}(t)$$
$$- (\kappa + \alpha)x\alpha p_x(t),$$
$$for\ x = 1,2, \ldots \ldots \ldots \ldots \ldots \ldots \ldots \ldots \ldots \ldots \ldots \ldots.8.4.9.1$$

and

$$\frac{dp_0(t)}{dt} = \alpha p_1(t) \ldots \ldots \ldots \ldots \ldots \ldots \ldots \ldots \ldots \ldots \ldots \ldots \ldots \ldots .8.4.9.2$$

The probability Generating Function F(S,t) of X(t), of the number of individuals alive at time t, is $F(S,t) = \sum_{x=0}^{\infty} p_x(t) S^x$. By multiplying the Kolmogorov Forward equation 8.4.9.1 by S^x and summing over x, we obtain a partial differential equation satisfied by F(S,t). The solution to this equation shows that F(S,t) satisfies the partial differential Equation

$$\frac{\partial F}{dt} =$$

$$\{\kappa S^2 + \alpha - (\kappa + \alpha)S\}\frac{\partial F}{\partial S}, \ldots \ldots \ldots \ldots \ldots \ldots \ldots \ldots \ldots ..8.4.9.3$$

This equation is linear in $\frac{\partial F}{\partial t}$ and $\frac{\partial F}{\partial S}$, so it can be solved by the Lagrangian method.
The solution can be found
(i) for $\kappa \neq \alpha$, birth and death rates are not equal and (ii) for $\kappa =$, birth and death rates are equal.

ASSIGNMENT 8.9 (FOR THE STUDENT).

(i)   Multiply the Kolmogorov forward equations 8.4.9.1 by S^x and sum over x to obtain the partial differential equation satisfied by F(S,t).
(ii)   By solving (i), the following partial differential equation is satisfied:

$$\frac{\partial F(S,t)}{\partial t} = \{\kappa S^2 + \alpha - (\kappa + \alpha)S\}\frac{\partial F(S,t)}{\partial S} \ldots \ldots \ldots \ldots \ldots .8.4.9.4$$

Use the Lagrangian method to solve equation 8.4.9.4

{Hint you may need to factorize}: $\alpha - [\kappa + \alpha]S + \kappa S^2 = [1 - S][\alpha - \kappa S]$.

For the case $\kappa \neq \alpha$, use the following result:

$$\frac{-1}{[1 - S][\alpha - \kappa S]} = \frac{1}{(\alpha - \kappa)}\left(\frac{\kappa}{\alpha - \kappa S} - \frac{1}{1 - S}\right)$$

(iii)

(a) For $\kappa \neq \alpha$, find the probability generating function F(S,t), given that at time t=0, the population size X(o)=1.

(b) Show that the probability generating function F(S,t), [given that at time t=0, the population size $X(0)=x_0$] is equal to:

$$F(S, t) = \left\{\frac{\alpha(1 - S) - (\alpha - \kappa S)\left(e^{(\alpha - \kappa)t}\right)}{\kappa(1 - S) - (\alpha - \kappa S)e^{(\alpha - \kappa)t}}\right\}^{x_0} \quad \dots \dots \dots \dots .8.4.9.5$$

(iv)

(a) For $\kappa = \alpha$, find the probability generating function F(S,t) [given that at time t=0, the population size X(0)=1.

(b) Show that the probability generating function F(S,t) [given that at time t=0, the population size $X(0) = x_0$] is equal to:

$$F(S, t) = \left\{\frac{\kappa t - S\kappa t + S}{\kappa t - S\kappa t + 1}\right\}^{x_0} \quad \dots \dots \dots \dots \dots \dots \dots \dots \dots \dots \dots .8.4.9.6$$

8.4.10 Some Properties of the Birth-Death Process

It has already been mentioned that the Birth and Death process is an example of a Markov Chain. One of the interesting characteristics of a Markov Chain is its behaviour in the long run. With specific reference to the Birth and Death process [in a population where births and deaths are occurring], we may wish to know what will be the state of this population as time progresses? What will happen if there are more deaths than births (or vice versa) in the population as time progresses? It is very clear that (i) if there are more deaths than births in the population, then it will be threatened and its size is bound to reduce with time. This situation is referred to as extinction. If this situation persists throughout time [i.e. if there are persistently more deaths than births] the population will die out, a situation referred to as ultimate extinction. (ii) If on the other hand, the birth rate exceeds the death rate, the population will survive with time, and depending on the magnitude of the difference between the two rates, the population may tend to explode with time.

ASSIGNMENT 8.10 (FOR THE STUDENT).

We assume that some population has size x at time t. We may write the random variable $X(t) = x$, and its probability is given as, $P[X(t) = x] = p_x(t)$. The probability generating function of this random variable is defined as,

$$F(S,t) = \sum_{x=0}^{\infty} p_x(t)S^x = p_0(t)S^0 + p_1(t)S^1 + p_2(t)S^2 + p_3(t)S^3 + \cdots \ldots \ldots \ldots \ldots \ldots \ldots \ldots \ldots .8.4.10.1$$

Case 1: death rate α not equal to death rate κ: $(\alpha \neq \kappa)$.
As time progresses, the probability that no individual is alive at time, t, which leads to extinction, is obtained by putting S=0 in F(S, t) given in 8.4.10.1 above, to give

$F(0,t) = p_0(t) = P[X(t) = 0] =$
probability of extinction at time t. Hence the probability of ultimate extinction is given as, $\lim_{t \to \infty} F(0,t)$.
(i)Looking at expression 8.4.9.4 in Assignment 8.9 above, find an expression for the probability of extinction when $\alpha \neq \kappa$.
(ii)From the expression obtained in (i), show that for ultimate extinction

$e^{(\alpha-\kappa)t} \to 0,$ *when* $\kappa > \alpha$ *and* $e^{(\alpha-\kappa)}$ *when* $\kappa < \alpha,$

hence find $\lim_{t \to \infty} F(0,t)$ *when* $\kappa > \alpha$ *and also when* $\kappa < \alpha.$

(iii)It can be seen from the solutions to (ii) that if the birth rate is greater than the death rate there is positive probability that the population does not die out. So if the population does not die out, show that for ultimate extinction

$\lim_{t \to \infty} F(S,t) = \lim_{t \to \infty} F(0,t) = 1,$ *when* $\kappa < \alpha.$

Case 2: death rate α *equal to birth rate* κ: $(\alpha = \kappa)$.
Looking at expression 8.4.9.5 of assignment 8.9 above,
(i) Obtain the probability of extinction by time t.
(ii) Obtain the probability of ultimate extinction.

8.5. Note: A special case of the Birth and Death process is the QUEUE, in which arrivals constitute 'births' and departures constitute 'deaths'. This approach will not be dealt with in this book, but you might find it useful to refer to other texts in Stochastic processes. Stochastic processes with continuous parameter time constitute a very broad area by itself, and a whole book can be produced on these topics alone. It is, therefore, not the intention of the authors to expand this book to that extent, and therefore topics like Queuing Theory, Branching Processes, Brownian motion, Diffusion Theory, Martingales and other related areas are beyond the scope of this simple text. Interested readers may refer, for instance, to the following texts: Bartlett (1960), Bharucha-Reid (1960), Chung (1967), Cox and Miller (1970), Cramer and Leadbetter (1967), Doob (1953), Feller (1957, 1970), Harris (1963), Hoel, Port and Stone (1972), Karlin and Taylor (1975), Kannan (1979), Parzen (1962) etc, where more detailed treatments of Stochastic processes are given.

APPENDICES

ANSWERS TO EXERCISE

ANSWERS TO EXERCISE 1

1.1   (a) both A and B (b) not (both A and B)   (c) A but not B   (d) Neither A nor B
(e)   At least one of A and B (f) Neither A nor B At least one of A and not B (h) Not both A and B.

1.2   (a) ϕ (phi-empty set) (b) A (c) U = Universal set
(d) A (e) A∩B (f) A (g) A (h) A∪B (I) A (j) A^c

1.3.(a) {5} (b) {1, 3, 4, 5, 6, 7, 8, 9, 10} (c) {2, 3, 4, 5}
(d) {1, 5, 6, 7, 8, 9, 10} (e) {1, 2, 5, 6, 7, 8, 9, 10}

1.4

$$(I) \; A \times A = \{(H,H), (H,T)(T,H), (T,T)\} : \underset{Combinations}{4}$$

$$(II) \; A \times A \times A = \left\{ \begin{array}{l} (H,H,H), (H,H,T), (H,T,H), (T,H,H), \\ (T,H,T), \;\; (T,T,H), (H,T,T), \;\; (T,T,T) \end{array} \right\} \underset{Combinations}{8}$$

(III)

$$B \times B = \left\{ \begin{array}{llllll} (1,1) & (2,1) & (3,1) & (4,1) & (5,1) & (6,1) \\ (1,2) & (2,2) & (3,2) & (4,2) & (5,2) & (6,2) \\ (1,3) & (2,3) & (3,3) & (4,3) & (5,3) & (6,3) \\ (1,4) & (2,4) & (3,4) & (4,4) & (5,4) & (6,4) \\ (1,5) & (2,5) & (3,5) & (4,5) & (5,5) & (6,5) \\ (1,6) & (2,6) & (3,6) & (4,6) & (5,6) & (6,6) \end{array} \right\} \underset{Combinations}{36}$$

(IV)

$$\left\{ \begin{array}{llllll} (H,1), & (H,2), & (H,3), & (H,4), & (H,5), & (H,6) \\ (T,1) & (T,2), & (T,3), & (T,4), & (T,5), & (T,6) \end{array} \right\} \underset{Combinations}{12}$$

1.5. (i) $A_1 \cup A_2 = \{n: 2 \le 155\}$ (ii) $A_1 \cap A_2 = \{n: 5 \le 15\}$
(iii) $A_1^c = \{n: 0 \le n \le 2\} \cup \{n: n \ge 15\}$
(iv) $A_2^c = \{n: n \le 5\} \cup \{n: n \ge 155\}$
(v) $A_2^c \cap A_1 = \{n: 2 \le n \le 5\}$

1.6. Only (iii) is a mapping. In (I) the member "a" has two images and in (iii)
d has no image.

1.7. The domain = {a, b} and the range = {x, y, z}. Note that C is not in the domain since it has no image and W is not in the range since it is not the image of any member.

1.8. The domain is {1, 2, 3} and the range is {2, 4, 6}.
Note: the domain need not be the set S but a subset of S whilst the range is a subset of T.

ANWERS TO EXERCISE 2

2.1. (a) The sample space S1 contains 12 ordered pairs
$$(i,j): i \neq j, 1 \leq i \leq 4, 1 \leq j \leq 4$$
where the first number indicates the first number drawn. Thus

$$S = \begin{cases} (1;\ 2)(1;\ 3)(1;\ 4) \\ (2;\ 1)(2;\ 3)(2;\ 4) \\ (3;\ 1)(3;\ 2)(3;\ 4) \\ (4;\ 1)\ (4;\ 2)\ (4;\ 3) \end{cases}$$

(a) The sample space S1 contains 16 ordered pairs
$$(i,j): i \neq j, 1 \leq i \leq 4,\ 1 \leq j \leq 4$$
where the first number indicates the first number drawn. Thus

$$S = \begin{cases} (1;\ 1)(1;\ 2)(1;\ 3)(1;\ 4) \\ (2;\ 1)(2;\ 2)(2;\ 3)(2;\ 4) \\ (3;\ 1)(3;\ 2)(3;\ 1)(3;\ 4) \\ (4;\ 1)\ (4;\ 2)\ (4;\ 3)\ (4;\ 4) \end{cases}$$

	Science Major	Non-Science Major	Total
Women	55	120	175
Men	115	110	225
Total	170	230	400

Now we can easily obtained the desired probabilities:

Let A denotes the event that a student is a science major. Realize that n(A) =

170; also let S denotes the total sample size, n(S) = 400.

(a)   There are 170 science majors out of 400 students, so

$$P\ (science\ major)$$
$$= P\ (A)$$
$$= \frac{n(A)}{n(S)}$$
$$= \frac{170}{400}$$

(b)   Let W represents the event a student is a woman, n(W) = 175 Since there are 175 women out of 400 student altogether, we have

$$P\ (Woman) = P\ (W) = \frac{n(w)}{n(S)} = \frac{175}{400}$$

(c)

There are 110 students who are neither women nor science majors so

$$P\ (neither\ woman\ nor\ science\ major) = \frac{110}{400} =$$

2.3 (a) (i) 0.1 (ii) 0.5 (iii) 0.7
(b) (i) 0.6 (ii) 0.9
(c) 0.9

2.4 (a) (i) 0.8 (ii) 0.3 (iii) 0.7
(b) 0.833 or $\frac{5}{6}$ (c) (i) 0.5 (ii) 0.05

2.5 (i) 0.26 (ii) 0.34 (iii) 0.20 (iv) 0.25 (v) 0.25

2.6 (i) $-\dfrac{4}{9}$ (ii) $\dfrac{5}{9}$

2.7 (i) $\dfrac{33}{95}$ (ii) $\dfrac{14}{95}$ (iii) $\dfrac{48}{95}$

2.8 (i) 0.512   (ii) 0.384

2.9 (i) 0.48   (ii) 0.36   (iii) 0.32   (iv) 0.16

2.10 (i) $\dfrac{7}{27}$ (ii) $\dfrac{1}{30}$

2.11 $\left(\dfrac{m}{m+n}\right)\left(\dfrac{p+1}{p+q+1}\right) + \left(\dfrac{n}{m+n}\right)\left(\dfrac{p}{p+q+1}\right)$

2.12 (i) 0.5   (ii) 0.5

2.13 <u>Ans:</u> $\dfrac{2}{3}$

2.14 (a) 0.202   (b) 0.9998   (c) 0.799

2.15 (a) 0.225   (b) (i) 0.111   (ii) 0.516

ANSWERS TO EXERCISE 3

3.1 (a) $\dfrac{1}{6}$ (b) $\dfrac{15}{36}$ (c) $\dfrac{15}{36}$ (d) $\dfrac{5}{9}$ (e) $\dfrac{1}{4}$ (f) $\dfrac{1}{4}$ (g) $\dfrac{1}{6}$ (h) $\dfrac{1}{6}$

3.2 (a) S = {H H, HT, TH, TH, TT}

P(H H) = P(H T) = P(T H) = P(T T) = $\dfrac{1}{4}$

(b)   (i) $\dfrac{1}{4}$ (ii) $\dfrac{3}{4}$ (iii) $\dfrac{3}{4}$ (iv) $\dfrac{1}{2}$

3.3 <u>Ans</u>: 120 ways

3.4 <u>Ans</u>: 720 ways

3.5 (a) $120 = 10c4$ (b) $5040 = 10P4$

3.6 (a) $(12_{C1})(10_{C1}) = 120$ (b) 2970

3.7 $15_{C3} = 455$ ways

3.9. (i) 495 (ii) 161,700 (iii) n

3.10 Denote the red, green, and black balls by R, G, and B respectively.
Therefore two balls of the same colour and the third different, there are six qualitative probabilities which are RRB; RRG; GGR; GGB; BBR; and BBG:
   a: The corresponding number of simple equally likely events are:

$$\binom{2}{2}\binom{4}{1} + \binom{2}{2}\binom{3}{1} + \binom{3}{2}\binom{2}{1} + \binom{3}{2}\binom{4}{1} + \binom{4}{2}\binom{2}{1}$$
$$+ \binom{4}{2}\binom{3}{1}$$
$$= 4 + 3 + 6 + 12 + 12 + 18$$
$$= 55$$

Number of equally likely events in the sample space is $\binom{9}{3} = 84$.
Hence the required probability is $\frac{55}{84}$

b: Let W be the event of drawing at least one red ball. Therefore W^c or $\overline{W}$ is the event of drawing no red ball. Number of elements in W^c or $\overline{W} = \binom{7}{3} = 35$
Hence $P(\overline{W}) = \frac{35}{84} = \frac{5}{12}$. Therefore $P(W) = 1 - P(\overline{W}) = 1 - \frac{5}{12} = \frac{7}{12}$

ANSWERS TO EXERCISE 4

4.1

x	0	1	10
$P(X = x)$	$\dfrac{1}{2}$	$\dfrac{1}{3}$	$\dfrac{1}{6}$

4.2 $\quad E(X) = (p_1 + p_2)$

4.3 (a)

u	3	4	5
$P(U = u)$	$\dfrac{1}{4}$	$\dfrac{1}{2}$	$\dfrac{1}{4}$

v	-2	-1	0
$P(V = v)$	$\dfrac{1}{4}$	$\dfrac{1}{2}$	$\dfrac{1}{4}$

(b)

u / v	3	4	5
-2	0	$\dfrac{1}{4}$	0
-1	$\dfrac{1}{4}$	0	$\dfrac{1}{4}$
0	0	$\dfrac{1}{4}$	0

u and v are not independent.

4.4 0.4

4.5 $E(Z) = 5,\ \sqrt{V(Z)} = 4$

4.6 $E(X - Y) = -1,\quad V(X - Y) = 1$

4.7 $E(X + Y) = 7,\quad V(X + Y) = 9$

4.8 (a) $E(XY) = 2.43,\qquad E(X/Y) = 1.89$

4.9

x	2	5
$P(X = x)$	0.4	0.6

4.10. X is a discrete random variable

(a)Remember the definition of $F(X)$: $F(X) = (X \leq x)$

i. $P(X \leq 4)\ F(4) = \frac{1}{6}$

ii. $P(-5 < X \leq 4) = P(x \leq 4\ ) - P(x \leq -5)$
$$= F(4) - F(-5)$$
$$= \frac{1}{6} - 0 = \frac{1}{6}$$

iii. $P(X = -3) = P(x \leq -3) - P(X \leq -3)$
$$= \frac{1}{6} - 0 = \frac{1}{6}$$

iv. $P(X = 4) = P(x \leq 4) - P(x < 4)$
$$= \frac{1}{6} - \frac{1}{6} = 0$$

c)   The probability function of X

$$f(x) = \begin{cases} \frac{1}{6}, for\ x = -3 \\ \frac{2}{6}, for\ x = 6 \\ \frac{3}{6}, for\ x = 10 \end{cases}$$

4.11 $E(X) = 4.5\ \sqrt{V(X)} = 0.866$

4.12 (a) $P(Y = 2) = P(Y = 3) = P(Y = 4) = \frac{1}{3}$

$$P(X = 2) = \frac{1}{2}, \ P(X = 1), P(X = 3) = \frac{1}{4}$$

(b) $\frac{1}{3}$ (c) $\frac{1}{2}$ (d) $\frac{5}{12}$

4.13 (a) [Hint: See example 4.3.8.1, section 4.3.8]

(b) X and Y are uncorrelated

(c) $\frac{3}{4}$

4.14 (a) $\rho_{XY} = 0$

(b) X and Y are not independent

4.15 (a) $\frac{2}{3}$ (b) $\frac{1}{3}$ (c) $\frac{1}{4}$

(d)

x	1	2	3
$p(x/2)$	$\frac{1}{4}$	$\frac{1}{2}$	$\frac{1}{4}$

(e)

y	2	3	4
$p(y/x = 1)$	$\frac{1}{3}$	$\frac{2}{3}$	0
$p(y/x = 2)$	$\frac{1}{3}$	0	$\frac{2}{3}$
$p(y/x = 3)$	$\frac{1}{3}$	$\frac{2}{3}$	0

Hence try and verify as the problem requires [Hint:see Example 4.3.6.1 of section 4.3.6]

4.16 $E(X) = 1, \quad V(X) = 0.5$

4.17 $\dfrac{19}{27} \approx 0.7$

4.18 0.99877

4.19 (a) 0.7734   (b) 0.04262

4.20 $\dfrac{77}{95}$

4.21 (a) $P(X = 0) = 0.264$ (b) $P(X = 4) = 0.09$

4.22 0.067

ANSWERS TO EXERCISE 5

5.1 (a) E(X) = 3.5 (b) $\sqrt{V(x)} = \sqrt{0.75} = 0.87$

5.2 (a) E(X) = 0   (b) $E(X^2) = \dfrac{1}{3}$ (c) E(X^3) = 0

   (d) $\sqrt{V(X)} = \sqrt{\dfrac{1}{3}} \approx 0.58$

5.3 (a) E(X) = 4.5   (b) $\sqrt{V(X)} = \sqrt{0.75} \approx 0.87$

5.4 (a) $E(X) = \dfrac{5}{3}$ (b) $V(X) = \dfrac{1}{18}$

   (c) $F_X(x) = \begin{cases} 0, & \text{if } x < 1 \\ (x-1)^2, & \text{if } 1 \le x < 2 \\ 1 & , \text{if } x \ge 2 \end{cases}$

   (d) 0.83

5.5 $V(X) = \dfrac{4}{3}a^2$

5.6 (a) $\dfrac{1}{4}$

(b) $F_X(x)=\begin{cases} 0 & , x<-1 \\ \dfrac{(x+1)}{2}, & -\le x<1 \\ 1 & , x\ge 1 \end{cases}$

5.7 (a) $P(X>1)=\dfrac{3}{4}$

(b) $E(X)=\dfrac{4}{3}$, $V(X)=\dfrac{2}{9}$

(c) $F_X(x)=\begin{cases} 0, & \text{if } x<0 \\ \dfrac{x^2}{4}, & \text{if } 0\le x<2 \\ 1 & , \text{if } x\ge 2 \end{cases}$

(d) $F_Y(y)=\begin{cases} 0 & , \text{if } y<0 \\ \dfrac{y}{4} & , \text{if } 0\le y<4 \\ 1 & , \text{if } y\ge 4 \end{cases}$

5.8 $f_Z(z)=\begin{cases} -\log z, & \text{if } 0<z<1 \\ 0, & \text{elsewhere} \end{cases}$

5.9 $f_Z(z)=\begin{cases} -\log z, & \text{if } 0<z<1 \\ 0, & \text{elsewhere} \end{cases}$

$$5.10 \quad f_z(z) = \begin{cases} 2 - \dfrac{1}{2Z^2} \,, & if \ \dfrac{1}{2} \le z \le 1 \\ \dfrac{2}{Z^2} - \dfrac{1}{2} \,, & if \ 1 \le z \le 2 \\ 0 & , \ otherwise \end{cases}$$

$$5.11 \ (a) \quad f_z(z) = \begin{cases} \dfrac{2}{Z^2} \,, & 2 < z < \infty \\ 0 \,, & otherwise \end{cases}$$

(b) median of $Z = 4$

$$5.12 \quad f_z(t) = \begin{cases} \dfrac{t+1}{2} \,, & -1 < t < 1 \\ 0 \,, & otherwise \end{cases}$$

$$5.13 \ (a) \quad F_Y(y) = \begin{cases} 1 - e^{-y2} \,, & if \ y \ge 0 \\ 0 & , \ if \ y < 0 \end{cases}$$

(b) median $\qquad \tilde{m} = \sqrt{\log 2} = 0.833$

$$5.18 \quad f_X(x) = f_Y(y) = e^{-\lambda}, \qquad \lambda > 0$$
$$f_{XY}(x,y) = (1 - e^{-x})(1 - e^{-y}),$$
$$x,\ y > 0$$

ANSWERS TO EXERCISE 6

6.1 (a) (i) 0.9332, (ii) 0.8413 (iii) 0.0227 (iv) 0.3085 (v) 0.0919
(v) 0.0919

(a) u = 18.12
(b) v = 15.68

6.2. 0.6528
6.3. 0.3413
6.4. 0.4672
6.5. 0.0227
6.6. (a) Y ~ N (50, 75) (b) $\overline{X}$ ~ N(2, 3)
6.7. 0.0287
6.8. (a) 0.7734 (b) 0.04262
6.9. $E(Y) = \sqrt{2/\pi}$, $V(Y) = (\pi - 2)/\pi$
6.10. 0.15
6.11. (a) Normal (b) Poisson (c) Direct (d) Not clear
6.12. $\dfrac{1}{4}$ (equal to the Chebyshev bound)
6.13. (a) n = 392 (b) n = 5000
6.14. (a) 0.083
6.15. 0.1814
6.16. (a) 0.1112 (b) 0.1915
6.17. (a) 120 (b) $\dfrac{945}{32}\sqrt{\pi}$ (c) $\dfrac{4}{15}$
6.18. $E(X) = \dfrac{\alpha}{\lambda}$, $Var\ X = \dfrac{\alpha}{\lambda^2}$
6.19. $E(W) = \dfrac{r}{s-1}$, $E(W^2) = \dfrac{r(r+1)}{(s-1)(s-2)}$
6.20. $(k-2)^{-1}$

ANSWERS TO EXERCISE 7

7.1 (a) $F(x) = \int_{-\infty}^{x} f(t)\,dt = \begin{cases} 1 - e^{-ax}, & , \le x < \infty \\ 0, & x < 0 \end{cases}$

(b) $M_X(t) = \dfrac{a}{a-t}\left(1 - t/a\right)^{-1}$, provided t < a

The n–th term in the Maclaurin series expansion is $\dfrac{n!}{a^n} \cdot \dfrac{t^n}{n!}$

$E(X) = \dfrac{1}{a}$, $Var(X) = \dfrac{1}{a^2}$

7.2 $C_Z(t) = \left(Pe^{it} + q\right)^{15}$

Hence Z ~ Bin (15, P) = Bin (n, P)

7.3 (a) $M_Y(t) = \left(\dfrac{1}{3} + \dfrac{2}{3}e^t\right)^8$, i.e. Y has a Binomial

distribution. Given by $Y \sim Bin\left(8, \dfrac{2}{3}\right) = Bin\,(n, P).$ (b)

$Var(Y) = \dfrac{16}{9}$

7.4 (a) $f_Z(z) = e^{-z}$

(b) $M_Z(t) = \dfrac{1}{1-t}$, $\left(for\ t < 1\right).$

7.5 (a) $F_X(t) = \dfrac{t}{2-t}$

(b) Var (X) = 2

7.6 $M_X(t) = e^{bt}\,M_X(a\,t)$

7.7 Var (X) = 2

7.8 (a) $F_X(t) = (2 - t)^{-1}$
(b) E(X) = 1, Var (X) = 2

7.9 $M_Y(t) = (1 - t)^{-n}$

$$f_y(y) = \frac{y^{n-1}\, e^{-y}}{(n-1)!}$$

7.10 $\quad M_X(t) = \left(\dfrac{\alpha}{\alpha - t}\right)^r$

7.11 $\quad$ (a) $\quad M_X(t) = \left(\dfrac{2}{t^2}\right)\left[e^t\,(t-1)+1\right]$

$\qquad$ (b) $\quad E(X) = \dfrac{2}{3},\, V(X) = \dfrac{1}{18}$

7.12 $\qquad$ (a) $\quad M_X(t) = \dfrac{\lambda\, e^{ta}}{(\lambda - t)}$

$\qquad\quad$ (b) $\quad E(X) = \dfrac{(a\lambda + 1)}{\lambda},\, V(X) = \dfrac{1}{\lambda^2}$

7.13 $\quad M_X(t) = (1 - t^2)^{-1}$

7.14 $\quad E(X) = 3.2$

7.15 $\quad Var\,(X) = \dfrac{1}{9}$

7.16 $\quad \left(e^{3t} - 1\right)\big/ 3t\, e^t$

7.17 $\quad \left(1 - i\,t\big/\lambda\right)^{-1}$

7.18 $\quad \exp\left(-\dfrac{t^2}{2}\right)$

7.20 $\quad$ Ans: 0.0515
7.21 $\quad$ Ans: 0.0651

7.22 $\quad$ Ans: 0.1172

7.23 $\quad$ Ans: (a) 0.0630 (b) 0.9730

7.24 $\quad$ 0.0025

7.25 $\dfrac{21}{256}$

7.26 $\dfrac{81}{1024}$

7.28 (a) $F(x) = \int_{-\infty}^{x} f(t)dt = \begin{cases} 1 - e^{-ax} & 0 < x < \infty \\ 0 & \text{otherwise} \end{cases}$

(b) $M_X(t) = \dfrac{a}{a-t}\left(1 - \dfrac{t}{a}\right)^{-1}$ provided $t < a$

[Hint: the nth term in the Maclaurin series expansion is $\left[\dfrac{n!\ t^n}{a^n\ n!}\right]$

(c) $E(X) = \dfrac{1}{a}, \quad V(X) = \dfrac{1}{a^2}$

BIBLIOGRAPHY

Applebaum, D., (1996), Probability and Information: An Integrated Approach, Cambridge University press, Cambridge.

Bajpal, A.C., Calus, L.M. and Fairley, J.A.., (1978), Statistical Methods for Engineers and Scientists, John Wiley and Sons, New York.

Barber, M.N., and Ninham, B.W., (1970), Random and Restricted Walks, Gordon and Breach, New York.

Barlow, R. (1989), Statistics. A Guide to the Use of Statistical Methods in the Physical Sciences, John Wiley and Sons, New York.

Bartlett, M.S., (1960), Stochastic Population Models in Ecology and Epidemiology, John Wiley and sons, New York.

Bartoszynski, R and Niewiadomska-Bugay, M. (1996), Probability and Statistical Inference, John Wiley and Sons, New York.

Beaumont, G.P. (1986), Probability and Random Variables, Ellis Norwood Ltd, Chichester.

Berger, M.A. (1993) An Introduction to Probability and Stochastic Processes, Springer-Verlag, New York.

Bharucha-Reid, A., (1960), Elements of the Theory of Markov Processes and Their Applications, McGraw-Hill Book Company, New York.

Billinsley, P. (1979) Probability and Measure, John Wiley and Sons, New York.

Birnbaum, Z.W. (1962) Introduction to probability and Mathematical Statistics, Harper and Brothers Publishers, New York.

Blake, Ian F.B. (1979), An Introduction to Applied Probability, John Wiley and Sons, New York.

Bradley, J., (1988) Introduction to Discrete Mathematics, Addison-Wesley Publishing Company, Reading.

Breiman L., (1972) Society for Industrial and Applied Maths, Siam, Philadelphia.
Bremaud, Pierre (1994), An Introduction to Probabilistic Modeline, Springer - Verlag, New York.

Chow, Y.S. Teicher, H. (1988), Probability Theory, Springer - Verlag, New York.

Chung, K.L., (1967), Markov Chains with Stationary Transition Probabilities, 2nd ed., Springer-Verlag, New York.

Cox, D.R., and Miller, H.D., (1970), The Theory of Stochastic Processes, Methuen, London.

Cramer, H., and Leadbetter, M.R., (1967), Stationary and Relted Stochastic Processes, John Wiley and sons, New York.

David F.N. (1951), Probability Theory for Statistical Methods, Cambridge University Press, Cambridge.

Doob, J.L., (1953), Stochastic Processes, John Wiley and Sons, New York.

Drake, A.W. (1967), Fundamentals of Applied Probability Theory, McGraw-Hill, New York.

Dudewicz, E.J. and Mishra, S.N. (1988), Modern Mathematical Statistics, John Wiley and Sons, New York.

Feller, W., (1957), An Introduction to Probability Theory and Its Applications, vol. I, 2nd ed., John Wiley and Sons, New York.

Feller, W., (1971), An Introduction to Probability Theory and its Applications Vol. II, Second Edition, JohnWiley and Sons, New York.

Freund, J.E. and Walpole, R.E., (1971), Mathematical Statistics, Prentice - Hall, Inc. Englewood Cliffs, New Jersey.

Fristedt, B. Gray, L. (1977), Modern Approach to Probability Theory, Birhhauser, Boston.

Galambos, J. (1984), Introductory Probability Theory, Marcel Dekker, Inc., New York and Basel.

Gnedenko, B.V. and Khinchin, A. Ya. (1962) An Elementary Introduction to the Theory of Probability, Dover Publications, Inc., New York.

Goldberg, S. (1960), Probability, An Introduction, Prentice-Hall, Inc., Englewood Cliffe, New Jersey.
Grimmeth, G. and Weslsh, D. (1986), Probability, An Introduction, Clarendon Press, Oxford.

Guttman, I., Wilks, S.S. and Hunter, J.S. (1982), Introduction Engineering Statistics, John Wiley and Sons, New York.

Haln, G.J., and Shapiro S.S. (1967) Statistical Models in Engineering, John Wiley and Sons.

Hoel, P.G., Port, S.C., and Stone, C.J., (1972), Introduction to Stochastic Processes, Houghton Mifflin, New York.

Hoel, P.G. (1984) Mathematical Statistics, Fifth ed., John Wily and Sons, New York

Hoel, P.G. Port, S.C., and Stone, `J.S. (1971), Introduction to Probability Theory, Houghton Mifflin.

Hogg, R.V. and Cragg, A.T. (1978), Introduction to Mathematical Statistics, Fourth Edition, Macmilian, New York.

Johnson, N.L. and Kotz, S. (1969) Distributions in Statistics: Discrete Distributions, John Wiley and Sons, New York.

Johnson, N.L. and Kotz, S. (1970) Distributions in Statistics: continuous Multivariate Distributions, John Wiley and Sons, New York.

Johnson, N.L. and Kotz, S. (1970) Distributions in Statistics: Continuous Univariate Distribution 1 & 2 John Wiley and Sons, New York.

Kannan, D, (1979). An introduction to Stochastic Processes. North Holland Series. New York.

Karlin, S., and Taylor, H.M., (1975), A First Course in Stochastic Processes, 2nd ed., Academic Press, New York.

Kendal, M.G. and Stuart, A. (1963) The Advanced Theory of Statistics, Vol. 1, 2nd ed., Charles Griffin & Company Ltd., London.

Kendal, M.G. and Stuart, A. (1976) The Advanced Theory of Statistics, Vol. 3, 4th ed., Charles Griffin & Company Ltd., London.

Kendal, M.G. and Stuart, A. (1979) The Advanced Theory of Statistics, Vol. 2, 4th ed., Charles Griffin & Company Ltd., London.

Kolmogorov, A.N. (1956), Foundations of the Theory of Probability, Chelsea Publishing Company, New York.
Lai, Chung Kai (1975), Elementary Probability: Theory with Stochastic Processes, Springer - Verlag, New York.

Lamperti, J. (1966), Probability: A Survey of the Mathematical Theory, W.A. Benjamin, INC, New York.

Lindgren, B.W, (1976). Statistical Theory. Third Edition, Macmillan Publishing Co. Inc. New York.

Lindgren, B.W. (1968) Statistical Theory, McMillian Company, New York.

Loeve, M., (1963) Probability Theory, 3rd ed., Van Nostrand, New York.

Lupton R., (1993), Statistics in Theory and Practice, Princeton, Princeton University Press.

McShane, E.J., (1974), Stochastic Calculus and Stochastic Models, Academic Press, New York.

Mendenhall, W. and Scheaffer, R.L. (1990), Mathematical Statistics with Applications, PWS-KENT Publishing Company, Mass.

Meyer, D. (1970), Probability, W.A. Benjamin, Inc. New York.

Meyer, P.L. (1970), Introductory Probability and Its Applications, 2nd Edition, Addison Wesley Publishing Company.

Miller, I. and Freund, J .E. (1987), Probability and Statistics for Engineers, Third Edition, Prentice-Hall, Englewood Cliffs, New Delhi.

Mood, A.M. Graybill, F.A. and Boes, D.C. (1974), Introduction to the Theory of Statistics, Third Edition, McGraw-Hill, New York.

Neveu, J., (1965), Mathematical Foundations of the Calculus of Probability, Holden- Day, San Francisco.

Nuamah, N.N.N.N., (1994), Statistical and Demographic Measures for Population Studies, Regional Institute for Population Studies, Legon.

Nuamah, N.N.NN., (1998), A first course in Probability Theory(Volumes 1,. Ghana Universities Press Accra.

Nuamah, N.N.NN., (1999), A first course in Probability Theory(Volumes 2,. Ghana Universities Press Accra.

Page, L.B. (1989). Probability for Engineering with Applications to Reliability, Computer Science Press.

Papoulis, A., (1965), Probability Random Variables, and Stochastic Processes, McGraw-Hill Book Company, New York.

Parzen, E., (1962), Stochastic Processes, Holden-Day, San Francisco.
Prabhu, N.U., (1966), Stochastic Processes, Basic Theory and its Applications, Mcmillan, New York.

Prohorov, Yu. V. and Rozanov, Yu. A. (1969) Probability Theory, Springer-Verlag, New York.

Robinson, E.A. (1985), Probability Theory and Applications International Human Resources Dev. Corporation.

Rosenblatt, M., (1974), Random Processes, 2nd ed., Springer-Verlag, New York.

Ross, S. (1984), A First Course in Probability, 2nd Edition, Macmillan Publishing Company.

Snyder, D.L., (1975), Random point Processes, John Wiley and Sons, New York.

Spiegel, M.R. (1980), Probability and Statistics, McGraw-Hill Book Company, New York.

Spitzer, F., (1964), Principles of Random Walk, Van Nostrand, New York.

Stirzaker, D. (1994), Elementary Probability, Cambridge University Press, Cambridge.

Stoyanov. J. et. al. (1989), Exercise Manual in Probability Theory, Kluwer Academic Publiishers, London.

Studies in the History of Statistics and Probability, Vol 1, edited by Pearson, E.S., and Kendal, M. (1970), Charles Griffin & Company Ltd., London.

Studies in the History of Statistics and Probability, Vol. 1, edited by Pearson, E.S., and Kendal, M. (1977), Charles Griffin & Company Ltd, London.

Trivedi, K.S. (1988), Probability and Statistics with Reliability, Queuing, and Computer Science Applications, Prentice-Hall of India, New Delhi.

Tucker, H.G. (1963), An Introduction to Probability and Mathematical Statistics Academic Press New York.

Wayne, W.D. (1991) Biostatistics: A Foundation for Analysis in the Health Sciences, Fifth Edition, John Wiley and Sons, Singapore.

Wilks, S.S. (1962), Mathematical Statistics, John Wiley and sons, New York.
Wonnaccott H. Thomas, Wonnacott Ronald J. (1990): Introductory Statistics. Fifth Edition.

INDEX

Forgetfulness Property, 117, 200
Forward Transformation, 162
Function, 8, 62
Of a discrete random variable, 62, 63, 72

G

Gamma Density Function, 202
Gamma Distribution, 201, 202
General Birth and Death Process, 315
Generating Functions, 218
Geometric Distribution, 115, 239
Expected Value(or Mean) of, 116
Probability Distribution of, 116
Variance of, 116

H

Hypergeometric Distribution, 113
Expected Value(or Mean) of, 114
Probability distribution of, 113
Variance of, 114

I

Immigration-Birth Process, 311
Inversion Formula, 236

J

Jacobian of Inverse Transformation, 155,156

Jacobian of Transformation, 155, 156, 181, 208

L

Law of Large Numbers, 195
Law of Total Probability, 29, 95

M

Maclaurin Series, 234
Marginal Distributions, 92, 198
Markov Chains, 255, 260
    Discrete Time, 255
Absorbing State of, 260, 261, 274, 275, 285
    Aperiodic, 279
    Communicating States 271,
Closed absorbing states, 275, 276
Closed Communicating Class, 272, 273, 278, 285
Closed communicating States of, 272
Equivalence Relation of, 271
Reflexive property of, 272
Symmetric Property of, 272
Transitivity Property of, 272
Initial Probability Vector of, 260, 266, 267
Irreducible , 272, 274, 281, 285, 286, 290
Non-irreducible, 260, 285
n-step Transition Probabilities of, 262
n-th Step Transition Probability, 266
One-Step Transition Probability
    of, 256
Periodicity of, 278, 281, 290